SCOPE 61

Interactions of the Major Biogeochemical Cycles

The Scientific Committee on Problems of the Environment (SCOPE)

SCOPE Series

SCOPE 1–59 in the series were published by John Wiley & Sons, Ltd., U.K. Island Press is the publisher for SCOPE 60 as well as subsequent titles in the series.

SCOPE 60: *Resilience and the Behavior of Large-Scale Systems,* edited by Lance H. Gunderson and Lowell Pritchard Jr.

SCOPE 61: *Interactions of the Major Biogeochemical Cycles: Global Change and Human Impacts,* edited by Jerry M. Melillo, Christopher B. Field, and Bedrich Moldan

SCOPE 61

Interactions of the Major Biogeochemical Cycles

Global Change and Human Impacts

Edited by
Jerry M. Melillo, Christopher B. Field,
and Bedrich Moldan

A project of SCOPE, the Scientific Committee on Problems of the Environment, of the International Council for Science

ISLAND PRESS

Washington • Covelo • London

2003

Library of Congress Cataloging-in-Publication Data

Interactions of the major biogeochemical cycles : global change and human impacts/ edited by Jerry M. Melillo, Christopher B. Field, Bedrich Moldan.
p. cm.—(SCOPE ; 61)
"A project of SCOPE, the Scientific Committee on Problems of the Environment, of the International Council for Science." Includes bibliographical references and index.
ISBN 1-55963-066-3 (pbk. : alk. paper)—ISBN 1-55963-065-5 (cloth : alk. paper)
1. Human ecology—Congresses. 2. Nature—Effect of human beings on—Congresses. 3. Biogeochemical cycles—Congresses. I. Melillo, Jerry M. II. Field, Christopher B. III. Moldan, Bedrich. IV. International Council for Science. Scientific Committee on Problems of the Environment. V. Series: SCOPE report ; 61.
GF3.I28 2003
577'.14—dc22 2003016784

British Cataloguing-in-Publication Data available.

Printed on recycled, acid-free paper

Manufactured in the United States of America
10 9 8 7 6 5 4 3 2 1

Contents

List of Colorplates, Figures, Tables, and Boxes

Colorplates

Colorplate section begins following page 138

Figures

Tables

Boxes

Foreword

The Scientific Committee on Problems of the Environment (SCOPE) is one of twenty-six interdisciplinary bodies established by the International Council for Science (ICSU) to address cross-disciplinary issues. SCOPE was established by ICSU in 1969 in response to environmental concerns emerging at that time. When establishing SCOPE, ICSU recognized that many of these concerns required scientific input spanning several disciplines represented within its membership. Today, representatives of forty member countries and twenty-two international, disciplinary-specific unions, scientific committees, and associates currently participate in the work of SCOPE, which directs particular attention to environmental issues in developing countries. The mandate of SCOPE has four parts: to assemble, review, and synthesize the information available on environmental changes attributable to human activity and the effects of these changes on humans; to assess and evaluate methodologies for measuring environmental parameters; to provide an intelligence service on current research; and to provide informed advice to agencies engaged in studies of the environment.

This synthesis volume continues SCOPE's discourse on the important biogeochemical cycles that are essential to life on this planet. It discusses our understanding of the major biogeochemical life cycles with special reference to the advances made in the past decade. It should provide a timely examination of the practical consequences of this knowledge for the sustainability of ecosystems affected by humans.

SCOPE publishes this book as the first of a series of rapid assessments of environmental issues. Our aim is to make sure that experts meet on a regular basis, summarize recent advances in related disciplines, and discuss their possible significance in understanding environmental problems and potential solutions. We aim to make this information available within six to nine months of an assessment's synthesis meeting. We hope that these assessments provide an important service to younger environmental scientists who want to remain informed about new developments and their significance across disciplines.

John W. B. Stewart, Editor-in-Chief

SCOPE Secretariat
51 Boulevard de Montmorency, 75016 Paris, France
Véronique Plocq Fichelet, Executive Director

Preface

Almost three decades ago SCOPE launched a major series of projects on the biogeochemical cycles. Initially these projects focused on the study of carbon, nitrogen, sulfur, and phosphorus separately, with much of the work on individual cycles being coordinated by separate SCOPE/UNEP (Scientific Committee on Problems of the Environment/United Nations Environmental Programme) units. Participants understood that none of these cycles operates independently and made attempts to include studies of element interactions in each of the unit's activities. In 1983 SCOPE produced its first major scientific assessment on interactions among biogeochemical cycles—*The Major Biogeochemical Cycles and Their Interactions,* edited by B. Bolin and R. B. Cook—a book that has guided biogeochemical research since its publication. The assessment's foundation was the basic stoichiometric model of life—the formation of the organic compounds in life processes requires the availability of the elements C, O, H, N, P, S, and a number of trace elements in distinct proportions.

Although SCOPE 21 challenged the stoichiometric model, the model stood up well as a basic paradigm for thinking about element interactions. After two decades of biogeochemical research using radioactive and stable isotopes, applying ever more sophisticated mathematical modeling approaches, and conducting a host of field manipulations, SCOPE delegates meeting at the XIth General Assembly in Bremen, Germany, decided that it was time to reevaluate how knowledge about element interactions had progressed. How insightful is the stoichiometric model? Have fundamental exceptions to it been found? What are they? What basic insights into biogeochemistry do these exceptions give us? Where applicable, does the stoichiometric model move easily across spatial and temporal scales? These and related questions form the basis for a new scientific assessment of element interactions in the biosphere.

The meeting on element interactions held in Prague, Czech Republic, in early October 2002 examined progress in several areas, including theory, measurements, design and interpretation of observation studies and manipulative experiments, and diagnostic and prognostic modeling. In the future sound management of element cycles and their interactions will be essential to fostering the transition to sustainable use of our planet's environment.

Acknowledgments

SCOPE acknowledges with thanks the financial support from the Andrew W. Mellon Foundation, the International Council for Science (ICSU), the US National Academy of Sciences (NAS), and the United Nations Educational, Scientific, and Cultural Organization (UNESCO) that allowed this assessment to be undertaken. The synthesis meeting for the assessment was held in the Congress Center in the historical complex of the Charles University "Karolinum" in Prague, Czech Republic. This complex is located in the Old Town, only 300 meters from the historic Old Town Square. SCOPE is indebted to Professor Bedrich Moldan and Charles University for hosting the meeting in such a wonderful place. Special thanks are given to Ms. Susan Greenwood Etienne of the SCOPE secretariat and Dr. Jiri Dlouhy of the Environment Center, Charles University, for their extraordinary efforts to make the Prague meeting a great success.

1

Element Interactions and the Cycles of Life: An Overview

Jerry M. Melillo, Christopher B. Field, and Bedrich Moldan

The grand cycles of C, H, O, N, P, S, and perhaps as many as 25 other elements sustain life on earth (Box 1.1). As these elements move through the environment, sometimes in inorganic forms and sometimes in organic forms, they interact in a variety of ways. These interactions can be direct, as in situations where one element serves to chelate, immobilize, or catalyze a reaction involving another, or they can be indirect, as in cases where a nutrient limitation imposed by one element influences the rates at which other elements cycle within and among ecosystems. Many of these interactions have consequences for basic ecosystem processes, such as organic matter production by plants and decay of organic materials by microbes, and so affect the way the world works in fundamental ways.

Human activities such as agricultural intensification, urbanization, and industrialization alter element interactions. These alterations contribute to major environmental problems such as climate change, acid precipitation, photochemical smog, and anoxic areas in the coastal zone. Looking to the future, the human race will have to manage element cycles and their interactions if it is to be successful in fostering the transition to sustainable use of the planet's environment.

Scientists have been aware of element interactions since the middle of the nineteenth century. The eminent German chemist Justus von Liebig was one of the first scientists to discuss element interactions as he investigated the role of soil in crop nutrition in the 1840s. In what has become known as the law of the minimum, Liebig stated that "by the deficiency or absence of one necessary constituent, all others being present, the soil is rendered barren for all those crops to

Box 1.1. Elements known or believed to be essential for animals, microbes, or plants

The elements shown below are known or believed to be essential for animals, microbes, or plants (Sterner and Elser 2002). The elements required in large quantities are referred to as macronutrients. Six elements—carbon, hydrogen, oxygen, nitrogen, phosphorus, and sulfur—are the major constituents of living tissue and comprise 95 percent of the biosphere (Schlesinger 1997).

Macronutrients	**Micronutrients**	
Carbon	Arsenic	Manganese
Hydrogen	Barium	Molybdenum
Oxygen	Boron	Nickel
Nitrogen	Bromine	Selenium
Phosphorus	Chlorine	Silicon
Sulfur	Chromium	Sodium
Calcium	Cobalt	Strontium
Magnesium	Copper	Tin
Potassium	Fluorine	Tungsten
	Iodine	Vanadium
	Iron	Zinc

the life of which that one constituent is indispensable" (cited in Brady 1974). This concept served as a cornerstone of soil science and is still considered basically sound, albeit with the added awareness that multiple resource limitations may also be important for processes that integrate many biological reactions over long time scales (e.g., net primary production, or NPP).

About a century after Liebig completed his pioneering work, an American biological oceanographer made a major advance in the conceptualization of element interactions. In 1958 Alfred Redfield proposed a stoichiometric model of life; that is, the formation of the organic compounds in biological processes requires the availability of the elements C, N, P, S, and a number of other elements in distinct proportions. Using this model, Redfield went on to synthesize a complex set of ideas regarding the chemistry of planktonic organisms and the chemistry of the marine environment. He put forth a scheme that suggested extensive control by the biota over the chemistry of some elements in seawater and oxygen in the atmosphere (Redfield 1958).

Since Redfield's time the stoichiometric model has been extended to freshwater and terrestrial ecosystems to explain the basic principles of element interactions and to explore the consequences of human perturbations for the major element cycles. In *Scientific American's* classic September 1970 issue on the biosphere, Edward Deevey Jr., an American geobiologist, argued that human disruption of the global carbon and nitrogen cycles "encourages the bloom on dry land." He conjectured that increases in "the input of industrial carbon dioxide" and the deposition of nitrate associated with acid rain might promote carbon storage in terrestrial ecosystems, especially forests (Deevey 1970).

In the early 1980s a team of scientists from around the world, working under the auspices of the Scientific Committee on Problems of the Environment (SCOPE), produced a scientific assessment on interactions among biogeochemical cycles (SCOPE's volume 21, Bolin and Cook 1983). The assessment's foundation was the basic stoichiometric model of life. Although SCOPE 21 challenged the stoichiometric model, it stood up well as a basic approach for thinking about element interactions.

Over the past two decades scientists have gained new insights into element interactions and ecological stoichiometry in water and land ecosystems through biogeochemical research involving a variety of field and laboratory studies and the development and application of ever more sophisticated mathematical models.

The understanding of element interactions in marine biogeochemistry has recently undergone revolutionary change with the recognition of iron as a key limiting nutrient for photosynthesis in parts of the Earth's ocean system. For decades biological oceanographers have been faced with a paradox—vast expanses of the Southern Ocean, equatorial Pacific, and subarctic North Pacific exhibit low phytoplankton productivity even though these waters have high levels of N and P, the nutrients that often control this key ecosystem process in ocean waters. Through a series of iron-addition experiments in the laboratory (Martin and Fitzwater 1988) and in the open ocean (Kolber et al. 1994; Coale et al. 1996), it has become clear that this micronutrient limits phytoplankton productivity in N- and P-rich marine waters. The productivity increase is generally greatest in large phytoplankton and is accompanied by reductions in CO_2, N, and P in the surface waters. Details of this "iron fertilization" story are given in Ducklow, Oliver, and Smith (Chapter 16, this volume), including implications for "managing" the global carbon cycle.

Scientists' understanding of element interactions in terrestrial biogeochemistry has developed in many ways over the past two decades. Much of this work has focused on carbon-nitrogen interactions. For example, Field and Mooney (1986) highlighted the role of nitrogen in controlling photosynthesis. Shaver et al. (1992) introduced the idea that moving nitrogen from soil into plants could enhance carbon storage in an ecosystem. Through modeling, Rastetter et al. (1992) explored how C:N interactions could affect an ecosystem's capacity to store carbon in response to increases in atmospheric CO_2 and temperature. Many more examples are given in the chapters in this volume.

Galloway (Chapter 14, this volume) explores the variety of mechanisms through which human disruptions of the global nitrogen cycle can affect carbon storage on land, with some disruptions causing increases and others causing decreases. Since the beginning of the Industrial Revolution, human activity has more than doubled the amount of reactive nitrogen that cycles through the atmosphere, land, and fresh and marine waters. This acceleration has many biogeochemical consequences, including increases in the nitrogen content of precipitation and increases in tropospheric ozone, an air pollutant formed when oxides of nitrogen react with volatile organic carbon compounds. On the one hand, nitrogen deposition in precipitation acts as a fertilizer in nitrogen-limited ecosystems, such as many extra-tropical forests, and so increases carbon storage in these ecosystems. On the other hand, tropospheric ozone reduces productivity and carbon storage in many land ecosystems, including extra-tropical forests. Some ecosystems, such as the forests of western and central Europe and the eastern United States, are subject to both increased nitrogen deposition and tropospheric ozone. Changes in carbon storage in these ecosystems must be calculated as the net of positives and negatives (Austin et al., Chapter 2, and Holland and Carroll, Chapter 15, both this volume).

Design of the Assessment

Recognizing the progress in understanding element interactions over the past two decades, SCOPE's Executive Committee concluded that it was time to revisit the topic of element interactions. SCOPE's second assessment of element interactions was designed at its triennial general assembly in Bremen, Germany, in September 2001. The assessment's design is a modified version of the format developed for the Dahlem Konferenzen, the series of interdisciplinary scientific conferences held in Berlin over the past three decades.

- a set of background papers written before a week-long workshop;
- a keynote address and four crosscutting discussion groups at the workshop; and
- an assessment volume with four parts (an overview chapter, four synthesis papers developed from the crosscutting discussions at the workshop, the keynote paper, and the background papers).

The keynote paper (Vitousek, Chapter 6, this volume) set the stage for the workshop by painting a broad picture of element interactions and illustrating key points with pioneering biogeochemistry research done along a chronosequence of forest ecosystems across the Hawaiian island chain.

The background papers were commissioned to provide a common base of reference in support of the crosscutting discussions at the workshop. The papers span the continuum from theoretical (Ågren et al., Chapter 7) to applied (Martinelli, Chapter 10; Tiessen and Shang, Chapter 12). Many include discussions of element exchanges among the land, fresh and marine waters, and the atmosphere (e.g., Lerdau, Chapter 9; Christensen and Keller, Chapter 13; Galloway, Chapter 14; Holland and Carroll, Chapter 15; Ducklow, Oliver, and Smith, Chapter 16; Ittekkot et al., Chapter 17). Two papers explore soil and sediment microbial activity in the context of element interactions (Scholes and Woghiren, Chapter 11; Ivanov and Lein, Chapter 18). And another considers biogeochemical interactions and biodiversity (Eviner and Chapin, Chapter 8).

The topics for the four crosscutting discussion groups were selected to maximize interactions among the workshop participants within, and especially across, scientific disciplines. The topics are given below:

- human disruption of element interactions: drivers, consequences, and trends for the twenty-first century
- disturbance and elemental interactions
- new frontiers in the study of element interactions
- potential for deliberate management of element interactions to address major environmental issues

Many new ideas have been developed in this assessment. We expect some of them to reshape current thinking about element interactions. In the rest of this chapter, we highlight just a few of these ideas to give the reader a sense of the breadth and innovative character of this assessment.

The Concept of Ecological Stoichiometry Revisited

In his keynote paper, Peter Vitousek (Chapter 6, this volume) updates the concept of ecological stoichiometry, building on the work of Redfield (1958), Reiners (1986), Sterner and Elser (2002), and others. Vitousek integrates three ideas into his broad view of ecological stoichiometry: biological stoichiometry, geochemical stoichiometry, and flexibility. Concerning biological stoichiometry, he makes the key point that element ratios are variable within as well as among groups of organisms. Several factors are responsible for this variability within groups of organisms; two important ones are, first, the quantity and biochemistry of structural tissues and, second, the uptake of an element in excess of immediate requirement when that element is abundant.

Geochemical stoichiometry refers to the element ratios of inputs to and outputs from terrestrial ecosystems. Vitousek argues that when the concepts of geochemical and biological stoichiometry are combined, they provide a common language for integrating biogeochemistry as a whole.

Finally, he proposes that it is important to recognize that some elements can cycle more rapidly within or between ecosystems than do other elements undergoing similar transformations. Vitousek refers to this phenomenon as "flexibility" and posits that in an ecosystem the element with the least flexible cycle will ultimately limit biological processes when viewed over geological time scales.

Changes in Community Composition and Element Interactions

The changes in community composition that have the greatest impacts on ecosystems are often those that have the largest effects on element cycling and element interactions. Furthermore, the resulting changes in element interactions can feed back to cause subsequent changes in community composition (Austin et al., Chapter 2; Eviner and Chapin, Chapter 8; both this volume).

Additions of species can have major impacts on element cycling and element interactions in terrestrial and aquatic ecosystems. The introduction of an exotic N fixer, *Myrica faya,* to Hawaii, where N_2 fixers had been uncommon, dramatically altered element cycling and element interactions (Vitousek et al. 1987). Increases in N inputs led to a series of linked effects including reduced N limitation for plants, narrowed C:N ratios in leaf tissue and litter, accelerated rates of litter decay, and increased leaching losses of inorganic and organic nitrogen. The introduction of zebra mussels *(Dresenna polymea)* to the Hudson River accelerated biological filtration by a factor of more than ten. Increased filtration caused removal of most of the algal biomass, a decline in summertime dissolved oxygen concentrations, and narrowing of the N:P ratio in the water column (Caraco et al. 1997, 2000).

Species removals can also affect element cycling and element interactions. For example, the reduction of fish stocks in aquatic ecosystems can result in high zooplankton populations. Their grazing on cyanobacteria can reduce the number of these N_2-fixing organisms, suppressing dinitrogen fixation and potentially keeping N:P ratios low and prolonging nitrogen limitation (Howarth et al. 1999; Marino et al. 2002).

Disturbance and Element Interactions

Disturbances such as floods, fires, and insect outbreaks have long been recognized as shapers of ecosystem structure and function. In the context of biogeochemistry, disturbances mobilize and redistribute elements in novel ratios and mediate element interactions on large spatial and long temporal scales. Ecosystem responses to the biogeochemical aspects of disturbances vary, sometimes amplifying the changes caused by disturbance (positive feedbacks) and sometimes dampening them (negative feedbacks). If the amplification is large enough, altered element distribution may exceed a threshold and force an ecosystem to reorganize in a new state with a new stoichiometry (Hungate et al., Chapter 3, this volume).

Human-induced environmental changes are likely to amplify ecosystem responses supporting positive feedbacks to element redistribution. Forests experiencing chronic acid deposition and estuaries enduring eutrophication are likely to reorganize along some new trajectory of element distribution following disturbance. Exotic species could become invasive following disturbance, with the results being alterations in element distributions and changes in the frequency and intensity of subsequent disturbances. Land-use changes may act in a similar manner, with a history of intensive land use creating a legacy favoring positive feedbacks that push the system along a new trajectory toward a new state.

Element Interactions and Managing the Biosphere Now and in the Future

Solutions to many of today's major environmental problems require that humans manage element cycles and element interactions. Oxygen depletion and the creation of "dead zones" in estuarine and coastal waters are good examples (Diaz and Rosenberg 1995; Nixon 1995). Nitrogen inputs to estuaries and coastal waters can result in excessive aquatic plant growth, including algal growth. When the plants die and sink to the sediment surface, they decompose and consume oxygen in the process. If oxygen is consumed during decomposition faster than it can be replenished through biological and physical processes, hypoxia results. Hypoxia occurs when dissolved oxygen concentrations are below those necessary to sustain most animal life.

Since 1993 the area of midsummer bottom-water hypoxia in the northern Gulf of Mexico along the Louisiana coast has been larger than 4,000 square miles. In 1999 it was 8,000 square miles, which is about the size of the state of New Jersey. A recent integrated assessment found that this hypoxia is caused primarily by excess nitrogen delivered from the Mississippi-Atchafalaya river basin in combination with stratification of Gulf waters (CENR 2000). To address the hypoxia problem in the Gulf of Mexico, the assessment team recommended two primary approaches to reduce, mitigate, and control the problem. First, they proposed reducing nitrogen loads to rivers and streams in the basin, including losses of nitrogen from farmlands, loads from point sources, and atmospheric deposition. Second, they suggested restoring and enhancing natural denitrification and nitrogen retention processes to reduce nitrogen loads from the Mississippi and Atchafalaya Rivers to the Gulf.

Two things helped the assessment team to relate cause and effect and identify solutions. To begin with, they had a fundamental understanding of the interactions among carbon, nitrogen, and oxygen in estuarine and coastal marine ecosystems. In addition, they took a "systems approach" to the problem that forced them to look beyond the estuary and the coastal waters and to include the drainage basin. The systems approach can lead to a fuller accounting of the environmental costs and benefits of management actions.

System-level environmental assessments have been conducted for several other ecosystems in the world. Jansson and Dahlberg (1999) report on one such effort for the Baltic Sea. In this assessment a team of scientists considered the environmental status of the Baltic Sea in the 1940s and today. This assessment also set out some alternative futures and considered environmental costs and benefits of various management options. Assessments like the ones done for the Mississippi drainage/northern Gulf of Mexico and the Baltic Sea are the basis for informed sustainable management of element cycles, their interactions, and, more generally, the biosphere.

Future Research Directions

To advance understanding of element interactions, we need to make progress in several areas, including theory, measurements, design and interpretation of observation studies and manipulative experiments, and diagnostic and prognostic modeling. These issues were dealt with in detail by one of the crosscutting discussion groups at the workshop (Ollinger et al., Chapter 4, this volume). Here we briefly discuss some of the ways forward.

Theory

There was a clear call for a comprehensive theory that offers a unified perspective on element interactions in ecosystems. Vitousek (Chapter 6, this volume) has proposed that this theory be developed initially through the linkage of the multiple-resource-limitation paradigm from plant and ecosystem physiology (Field et al. 1992; Rastetter, Ågren, and Shaver 1997) with the stoichiometric perspective from biogeochemistry. Yet it is also important to address the contrast that still exists between ecosystem stoichiometry as a guiding principal and the numerous sources of variability that limit the applicability of simple element ratios. What will ultimately be needed is a more basic understanding of the degree to which multiple element interactions follow globally generalizable patterns—across elements, species, and environmental conditions—versus local sources of complexity such as variation in carbon quality and nutrient use among species. One of the great challenges in developing this understanding will be to structure future research so that it scales in both time and space.

Measurements

The development of new measurement techniques often leads to major advances in science. For biogeochemistry the second half of the twentieth century was a period of rapid development of a host of techniques, including gas chromatography, accelerator mass spectrometry, and air-borne and satellite-base scanning near infrared reflectance. These and many other extant measurement techniques have begun to give scientists new insights into element interactions. In the early years of the twenty-first century, scientists are looking for better applications of extant measurement techniques in combination with emerging molecular techniques and nanotechnologies. These combinations should help to link organisms with their effects on element interactions across space scales, from the fine-grain world of microbes and plankton to the coarse-grain world of landscapes, drainage basins, and beyond.

Design and Interpretation of Observation Studies and Manipulative Experiments

Although they have been very useful, simple factorial experiments, statistically interpreted with the aid of analysis of variance techniques, allow only partial understanding of element interactions. For a world in which heterogeneity and complex responses are the norm, it is appropriate to consider experimental studies with a focus on multifactorial/multitreatment manipulations that more closely represent the range of conditions present in natural systems. Results from these studies would enable researchers to derive more complete response surfaces and perhaps give better insights about where there are thresholds for element interactions. These "grand experiments" should be long term (five years to a decade) because biogeochemical responses to manipulations can involve long time lags and dramatically changing patterns over time.

"Grand experiments" cannot and should not be conducted everywhere since both human and financial resources are limited. To facilitate spatial scaling and paradigm confirmation, the intensive "grand experiments" should be combined with less-intensive manipulation experiments and observation sites distributed along broad gradients of the major biogeochemical drivers.

Diagnostic and Prognostic Models

Despite the well-known roles of multiple element interactions in biogeochemical processes in terrestrial and aquatic ecosystems, few biogeochemistry models, either diagnostic or prognostic, include interactions beyond those that involve carbon and nitrogen. As many of the background papers have noted, other macronutrients and micronutrients can also be important to element interactions and the control of ecosystem processes. Clearly this shortcoming in biogeochemistry models must be addressed.

The background paper by Eviner and Chapin (Chapter 8, this volume) and the crosscutting discussion syntheses by Austin et al. (Chapter 2) and Hungate et al. (Chapter 3) sharpen the focus on the linkages between community composition and element interactions. To accurately simulate and project biogeochemcial behavior, ecosystem-level models will have to incorporate these linkages. This task will certainly be aided by the development of a comprehensive theory that offers a unified perspective on energy, water, and elements in ecosystems.

Closing Thoughts

Three important messages kept surfacing at the element interactions workshop discussions at Prague's Charles University: "be careful," "expect surprises," and "take advantage of new interdisciplinary links." The essence of the caution is as follows: When we manage element interactions, we need to place a priority on not making new problems in attempts to solve old ones. Extensive systems-level knowledge used with a unified perspective on energy, water, and elements will help to make us problem solvers rather than problem makers.

Scientists' knowledge of element interactions is far from complete. To date, much of the thinking about element interactions has been based on systems at or near the natural state. As human actions push ecosystems farther and farther from natural conditions, we must expect surprises—new interactions and elements not often considered today could become important.

Finally, many of the workshop's discussions about new approaches to the study of element interactions called for new intellectual partnerships—between molecular biologists and biogeochemists, between geologists and biogeochemists, between engineers and biogeochemists, and between social scientists and biogeochemists. We look to these new partnerships for innovative approaches to some of society's most pressing environmental issues.

Literature Cited

Bolin, B., and R. B. Cook, eds. 1983. *The major biogeochemical cycles and their interactions.* SCOPE 21. Chichester, U.K.: John Wiley and Sons.

Brady, N. C. 1974. *The nature and properties of soils.* 8th ed. New York: Macmillan.

Caraco, N. F., J. J. Cole, P. A. Raymond, D. L. Strayer, M. L. Pace, S. E. G. Findley, and D. T. Fischer. 1997. Zebra mussel invasion in a large turbid river: Phytoplankton response to increased grazing. *Ecology* 78:588–602.

Caraco, N. F., S. E. G. Findley, D. T. Fischer, M. L. Pace, and D. L. Strayer. 2000. Dissolved oxygen declines in the Hudson River associated with the invasion of the zebra mussel *(Dressena polymorpha). Environmental Science and Technology* 34:1204–1210.

CENR (Committee on Environment and Natural Resources). 2000. Integrated assessment of hypoxia in the northern Gulf of Mexico. National Science and Technology Council Committee on Environment and Natural Resources, Washington, D.C.

Coale, K. H., S. E. Fitzwater, R. M. Gordon, K. S. Johnson, and R. T. Barber. 1996. Control of community growth and export production by upwelled iron in the Equatorial Pacific Ocean. *Nature* 379:621–624.

Deevey, E. S., Jr. 1970. Mineral cycles. *Scientific American* 223:148–158.

Diaz, R. J., and R. Rosenberg. 1995. Marine benthic hypoxia: A review of its ecological effects and the behavioral responses of benthic macrofauna. *Oceanography and Marine Biology: An Annual Review* 33:245–303.

Field, C. B., and H. A. Mooney. 1986. The photosynthesis-nitrogen relationship in wild plants. Pp. 25–55 in *On the economy of plant form and function,* edited by T. J. Givnish. Cambridge: Cambridge University Press.

Field, C. B., F. S. Chapin III, P. A. Matson, and H. A. Mooney. 1992. Responses of terrestrial ecosystems to the changing atmosphere: A resource-based approach. *Annual Review of Ecology and Systematics* 23:201–235.

Howarth, R. W., F. Chan, and R. Marino. 1999. Do top-down and bottom-up controls interact to exclude nitrogen-fixing cyanobacteria from the plankton estuaries: Explorations with a simulation model. *Biogeochemistry* 46:203–231.

Jansson, B.-O., and K. Dahlberg. 1999. The environmental status of the Baltic Sea in the 1940s, today and in the future. *Ambio* 28:312–319.

Kolber, Z. S., R. T. Barber, K. H. Coale, S. E. Fitzwater, R. M. Greene, K. S. Johnson, S. Lindley, and P. G. Falkowski. 1994. Iron limitation of phytoplankton photosynthesis in the Equatorial Pacific Ocean. *Nature* 371:145–149.

Marino, R. F., F. Chan, R. W. Howarth, M. Pace, and G. Likens. 2002. Ecological and biogeochemical interactions constrain planktonic nitrogen fixation in estuaries. *Ecosystems* 5:719–725.

Martin, J. H., and S. Fitzwater. 1988. Iron deficiency limits phytoplankton growth in the northeast Pacific subarctic. *Nature* 331:341–343.

Nixon, S. W. 1995. Coastal marine eutrophication: A definition, social causes, and future concerns. *Ophelia* 41:199–219.

Rastetter, E. B., R. B. McKane, G. R. Shaver, and J. M. Melillo. 1992. Changes in C storage by terrestrial ecosystems: How C-N interactions restrict responses to CO_2 and temperature. *Water, Air, and Soil Pollution* 64:327–344.

Rastetter, E. B., G. I. Ågren, and G. R. Shaver. 1997. Responses of N-limited ecosystems to increased CO_2: A balanced nutrition, coupled-element-cycles model. *Ecological Applications* 7:444–460.

Redfield, A. C. 1958. The biological control of chemical factors in the environment. *American Scientist* 46:205–221.

Reiners, W. A. 1986. Complementary models for ecosystems. *American Naturalist* 127:59–73.

Schlesinger, W. H. 1997. *Biogeochemistry: An analysis of global change.* 2nd ed. New York: Academic Press.

Shaver, G. R., W. D. Billings, F. S. Chapin III, A. E. Giblin, K. J. Nadelhoffer, W. C. Oechel, and E. B. Rastetter. 1992. Global change and the carbon balance of Arctic ecosystems. *BioScience* 42:433–441.

Sterner, R., and J. J. Elser. 2002. *Ecological stoichiometry: The biology of elements from molecules to the biosphere.* Princeton: Princeton University Press.

Vitousek, P. M., L. R. Walker, L. D. Whiteaker, D. Mueller-Dombois, and P. A. Matson. 1987. Biological invasion by *Myrica faya* alters ecosystem development in Hawaii. *Science* 238:802–804.

PART I
Crosscutting Issues

2

Human Disruption of Element Interactions: Drivers, Consequences, and Trends for the Twenty-first Century

Amy T. Austin, Robert W. Howarth, Jill S. Baron, F. Stuart Chapin III, Torben R. Christensen, Elisabeth A. Holland, Mikhail V. Ivanov, Alla Y. Lein, Luiz A. Martinelli, Jerry M. Melillo, and Chao Shang

Human activity is having dramatic and unprecedented impacts on global biogeochemical cycles, and these impacts will continue well into the twenty-first century and beyond (Vitousek et al. 1997; Falkowski et al. 2000). These changes are due, in large part, to activities related to food production, urbanization, industrialization, and water management. Changes in the availability of elements coopted for human use result from the introduction of novel forms of elements (i.e., synthetic nitrogen fertilizer), mineralization of organic elements in unavailable forms (i.e., burning of fossil fuel to form CO_2), or changes in rates of natural biogeochemical processes. The effects of these alterations go beyond changes in the availability of a single element due to increased loading of nutrients in the environment; these disruptions radically alter the relationships among elements. Natural variation in element ratios, determined by both biological and geochemical factors including climate, parent material, and time (Reiners 1986; Vitousek, Chapter 6, this volume), combine with the geographical variation in human activity, resulting in a heterogeneous human impact on element interactions in both aquatic and terrestrial ecosystems. Anthropogenic impacts do not operate in isolation, and many of the consequences on element interactions result from the net effect of multiple drivers occurring simultaneously. Because much of the global economy is driven by these alterations of element cycles, the understanding of how human activities affect element interactions at all scales is critical to our ability to evaluate and mitigate human impacts on the landscape in the next century.

In this chapter we review some of the major human drivers affecting element interactions in the next century, focusing on the effects of agricultural intensification and extensification, urbanization, changes in atmospheric chemistry, and changes in species composition. It is not possible to comprehensively review projections for all human-induced global change here, but we focus on those drivers that we expect will have large effects on element interactions. In addition, we explore the effects of the interactions of these drivers, which result in unforeseen consequences for element interactions (Figure 2.1). These interactions result in changes in the net carbon balance of terrestrial ecosystems, impacts from human alteration on transport of particulate matter, climate warming, and changes caused by shifts in species composition.

Anthropogenic Drivers of Change in Element Interactions

Agricultural Intensification and Extent

In the twenty-first century, food production will be of major concern to most areas of the world because of the growing human population and increasing consumption per capita (FAO 2001). In some areas, the amount of agricultural land is projected to increase dramatically in the next century (Tilman et al. 2001), with agricultural expansion primarily in the tropical regions of Africa and Latin America. In Europe and North America, more intensive management, including increased use of fertilizer and pesticides, and land degradation will reduce the land area devoted to agricultural activities (Alcamo et al. 1996; Matson et al. 1997). These changes in land use and technologies will alter the main biogeochemical cycles and consequently element interactions, but their effects will vary depending on the type of agriculture, the original fertility of the soil, and the economic circumstances that determine the level of inputs into the agricultural system (Tilman et al. 2001). The net effect of agricultural activity through disruption of element interactions leads to increased leakage of nutrients from the system, either in flow-through from excessive inputs or from disturbance and subsequent effects on ecosystem nutrient retention. Because of the economic dependence of agricultural activities on subsidy policies, world market pricing, and short-term responses to changes in markets, trends in industrial agriculture are more difficult to predict, but agricultural activity will clearly be one of the most prominent drivers of changes in element interactions in the coming century.

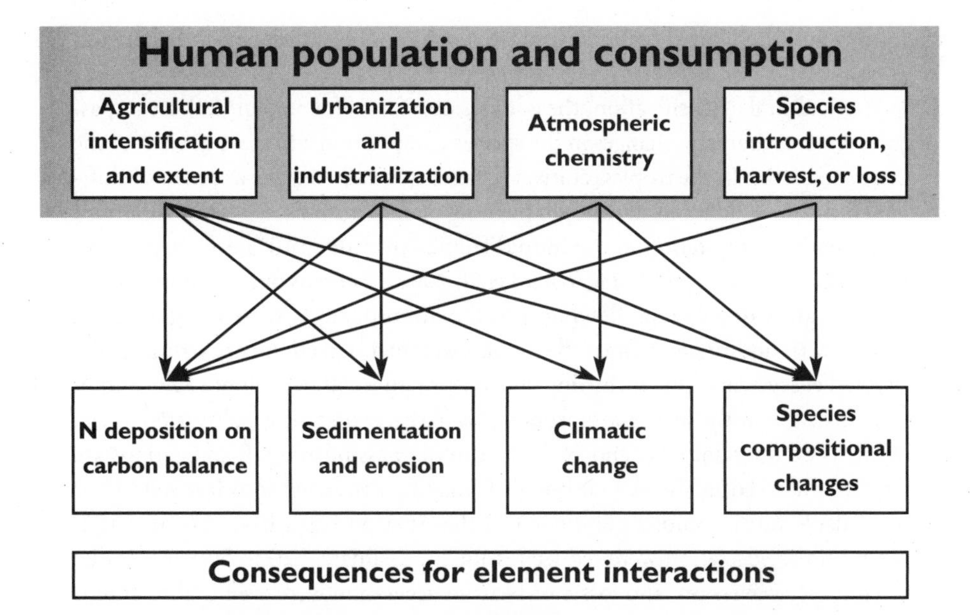

Figure 2.1. Schematic diagram showing anthropogenic drivers of change in element interactions and the resulting consequences from the interaction of multiple drivers. Human population growth and increased consumption per capita in the next century are the basis for change in anthropogenic drivers. The consequences on element interactions results from the interaction of multiple drivers.

Agricultural intensification can cause major alterations in the interaction among element cycles. In the United States and Western Europe, many fields are heavily fertilized with both N and P. Agricultural subsidies and the need to predictably achieve high yields in capital-intensive agriculture result in a tendency to apply excessive fertilizer. In the United States this overfertilization is dramatic, and on average farmers could use 20–30 percent less N fertilizer with little or no decrease in crop yields (Howarth et al. 2002). N and P behave quite differently in the field. Phosphate is highly adsorbed onto soils, unless extremely high fertilization with animal wastes results in P saturation, and so P loss is primarily associated with soil erosion (NRC 2000; Howarth, Walker, and Sharpley 2002). On the other hand, nitrogen readily leaves agricultural fields as dissolved nitrate in both surface and subsurface flows. Globally, the P content of agricultural products has increased by an estimated 25 percent over the past fifty years (Carpenter et al. 1998), with most of this increase occurring in Europe and North America. In contrast, there is little evidence of nitrogen accumulation in agricultural soils (McIsaac et al. 2001; van Breemen et al. 2002), probably because of its greater mobility. Overall, nutrients exported from agricultural fields to surface waters in

Europe and the United States have relatively high N:P ratios, generally between 30:1 and 300:1 (Billen, Lancelot, and Meybeck 1991).

Agricultural extensification, the conversion of land for agricultural use or pasture, involves dramatic changes in the species composition and management of the converted areas. In the tropics, cultivated lands generally require a system of shifting cultivation and removal through burning of aboveground vegetation. Current estimates of deforestation in the humid tropics are 3.09×10^6 hectares per year, primarily for use in agriculture (Achard et al. 2002). Biomass burning is one of the major sources of elements flowing into the atmosphere (Crutzen and Andreae 1990), with additionally large effects on element interactions occurring in the affected ecosystems. Under shifting cultivation, up to 50 percent of soil C and N can be lost, mostly to the atmosphere, over the course of a cultivation cycle. While large fractions of C and N are lost in biomass burning, P and cations are largely conserved in the soil, although P may be converted into less soluble or unavailable forms. Studies conducted in the Amazon basin have shown that C stocks in the vegetation were reduced from approximately 20 to 56 percent due to land use conversion, and losses of N from vegetation were even higher, reaching 70 percent (Kauffman et al. 1995; Hughes, Kauffman, and Cummings 2002). Additionally, soil C and inorganic N cycling (mineralization, nitrification, and N_2O emissions) are reduced (Martinelli, Chapter 10, this volume). Movement from shifting cultivation, where land is burned, converted, and then abandoned, to agroforestry, using woody species that are climatically appropriate for the region and residue management alternatives, is aimed at reducing nutrient leakage from low-input agricultural systems (Moldan et al., Chapter 5, this volume). In addition, leguminous-based cropping systems show great potential to restore the coupling of C and N cycling and are increasingly being used in soil conservation efforts (Drinkwater, Wagoner, and Sarrantonio 1998).

Agricultural activity also alters other element cycles and interactions in ways that are regionally variable and depend on local climate, soils, and agricultural practices. Low-input agriculture in tropical soils decouples element ratios inherited from native vegetation and soil organic matter, primarily by extracting elements with low availability, which are then removed in the harvested biomass. In contrast, rice cultivation in anaerobic soils leads to a different pattern of change in element interactions. Paddy soils are flooded and fully waterlogged for four months a year on average. During this time the anaerobic conditions in the soil increase the availability of P to rice plants, reducing the need to add P fertilizer (these soils are generally P-deficient under aerobic conditions), but conversely release both N and CH_4 to the atmosphere. The development of high-yielding rice varieties in the 1970s and the commercial availability of N fertilizers have led to a large increase in rice production over the past decades. Rice production is expected to grow by an additional 60 percent by the year 2020, but the rate of increase in yield is not expected to keep pace with projected population growth and demand (Cassman 1999). Thus the need to increase N fertilizer use will further impact element cycles, as a doubling of NH_3 emissions is predicted by 2030 for East Asia (Klimont et al. 2001; Zheng et al. 2002).

Urbanization and Industrialization

At the beginning of the twenty-first century, urban areas cover about 2 percent of the Earth's land surface and are home to about 50 percent of the world's 6 billion people (UNEP 2002). Over the next century the world's urban population is expected to grow by 2 percent per year (United Nations Population Division 2002). In much of the developing world over the past decade, there has been a net migration from rural areas into cities, and this trend is likely to continue into the future (Tibbets 2002). Conservative estimates are that within thirty years, cities will cover 3–4 percent of the land surface of the planet, due to both the expansion of existing cities and the creation of new urban areas (WRI, UNEP, and World Bank 1996). Over this time period the extent of urban areas is expected to be even higher in Asia and the Pacific, with up to 6 percent of the land area being urban (UNEP 2002). The spatial "footprint" of this urbanization will extend far beyond the actual area occupied by urban zones.

Cities and industries emit large amounts of many elements to the atmosphere and to surface waters, and these releases occur in element ratios that are far different from those that flow through natural ecosystems or from agricultural runoff. For example, compared with natural systems, releases from urban and industrial environments are enriched in a variety of metals instead of nitrogen or phosphorus. These element releases are under technological control and will vary depending upon the technologies employed.

The major releases of nutrients from cities to surface waters come from sewer systems and are dominated by human waste. In New York City, for example, the nitrogen and phosphorus loads to the Hudson River from sewage treatment plants are some thirtyfold greater than are the inputs from street runoff and storm sewers (Howarth et al. 2003). The approach to sewage treatment, therefore, is a prime determinant in the stoichiometry of element releases from cities to surface waters. Whereas 90 percent of urban sewage is treated in North America, only 66 percent is treated in Europe, 35 percent in Asia, 14 percent in Latin America and the Caribbean, and 0 percent in Africa (World Health Organization 2000; Martinelli, Chapter 10, this volume). Untreated sewage has on average a relative composition of organic carbon and nutrients of C:N:P = 12:5:1 (NRC 1993). Many cities in developed countries provide secondary sewage treatment, which is designed to reduce the organic carbon loading by an order of magnitude but generally reduces nitrogen by only one-third and phosphorus by one-half. The resulting C:N:P stoichiometry is 2:6:1. Further treatment for nutrient removal occurs in some cities but is far less common. This tertiary treatment can result in a C:N:P stoichiometry of 4:2:1 (NRC 1993).

In addition to altering the element ratios of carbon and nutrients released to surface waters, the extent of sewage treatment can have major consequences for the processing of materials in the receiving surface waters in ways that further alter the element ratios (Daniel et al. 2002). Where cities release untreated sewage into rivers, as occurs in most of the developing world, the high organic carbon loads stimulate respiration. Most of the dissolved oxygen is consumed, leading to anoxia and promoting production of CH_4 and N_2O (Ballester et al. 1999). In addition, under anoxic conditions, nitrification is inhibited and direct denitrification of nitrate is enhanced. Therefore, nitrate concentrations are lowered, and ammonia accumulates in the water (Martinelli et al. 1999). Because the nitrification-denitrification couple is broken, total N levels may remain higher than in rivers receiving sewage with more advanced forms of treatment, where oxygen concentrations typically remain much higher (NRC 1993).

Atmospheric Chemistry

The projections of human-driven emissions to the atmosphere over the next 100 years follow quite different trajectories depending on the predicted changes in socioeconomic factors affecting the magnitude of human impact (IPCC 2000). These socioeconomic factors include the degree of globalization versus regionalization and the relative importance of material versus social and ecological values (Table 2.1). The *Special Report on Emissions Scenarios* (SRES) of the Intergovernmental Panel on Climate Change (IPCC 2000) reflects the state of our knowledge on these projections, providing estimates of how the emissions and their elemental balance will change in the next century. The increased emissions to the atmosphere from human alterations of C, S, and N cycles have multiple effects on terrestrial and aquatic ecosystems, including acidification, soil fertility loss, and downstream eutrophication (Galloway, Chapter 14, this volume). These scenarios suggest that there will be substantial changes in the elemental composition of emissions due to human activity but highlight that the nature and extent of change will be highly dependent on economic development, population growth, public policy, and technological progress.

Predictions in all scenarios indicate that SO_2 emissions will decline in the next century, an outcome that has important implications for element interactions. Acidification due to sulfate deposition has been a concern over the past several

Table 2.1. Projected percentage increases in human-driven C, N, and S emissions between 2000 and 2100

	SRES scenario					
Compound	*A1FI*	*A1B*	*A1T*	*A2*	*B1*	*B2*
CO_2-fossil fuel	405	118	-28	382	-13	130
CO	192	89	136	165	-59	128
NMVOCs	202	40	-8	146	-37	22
CH_4	137	-7	-12	187	-24	93
NO_x	256	29	-9	253	-39	97
N_2O	148	4	-19	146	-15	3
SO_2	-44	-61	-72	-15	-65	-32

Source: Calculated from SRES scenarios (IPCC 2000).

Note: Compounds are carbon dioxide (CO_2), carbon monoxide (CO), nonmethane volatile organic carbon (NMVOC), methane (CH_4), nitrogen oxide (NO_x = NO + NO_2), nitrous oxide (N_2O), and sulfur dioxide (SO_2). The A1 family of scenarios has rapid economic growth, a population stabilizing in midcentury, and alternative developments of energy technologies: A1FI (fossil fuel intensive), A1B (balanced), and A1T (predominantly non–fossil fuel). The A2 scenario includes projections of high population growth in a heterogenous world of regional economic development and slower technological change than other scenarios. The B1 scenario follows the same population growth patterns as A1, with a service and information economy, reductions in intensity of material use with the introduction of clean, resource efficient technologies, and sustainability. The B2 scenario has a slower population growth and relies on local and regional balancing of economic growth and environmental solutions.

decades, but acid precipitation inputs will shift from being dominated by sulfate acids to nitric and organic acids. The reduction in sulfate aerosols and increase in organic aerosols will reduce reflectance, which may enhance climatic warming, but the net change in acidification is expected to be relatively small. Although the SRES scenarios do not include NH_3 emissions, ammonia will become a larger contributor to increasing N deposition inputs due to the increases in animal production and agricultural intensification necessary to sustain a growing human population and the projected global increase in per capita meat consumption (FAO 2001; Howarth et al. 2002). This shift in the dominant form of anthropogenic input has important implications for element interactions, as S and N play fundamentally different roles in terrestrial and aquatic biogeochemistry.

Differences in fuel source emissions of NO_x from human activity result in atmospheric chemistry changes that have regional and global implications for element interactions. NO_x emissions from Europe and North America are largely fossil-fuel based and increase the presence of OH^- radicals in the troposphere. These emissions serve as a precursor for the production of tropospheric O_3. In contrast, NO_x emissions in South and Southeast Asia are proportionally much smaller than the CO emissions that are produced primarily by biofuel and biomass burning. CO emissions have the opposite effect from NO_x emissions, actually reducing the oxidative power of the atmosphere (Lelieveld et al. 2001). Tropospheric O_3 has a high estimated global warming potential (0.35 +/- 0.15 watts per meter squared [Wm^{-2}]), and ranks among the top five greenhouse gases. Nonetheless, reducing the quantities of atmospheric OH^- as a result of biomass burning severely constrains the potential to scrub gases from the atmosphere, thereby increasing the lifetime of methane (CH_4), another important greenhouse gas. The nature of these regional feedbacks (production of O_3 and residence time of CH_4) reflects the changing balance in emissions due to regional differences in anthropogenic activities (Prather et al. 2001; Holland and Carroll, Chapter 15, this volume).

Species Introduction and Removal

Human activities have profoundly altered the species composition of the biosphere by transporting species to new locations and by altering the global environment in ways that eliminate or change the abundance of naturally occurring species (Chapin, Sala, and Huber-Sannwald 2001). Many of these species changes have effects on element interactions that are just as large as direct effects of human actions on biogeochemical cycles. In addition, extensive harvest of natural populations in the ocean and on land has greatly altered the species composition of ecosystems and increased rates of species loss (Chapin et al. 2000). Humans are responsible for the extinction of 5–20 percent of many groups of organisms, and current rates of extinction are 100–1,000 times greater than before the Industrial Revolution (Pimm et al. 1995). The rates of species change are projected to continue at a high rate through the twenty-first century, with the relative importance of the drivers of change differing among biomes (Sala et al. 2000; Chapin, Sala, and Huber-Sannwald 2001). The effects of these changes on element interactions are idiosyncratic depending on the system and the type of transformation. The effects and feedbacks from element cycling are discussed more fully in the following section on consequences of interacting anthropogenic drivers.

Consequences of Interacting Anthropogenic Drivers on Element Interactions

Many of the most important environmental consequences of human activities expected in the twenty-first century will be mediated by changes in the interactions of element cycles. In this section we briefly summarize some of these changes that cannot be predicted without an understanding of how multiple anthropogenic drivers together result in consequences that may not have been predicted when considering a single driver alone.

Effects of N Deposition on Net Carbon Balance

Net carbon balance of ecosystems has attracted considerable attention in recent years because of the potential for ecosystems to act as sinks or sources for atmospheric CO_2. The net balance of carbon, however, is tightly controlled by a complex set of processes involving numerous element interactions. Recent research on the effect of nitrogen deposition on ecosystems has shown that the responses are much more complex than the simple stimulation of carbon accumulation by N fertilization that was initially hypothesized (Nadelhoffer et al. 1999; Matson, Lohse, and Hall 2002). Current views suggest that changes in element interactions play a key role in explaining the effects of N deposition on trace gas fluxes and feedbacks to climate (Figure 2.2). The direct effect of N deposition on net primary production (NPP) depends on the form of N deposited and its interaction with other pollutants. NO_y, which is taken up by the canopy and stimulates NPP in N-limited systems, has a geographic distribution almost identical to that of tropospheric ozone, which is also absorbed by the canopy and can negatively affect NPP (Ollinger et al. 2002). The parallel distribution of NO_y and O_3 results from the role that nitrogen oxides play in generating ozone. The net balance of these effects may be nearly neutral except at extremely high pollutant levels. The actual effect may depend on ecosystem type and environmental stress, both of which influence stomatal conductance and canopy uptake. N deposition as NH_3 acts primarily as an N fertilizer. This fertilization effect depends on the degree of N saturation of the system. In low-N ecosystems, NPP is stimulated; as N loading increases, nitrified N leaches from the ecosystem, carrying with it cations from soil surfaces and humic complexes; at high levels of N, the negative effects of cation loss exceed the positive effects of N fertilization, causing declines in NPP (Aber et al. 1998; Driscoll et al. 2001). In Europe, where both NO_y and NH_3 are important in deposition, the net effect appears to be positive (Ågren, pers. comm.), whereas in North America, where NO_y dominates, the net effect of deposition on NPP is often negative (Aber et al. 1998; Driscoll et al. 2001).

The effect of N deposition on the net carbon balance depends on the relative effects on NPP and the decomposition of litter and soil organic matter (SOM). The rate of buildup of a soil organic matter layer depends not only on the amount and quality of the litter formed but also on the rate and pattern of decomposition as constrained by climatic factors (Schlesinger 1977; Austin 2002). The differential effects of increased N inputs on ecosystem processes may be complex. Comparisons of litter decomposition across and within ecosystems generally show a positive relationship between litter N content and decomposition rate (Vitousek 1982; Coûteaux, Bottner, and Berg 1995), although carbon chemistry and its interaction with litter N may play an equally important role in determining decomposition rates (Melillo et al. 1982; Austin and Vitousek 2000). In addition, the effect of litter N content on decomposition may depend on the time scale considered. In some studies, litter with high N concentrations decomposes rapidly at first and then approaches an upper limit because of an accumulation of a large recalcitrant fraction of SOM (Berg et al. 2001). In contrast, litter with lower N concentrations decomposes slowly at first but eventually loses more mass (Berg and Ekbohm 1991; Berg et al. 1996; Coûteaux et al. 1998).

The effects of N fertilization on litter decomposition are also mixed. N fertilization led to increased decomposition in prairie, pine forests, and alpine meadows (Hunt et al. 1988; Arnone and Hirschel 1997) but had no effect on decomposition in jack pine forests (Prescott, Kishchuk, and Weetman 1995), tropical forests (Hobbie and Vitousek 2000), or peatlands (Aerts et al. 1995). In contrast, N fertilization decreased decomposition in temperate forests (Magill and Aber 1998; Carreiro et al. 2000). The complexity of response in litter decomposition to increased N deposition may be tied to the carbon chemistry of the litter and to changes in the efficiency in the microbial community (Ågren, Bosatta, and Magill 2001). For example, in tropical soils N stimulates the decomposition of low-lignin litter, but this N effect weakens as the litter lignin concentration increases (Hobbie 2000). This result is consistent with enzymatic studies showing that N addition decreases the activity of lignin-degrading enzymes (phenol oxidase) while stimulating the activity of some enzymes (e.g., cellulase) that degrade more labile compounds (Carreiro et al. 2000).

Taken together, these results suggest that N deposition may either stimulate or have complex effects on decomposition, and the effects on litter decomposition may not translate directly to the accumulation of soil organic matter because of the changes in decomposition dynamics over time. Together with the uncertain effects of N deposition on NPP, this finding indicates an insufficient understanding of C:N interactions in plants and microbes in terrestrial ecosystems. Consequently, firm conclusions about the net effect of N deposition on ecosystem carbon balance in recent decades or in the future are not yet possible.

In contrast to the uncertain effects of N deposition on net CO_2 flux, there is clear evidence that N deposition, in some ecosystems, decreases soil methane oxidation, augmenting CH_4 concentrations in the atmosphere. Dry forest and grassland soils globally take up about 30 teragrams (10^{12} g) CH_4 per year, and N deposition decreases atmospheric CH_4 consumption rates, particularly in temperate

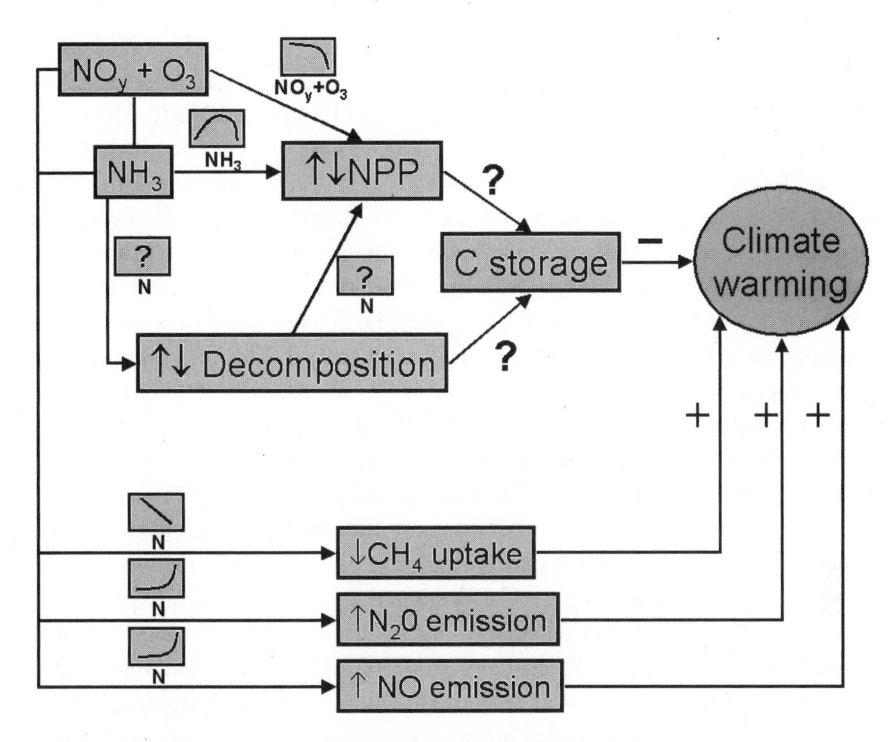

Figure 2.2. Schematic representation of the effects of nitrogen deposition on the carbon balance and the feedbacks to climate warming in terrestrial ecosystems. Deposition in the form of NH_y and NO_x are shown on the left with the hypothesized effects on carbon turnover and trace gas emissions. Question marks indicate a high level of uncertainty about the direction of change due to anthropogenic N deposition.

forests (e.g., Steudler et al. 1989; Schnell and King 1994; Fenn et al. 1998). Decreased CH_4 uptake contributes to atmospheric warming (Figure 2.2).

Further, N deposition also increases fluxes of N_2O, a greenhouse gas, and of NO, a gas that catalyzes the formation of tropospheric ozone (O_3), another greenhouse gas. The N_2O emissions depend on the initial nutrient status of the soils. N-saturated soils are most likely to increase their rates of nitrification and denitrification and hence gaseous losses of N. If nitrogen fertilizers were applied to forests at current agricultural rates, this would induce an emission on the order of 5 kilograms (kg) N_2O-N per hectare per year (Groffman et al. 2000), which has a greenhouse warming potential equivalent to 2,300 kg CO_2 per hectare per year. In perspective, these rates could change about half of the twenty-six measurement-years in the EUROFLUX project (Valentini et al. 2000) from sinks (if only CO_2 is considered) to sources of greenhouse warming (Christensen and Keller, Chapter 13, this volume).

Box 2.1. Fertilizing the forest may not fill the carbon sink

One possibility for increasing the carbon sink in Europe and North America is to fertilize large areas of temperate and boreal forest with nitrogen (Schultze et al. 2001) to enhance productivity (Bergh et al. 1999) and carbon storage through increased humus formation (Berg and Meentemeyer 2002). A simple model projection of the accumulated sink strength of carbon in a temperate Northern Hemisphere forest with and without nitrogen additions in the twenty-first century is shown in this figure. The figure is based on a gradually diminishing Northern Hemisphere sink strength of 0.7 Mg per hectare per year as a compromise between current estimates from Europe (Valentini et al. 2000; Schultze et al. 2001) and North America (Goodale et al. 2002). The estimated effect of trace gases on the radiative forcing is expressed as C equivalents (global warming potentials [GWP] used are CO_2 = 0.27; CH_4 = 5.7; N_2O = 84.5), allowing for a comparison of the combined effects of different elements (CO_2, CH_4, and N_2O) under background conditions and in a situation where fertilizer is added to increase the C sink strength by approximately 30 percent. Soil response is based on a generalized N_2O emission model that is nonlinear and includes a threshold effect, here estimated at 10 years, where increased N_2O emissions are initiated in response to soil N saturation (Moldan et al., Chapter 5, this volume). Following saturation, N_2O emissions are gradually increased to the documented level of 3–5 kg of N_2O-N per hectare per year for fertilized (or high deposition) forests (Papen and Butterbach-Bahl 1999).

(continued on next page)

The net effect of these changes in trace gas flux is uncertain. Changes in CH_4 and N_2O flux contribute positively to greenhouse warming (see Box 2.1), and the net changes in CO_2 flux, which were previously considered the dominant interaction between N deposition and climate feedbacks, are uncertain. Resolving this uncertainty requires an improved understanding of the effects of N deposition on decomposition and on various indirect effects, including grazing, fire, and water balance (Hungate et al., Chapter 3, this volume).

Sedimentation and Erosion

Human activity causes both the accumulation and the liberation of particulate matter, causing dramatic changes in element interactions. The disruption of the natural pathways of water movement has profound effects on element interactions in systems that are directly manipulated, as well as downstream, coastal, and ulti-

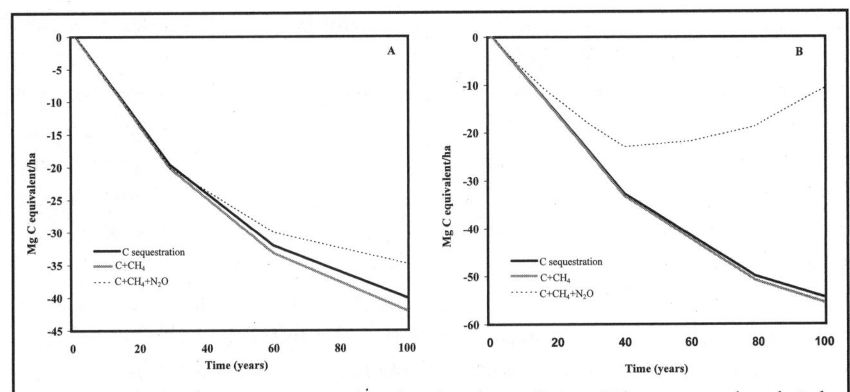

Panel A shows a seminatural unfertilized condition. The accumulated sink in 100 years is 40 Mg of C per hectare; this figure increases by 5 percent if CH_4 uptake is included but falls by 13 percent if both CH_4 and N_2O fluxes are included. Panel B shows the same forest under conditions of continued fertilization. Carbon uptake is increased by 35 percent and the increase by including CH_4 is reduced to 2 percent (since N addition generally reduces CH_4 uptake; Christensen and Keller, Chapter 13, this volume). The main effect, however, is the marked reduction in carbon uptake, by more than 70 percent, when the effect of N_2O emissions is included. This simple projection shows that the multiple element interactions cause fertilization to be a counterproductive measure for reducing greenhouse forcing in situations where the soil N_2O emissions respond as modeled here (Moldan et al., Chapter 5, this volume).

mately ocean systems (Naiman and Turner 2000; Jackson et al. 2001). Land use change has also increased the liberation of particulate matter through increased soil erosion (Schlesinger et al. 1990). Changes in movement of particulate matter result in important changes in element interactions due to the transport of nutrients to nonadjacent ecosystems, as well as to the limitations created when critical elements do not reach downstream ecosystems (Ittekkot et al., Chapter 17, this volume).

There are now more than 40,000 large dams on the Earth's rivers to store water for human needs, and reservoirs worldwide hold an estimated 6,600 cubic kilometers (km^3) per year (Postel 1999). Humans appropriate more than 50 percent of global freshwater for their use, 75 percent of which is used for irrigation in agriculture (Gleick 2000; Baron et al. 2002). The number of dams and reservoirs will increase in the next century because of the increasing need for freshwater from a growing human population and economic development (Vörösmarty et al.

2000), with a projected increase of 150–200 dams per year over at least the next two decades (Jackson et al. 2001).

The disruption of element ratios by these novel reservoir ecosystems has consequences for downstream zones. Sediment impoundment behind dams can also affect element interactions between silica, carbon, and nitrogen. Elevated concentrations of N, P, and Si from anthropogenic inputs create a nutrient-rich environment in reservoirs where diatom productivity is enhanced (Ittekkot et al., Chapter 17, this volume; Filoso et al. in press). Increased inputs of N and P both behind dams and in downstream ecosystems can cause algal blooms because of enhanced productivity, detrimentally affecting the fisheries in the coastal areas (Rabalais et al. 1996). Diatom blooms in lakes and reservoirs result in silica burial and sequestration behind dams, markedly restricting the silica supply to downstream and coastal systems. The silica flux down the Mississippi River has decreased by approximately 50 percent over the past half-century (Rabalais et al. 1996), and silica flux to the Rhine River and silica inputs to the Baltic Sea and Black Sea have also decreased (Billen, Lancelot, and Meybeck 1991; Ittekkot et al., Chapter 17, this volume). Farther downstream, the supply of silica in coastal ecosystems can decrease even more as a result of increased sedimentation of diatoms, as increased nitrogen loads stimulate production (Wulff, Stigebrandt, and Rahm 1990; Conley, Schelske, and Stoermer 1993). The depletion of silica relative to nitrogen and phosphorus can change the composition of the phytoplankton community, away from diatoms and toward green algae and dinoflagellates, altering the food webs and trophic structure of coastal ecosystems (NRC 2000).

The increased movement of soil particles due to wind erosion and transport creates a source-sink phenomenon, in which impacted ecosystems suffer a reduction in fertility due to loss of sediments, while adjacent or downwind ecosystems experience larger inputs of wind-borne particulate matter, primarily the rock-derived nutrients phosphorus, cations, and iron. This transport of material alters element interactions of both terrestrial and aquatic ecosystems (Schlesinger et al. 1990; Ducklow, Oliver, and Smith, Chapter 16, this volume). The Sahara Desert alone is estimated to emit about 1 billion tons of dust per year (D'Almeida 1986), and increased dust loads from this region over the past fifteen years have been linked to changes in the North Atlantic oscillation (Moulin et al. 1997).

Even small but persistent inputs of dust provide essential nutrients for plant growth (Chadwick et al. 1999), but alteration of dust inputs and losses are more likely to affect the element cycling of arid and semiarid ecosystems, where the impact on soil erosion will be largest. Land use and climate interact with soil stability to influence the potential for wind erosion. For example, cryptobiotic crusts, an association of lichens, mosses, and cyanobacteria, are common ground cover in unvegetated areas in arid regions of the world but are quite susceptible to disturbance by grazing, agriculture, and recreational use (West 1990). Once disturbed, these crusts can take decades to recover, and dust losses can be significant. As soil erosion increases, the degradation of the crusts decreases N input into these ecosystems, altering N availability and decreasing soil fertility (Evans and Ehleringer 1993). Although agricultural practices in the developed world have reduced erosion over the past century, development pressures continue to build in Africa and Asia, with resulting increased potential for wind erosion of surface soils.

Long-range transport and subsequent deposition of nitrogen, phosphorus, and iron in the surface layer of the ocean from terrestrial ecosystems can potentially stimulate productivity and affect element interactions in numerous ways (Falkowski, Barber, and Smetacek 1998). However, critical to the response of oceans to increased dust inputs from human activity is the natural geographic variation in element limitation in the oceans (Moore et al. 2001). For example, nitrogen availability is considered to limit productivity of diatoms and small phytoplankton in the North Atlantic Ocean. Present atmospheric inputs of around 250–500 milligrams (mg) N per square meter per year from North American emissions (Galloway and Cowling 2002) can therefore stimulate primary production in the North Atlantic. In contrast to the North Atlantic, the North Pacific Ocean (north of the Kuroshio Current) is iron-limited (Martin and Fitzwater 1988; Moore et al. 2001); addition of nitrogen without iron will likely have little immediate effect. Pacific aerosols of Chinese origin may indeed be enriched with both N and Fe as a result of eolian dust inputs, and the co-deposition of both elements may therefore enhance primary production by reducing both Fe and N limitation (Ducklow, Oliver, and Smith, Chapter 16, this volume).

Climate Change

Climate change due to anthropogenic forcing, particularly warming, will increase in the next century (Ramaswamy et al. 2001). Through several feedback processes including element interactions, the consequences on climate change may have important direct and indirect effects on atmospheric chemistry and energy exchange.

Climate Change and Nitrogen Fixation in the Oceans

Over geological time scales, nitrogen fixation in the oceans maintains a balance between nitrogen and phosphorus near the Redfield ratio of 16:1. Phosphorus is not believed to vary much in time compared with N, and the factors that regulate P availability in oceans are not well understood (Howarth et al. 1995).

Nevertheless, the quantity of phosphorus in the oceans thereby determines the quantity of nitrogen as well, and phosphorus is viewed as the master element regulating primary production. At shorter time scales, however, portions of the world's oceans can be slightly depleted in nitrogen relative to phosphorus, and primary production is therefore limited by nitrogen. Recent evidence has suggested that the rates of nitrogen fixation and the ability of nitrogen fixation to bring nitrogen levels up to the Redfield ratio can vary over the time scale of years to decades (Karl et al. 2002).

In both the Sargasso Sea of the North Atlantic (Michaels et al. 1996) and in the subtropical gyre of the North Pacific Ocean near Hawaii (Karl et al. 2001), rates of nitrogen fixation appear to have substantially increased over the past few decades. For the North Pacific the most likely driver behind this change may be climatic (Karl et al. 2002). Between the 1970s and the 1990s, stratification of the North Pacific gyre increased. The increased stratification may provide conditions more favorable for nitrogen fixation by changing the relative proportion of nitrate to iron in surface waters. Iron is supplied to surface waters from dust inputs and from upwelling or mixing of deep ocean waters. When the upwelling/mixing term is relatively large, significant quantities of nitrate are also supplied to the surface waters, and the relatively high nitrate-to-iron ratio may not be conducive for high rates of nitrogen fixation, since the nitrate can suppress fixation (Karl et al. 2002). As an ocean becomes more stratified, however, the upwelling/mixing term is decreased, and the atmospheric input of iron in dust becomes more important. Under these conditions, the nitrate-to-iron ratio is lower, and with less suppression from nitrate and more stimulation from iron, nitrogen fixation rates may increase substantially (Karl et al. 2002). Thus, increased stratification from a climatic oscillation or climatic warming, combined with the input of iron in dust, can "serve to decouple the otherwise linked C-N-P-Si cycles in the sea" (Karl et al. 2002).

Climate Change Effects on Tundra Ecosystem Element Interactions

Climatic warming is expected to be most pronounced at high latitudes, where there are large deposits of soil organic matter, which is currently a significant source of atmospheric CH_4 (Christensen and Keller, Chapter 13, this volume). Differential effects of climatic warming on the productivity and decomposition of these ecosystems could cause climatically important changes in carbon storage and trace gas exchange. The direction and timing of these responses to warming will depend on changes in the soil moisture regime and in element interactions. Long-term warming experiments show an increase in N availability that probably results from enhanced respiration and loss of soil C (Chapin et al. 1995). Absorption of this available N causes an increase in shrub growth (Chapin et al. 1995; Hobbie and Chapin 1998), which may enhance net ecosystem carbon gain because of the transfer of nitrogen from the soil, which has a low C:N ratio, to woody plants with a higher C:N ratio (Shaver et al. 1992).

Permafrost disintegration caused by climatic warming (Brown, Hinkel, and Nelson 2000) may also affect element interactions in tundra ecosystems. Changes in permafrost distribution in subarctic Sweden cause mires to shift from ombrotrophic moss and shrub-dominated systems to minerotrophic wet vascular plant-dominated systems (Malmer and Wallén 1996; Svensson et al. 1999). This shift, in turn, leads to a significant lowering of soil redox potentials, an increase in anaerobic decomposition, and, consequently, increased carbon emissions as methane. Wet minerotrophic soils and vegetation are associated with the highest methane emissions in subarctic and Arctic tundra environments (Joabsson and Christensen 1999; Christensen et al. 2000; Joabsson and Christensen 2001; Nykänen et al. in press). Discontinuous permafrost regions are considered some of the most vulnerable to climate warming, so with the predicted warming over the next 100 years, effects such as the one described on changes in element interactions are expected to be strong.

Changes in Species Composition

Changes in species diversity and composition can be triggered by many human-induced changes, including the introduction of exotic species, overharvest of selected species, and changes in land use, climate, atmospheric composition, and trophic interactions (Figure 2.3; Sala et al. 2000). Regardless of the cause of species change, those changes that alter the interactions among element cycles often have large impacts on ecosystems. Changes in element cycles, in turn, alter the availability and balance of resources available to organisms, and these resource changes cause further modifications in species composition (Figure 2.3). This feedback between species composition and element interaction causes ecosystem alterations that are often just as large as the direct effects of human action on biogeochemical cycles. Species changes have a broad array of effects on element interactions that are specific to the nature of the species that change in abundance (Eviner and Chapin, Chapter 8, this volume). The universal consequence, however, is that species changes with large effects on element cycles always have large ecosystem effects. In this section we present a few of the myriad examples of these effects of species composition on changes in element cycles, organizing our examples according to the driver that precipitated the species change (Table 2.2).

Table 2.2. Effects of species changes on element interactions in terrestrial and aquatic ecosystems

Human impact	*Example*	*Element interaction*	*Reference*
Species introduction	*Myrica faya*, Hawaii	↑ N input	Vitousek et al. 1987
	Zebra mussel, N and S America	↓ N:P ratios	Caraco et al. 1997
Land use change	Reducing fallow cycle, India	↓ N and K	Ramakrishnan et al. 1992
	Aquatic plants, S America	↑C:O ratios	Fearnside 2002
Nutrient enrichment	N deposition and mycorrhizae		Bowman and Steltzer 1998
Trophic interactions	Overgrazing by domestic livestock	↓N, cations, changes in C:N	Evans and Ehleringer 1993
	Overharvesting	Changes in N:P	Marino et al. 2002
Climate change	Shrub encroachment	Changes in C:N	Hobbie 1996

Species Introductions

Introduction of exotic N fixers to environments where N fixers were previously uncommon, such as the introduction of *Myrica faya* in Hawaii (Vitousek et al. 1987), can radically alter element cycling and interaction in ecosystems. Increases in N input to ecosystems reduces the degree of N limitation and shifts the ecosystem toward limitation by other elements such as P, causing a decline in the C:N ratio of live tissue. This augments herbivory and can cause declines in the litter C:N ratio, which enhances decomposition. Another example of the introduction of an N-fixing species results from the encroachment of woody N-fixing shrubs in semiarid grasslands with high domestic grazing intensity. These shrubs can increase N mineralization fivefold and increase storage of both C and N in soils (Hibbard et al. 2001).

Species introductions have caused equally large changes in element interactions in aquatic ecosystems. Before the introduction of the zebra mussel *(Dresenna polymea)* to the Hudson River, biological water filtration, primarily by zooplankton, filtered the water approximately every fifty days. After one year the zebra mussel filtered the entire water column every two to three days. This rapid filtration caused a removal of 80–90 percent of algal biomass, a 15 percent decline in summertime dissolved oxygen concentrations, and doubled PO_4 concentrations

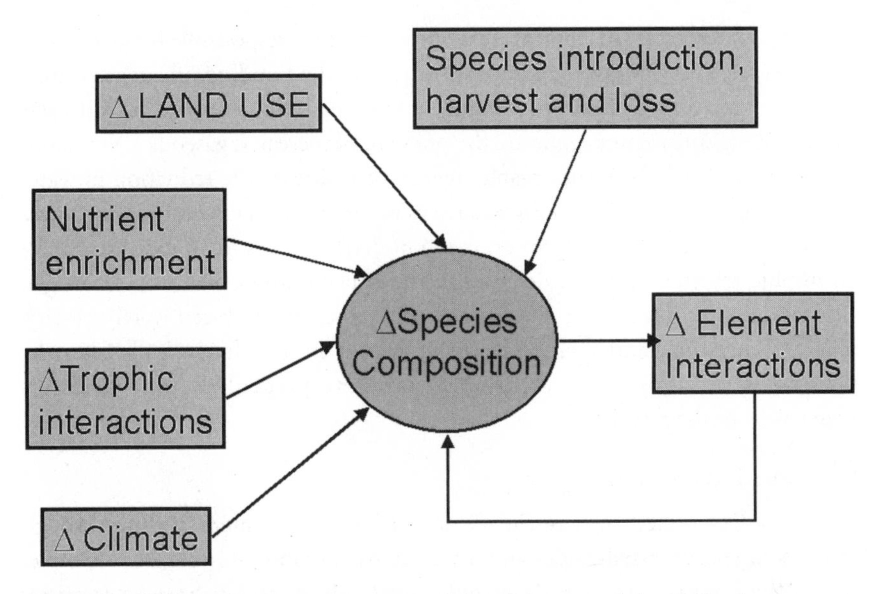

Figure 2.3. Drivers affecting species compositional change and the feedbacks to element interactions. Many human impacts have consequences for changes in species composition, which then affect element interactions. These changes in element ratios can then feedback to cause subsequent changes in species composition.

(Caraco et al. 1997; Caraco et al. 2000). Zebra mussels have had similar effects throughout North and South America.

Land Use Change

Changes in land use are projected to be the greatest cause of changes in species composition in the twenty-first century (Sala et al. 2000). In northeast India pressure from an increasing human population has reduced the length of the slash-and-burn cycle from thirty-plus years to less than ten years (Ramakrishnan 1992). In a thirty-year rotation the fallow period is long enough for N fixers to replenish N pools and for potassium accumulators such as the native bamboo and other species to accumulate enough nutrients to balance the erosional losses that occur during the agricultural phase. The resulting N and K accumulation reduces the nutrient limitation of NPP, leading to the accumulation of soil organic matter that is critical to water and nutrient retention. As the rotation period becomes shorter, the landscape is increasingly dominated by agricultural land and no longer has seed sources of element-accumulating species. With a rotation shorter than eight to ten years, soil fertility declines precipitously, causing a shift to intensive agriculture that requires external nutrient inputs.

Artificial reservoirs in tropical areas of the world are responsible for large emissions of CO_2 and CH_4 to the atmosphere, due to the flooding of extensive areas of dense vegetation. The constant inputs of untreated urban sewage and diffusive sources of organic matter aggravate the problem of increased gaseous C emissions due to these novel inputs of organic matter. In addition, the reduction in water velocity allows floating macrophytes such as water hyacinth to colonize large areas of these reservoirs. Their large production of detritus leads to an increase in heterotrophic respiration and a decrease in the concentration of dissolved oxygen (Fearnside 2002; Lima et al. 2002). Taken together, the reduced water velocity, plant colonization, and increased organic matter inputs dramatically alter the C:O ratios, resulting in increased CH_4 emissions, particularly in plant-dominated areas of the reservoir.

Nutrient Enrichment

Recent and projected increases in N deposition alter element cycling through changes in species assemblages and associations. In alpine tundra, for example, replacement of the dominant *Acomastylis rossii* by the grass *Deschampsia caespitosa* leads to an eightfold increase in net N mineralization and nitrification rates, which in turn increases groundwater loss of NO_3^- to aquatic ecosystems (Bowman and Steltzer 1998). The shift in species reduces inputs of phenols and lignin in soils, thus increasing microbial activity. Another effect of N deposition observed in several fertilization experiments is the reduction of arbuscular and ectomycorrhizal fungi, observed in the Czech Republic (Titus and Leps 2000), in transects away from an N point source (Lilleskov et al. 2002), in grasslands in the central United States (Corkidi et al. 2002), and in sage scrub communities of southern California (Egerton-Warburton, Graham, and Allen 2001). The loss of mycorrhizal associations affects species composition by selecting for plants with rapid N uptake and cycling, altering C:N ratios of plants and soil. In sage scrub these plants are nonnative annual grasses that encourage frequent fires that reduce soil organic matter and destroy native seed stocks.

Trophic Interactions

Human-induced changes in trophic interactions frequently alter element interactions. In aquatic ecosystems introduction or overharvesting of fish often initiates a dramatic restructuring of food webs, and this restructuring can have a profound effect on the interactions of N and P (Elser et al. 1998; Sterner and Elser 2002). For example, when fast-growing zooplankton species are favored, they can store N and P in their biomass at a relatively low N:P ratio, which increases the N:P ratio of available nutrients in the water column (Mackay and Elser 1998). Also, when zooplankton populations are high, their grazing on planktonic cyanobacteria can keep abundances of N-fixing species low and thereby suppress nitrogen fixation, potentially keeping N:P ratios low and prolonging nitrogen limitation (Howarth, Chan, and Marino 1999; Marino et al. 2002).

Climate Change

Changes in climate are likely to cause large changes in species abundances, especially in extreme environments such as Arctic tundra or boreal forest, where climatic warming is expected to be the largest. Many of these species changes will affect element cycling. Recent high-latitude warming has triggered a northward forest expansion of the latitudinal tree line (Lloyd and Fastie 2002) and an expansion of shrubs (Sturm, Racine, and Tape 2001) in northern Alaska. The resulting increase in woody biomass increases ecosystem level C:N ratio and reduces decomposition and N mineralization rates (Hobbie 1996). The increased woody biomass combined with warmer climate also increases the probability of fire, which massively alters the interaction of element cycles (Hungate et al., Chapter 3, this volume).

In summary, many of the largest human impacts on element cycles are mediated by changes in species diversity, species composition, and the effects of these species on element cycles (Table 2.2). Many of these species changes produce effects that are highly predictable (e.g., N fixers), whereas other species effects require improved understanding of the interface between community dynamics and biogeochemistry.

Conclusions and Synthesis

Human activity now and in the next century will have dramatic impacts on element interactions. Nevertheless, the complexity of anthropogenic drivers of global change and their interaction make it difficult to predict ecological outcomes with certainty, but a first step may be defining some of the most important questions that must be addressed in the coming years (Box 2.2). Our conclusions regarding human impact on element cycles can be summarized as follows:

- Human activities release elements into the environment with element ratios that tend to be very different from those found in natural ecosystems. The resulting new element ratios vary with technologies and with human practices and often therefore differ significantly between the developing and the developed world because of the different underlying economic drivers.
- Element ratios in regions on land and in the oceans exhibit natural heterogeneity, and this heterogeneity is further enhanced by element loading or redistribution caused by human activity. This heterogeneous effect of human activ-

Box 2.2. Key questions for understanding human impact on element interactions in the twenty-first century

- How does element residence time (balance between sequestration and liberation) of C and other elements determine human impact on natural ecosystems? What are the controls on velocity of turnover (mineralization rates) and residence time (storage of SOM), and how are they affected by human activity?
- How will water management in the twenty-first century affect element interactions along a continuum from uplands to oceans? What are the connections from upstream to downstream ecosystems? To coastal zones? To oceans?
- What is the general theoretical framework that links species interactions to element cycles? Can we say more than that there are strong effects of species changes on element interactions, but these effects are idiosyncratic?
- What is the effect of climatic forcing on element interactions? Will we see feedbacks from increased ocean stratification and/or conveyor belt shutdown? What could the impact on fisheries be?

ity may be due to differences in contemporary drivers that are heterogeneous or to ecological memory (i.e., land degradation, salinization, or previous nutrient loading), but the response due to human activities on the landscape will reflect a mixture of natural and imposed heterogeneity.

- A consideration of complex element interactions can lead to a fundamentally different conclusion than that from consideration of effects of single elements. For example, the hypothesis that nitrogen deposition has a negative feedback on climatic warming is based largely on consideration of the fact that nitrogen stimulates primary production and net carbon storage in forests. When one also considers the effect of ozone on primary production, the possibility that nitrogen deposition will stimulate litter decomposition while increasing storage of soil organic matter, the decrease in methane consumption by forest soils, and the increase in N_2O flux from forest soils, one reaches the opposite conclusion: nitrogen deposition and related factors may have a positive or neutral feedback on climatic warming.
- Human activity can both mobilize and sequester elements, disrupting element interactions. For example, the construction of dams and reservoirs, in combination with nutrient loading to the reservoirs, can result in storage of silica. Such effects have lowered the silica-to-nitrogen and silica-to-phosphorus ratios in many large rivers and coastal seas. This change, in turn, can result in loss of

diatoms from coastal marine ecosystems, with resulting alteration of food-web structure and increased incidences of harmful algal blooms.

- Climate change may have complicated influences on element interactions. As one example, stratification in the subtropical North Pacific Ocean has increased over the past few decades, which may have greatly increased nitrogen fixation because of an increase in the iron-to-nitrate ratio in surface waters. The increased fixation in turn has increased the nitrogen-to-phosphorus ratio and made primary production more limited by phosphorus and less by nitrogen.
- Species interactions that have large impacts on element interactions and element cycles such as N fixers, woody plant encroachment, and zebra mussels have dramatic impacts on ecosystems. The resulting changes in element interactions can then feed back to cause subsequent changes in species composition.

Acknowledgments

This document is a result of a working group from the Element Interactions Rapid Assessment Project Workshop organized by the Scientific Committee on Problems of the Environment (SCOPE) and the Andrew W. Mellon Foundation. We would like to acknowledge the contributions of Walker Smith, Jon Cole, Bob Sterner, Michael Keller, Mary Scholes, Jason Neff, Björn Berg, Goran Ågren, John Stewart, and Holm Tiessen, and we are grateful to Peter Vitousek, James Galloway, Bob Naiman, Paolo Nannipieri, and Osvaldo Sala for helpful comments on the manuscript. Amy T. Austin acknowledges support from the Inter-American Institute for Global Change Research (CRN-012) and the Fundación Antorchas of Argentina.

Literature Cited

Aber, J., J. W. McDowell, K. Nadelhoffer, A. Magill, G. Bernstson, M. Kamakea, S. McNulty, W. Currie, L. Rustad, and I. Fernandez. 1998. Nitrogen saturation in temperate forest ecosystems: Hypothesis revisted. *BioScience* 48:921–934.

Achard, F., H. D. Eva, H. J. Stibig, P. Mayaux, J. Gallego, T. Richards, and J. P. Malingreau. 2002. Determination of deforestation rates of the world's humid tropical forests. *Science* 297:999–1002.

Aerts, R., R. van Logtestijn, M. van Staalduinen, and S. Toet. 1995. Nitrogen supply effects on productivity and potential leaf litter decay of Carex species from fens differing in nutrient limitation. *Oecologia* 104:447–453.

Ågren, G., E. Bosatta, and A. H. Magill. 2001. Combining theory and experiment to understand effects of inorganic nitrogen on litter decomposition. *Oecologia* 128:94–98.

Alcamo, J., G. J. J. Kreileman, J. C. Bollen, G. J. van den Born, R. Gerlagh, M. S. Krol, A. M. C. Toet, and H. J. M. de Vries. 1996. Baseline scenarios of global environmental change. *Global Environmental Change* 6:261–303.

Arnone, J. A., and G. Hirschel. 1997. Does fertilizer application alter the effects of elevated CO_2 on Carex leaf litter quality and in situ decomposition in an alpine grassland? *Acta Oecologia* 18:201–206.
Austin, A. T., and P. M. Vitousek. 2000. Precipitation, decomposition, and litter decomposability of *Metrosideros polymorpha* on Hawai'i. *Journal of Ecology* 88:129–138.
Austin, A. T. 2002. Differential effects of precipitation on production and decomposition along a rainfall gradient in Hawai'i. *Ecology* 83:328–338.
Ballester, M. V., L. A. Martinelli, A. B. Krusche, R. L. Victoria, and P. B. Camargo. 1999. Effects of increasing organic matter loading on the dissolved O_2, free dissolved CO_2 and respiration rates in the Piracicaba River basin, Southeast Brazil. *Water Research* 33:2119–2129.
Baron, J. S., N. L. Poff, P. L. Angermeier, C. N. Dahm, P. H. Gleick, N. G. Hairston, R. B. Jackson, C. A. Johnston, B. D. Richter, and A. D. Steinman. 2002. Meeting ecological and societal needs for freshwater. *Ecological Applications* 12:1247–1260.
Berg, B., and G. Ekbohm. 1991. Litter mass-loss rates and decomposition patterns in some needle and leaf litter types: Long-term decomposition in a Scots pine forest. *Canadian Journal of Botany* 69:1449–1456.
Berg, B., G. Ekbohm, M. B. Johansson, C. McClaugherty, F. Rutigliano, and A. Virzo De Santo. 1996. Some foliar litter types do have a maximum limit for litter decomposition: A synthesis of data from forest systems. *Canadian Journal of Botany* 74:659–672.
Berg, B., C. McClaugherty, A. Virzo De Santo, and D. Johnson. 2001. Humus buildup in boreal forests: Effect of litterfall and N concentration. *Canadian Journal of Forest Research* 31:988–998.
Berg, B., and V. Meentemeyer. 2002. Litter quality in a north European transect versus carbon storage potential. *Plant and Soil* 242:83–92.
Bergh, J., S. Linder, T. Lundmark, and B. Elfving. 1999. The effect of water and nutrient availability on the productivity of Norway spruce in northern and southern Sweden. *Forest Ecology and Management* 119:51–62.
Billen, G., C. Lancelot, and M. Meybeck. 1991. N, P, Si retention along aquatic continuum from land to ocean. Pp. 19–44 in *Ocean margin processes in global change*, edited by R. F. C. Matoura, J. M. Martin, and R. Wollast. Chichester, U.K.: John Wiley and Sons.
Bowman, W. D., and H. Steltzer. 1998. Positive feedbacks to anthropogenic nitrogen deposition in Rocky Mountain alpine tundra. *Ambio* 27:514–517.
Brown, J., K. N. Hinkel, and P. E. Nelson. 2000. The circumpolar active layer monitoring (CALM) program: Research design and initial results. *Polar Geography* 24:165–258.
Caraco, N. F., J. J. Cole, P. A. Raymond, D. L. Strayer, M. L. Pace, S. E. G. Findlay, and D. T. Fischer. 1997. Zebra mussel invasion in a large turbid river: Phytoplankton response to increased grazing. *Ecology* 78:588–602.
Caraco, N. F., S. E. G. Findlay, D. T. Fischer, G. G. Lampman, M. L. Pace, and D. L. Strayer. 2000. Dissolved oxygen declines in the Hudson River associated with the invasion of the zebra mussel (*Dreissena polymorpha*). *Environmental Science and Technology* 34:1204–1210.
Carpenter, S. R., N. F. Caraco, D. L. Correll, R. W. Howarth, A. N. Sharpley, and V. H.

Smith. 1998. Nonpoint pollution of surface waters with phosphorus and nitrogen. *Ecological Applications* 8:559–568.
Carreiro, M. M., R. L. Sinsabaugh, D. A. Repert, and D. F. Parkhurst. 2000. Microbial enzyme shifts explain litter decay responses to simulated nitrogen deposition. *Ecology* 81:2359–2365.
Cassman, K. G. 1999. Ecological intensification of cereal production systems: Yield potential, soil quality, and precision agriculture. *Proceedings of the National Academy of Sciences* 96:5952–5959.
Chadwick, O. A., L. A. Derry, P. M. Vitousek, B. J. Huebert, and L. O. Hedin. 1999. Changing sources of nutrients during four million years of ecosystem development. *Nature* 397:491–497.
Chapin, F. S., G. R. Shaver, A. E. Giblin, K. G. Nadelhoffer, and J. A. Laundre. 1995. Response of arctic tundra to experimental and observed changes in climate. *Ecology* 76:694–711.
Chapin, F. S., E. S. Zavaleta, V. T. Eviner, R. L. Naylor, P. M. Vitousek, H. L. Reynolds, D. U. Hooper, S. Lavorel, O. E. Sala, S. E. Hobbie, M. C. Mack, and S. Díaz. 2000. Consequences of changing biodiversity. *Nature* 405:234–242.
Chapin, F. S., O. E. Sala, and E. Huber-Sannwald. 2001. *Global biodiversity in a changing environment.* New York: Springer.
Christensen, T. R., T. Friborg, M. Sommerkorn, J. Kaplan, I. Illeris, H. Søgaard, C. Nordstrøm, and S. Jonasson. 2000. Trace gas exchange in a high arctic valley 1: Variations in CO_2 and CH_4 flux between tundra vegetation types. *Global Biogeochemical Cycles* 14:701–714.
Conley, D. J., C. L. Schelske, and E. F. Stoermer. 1993. Modification of the biogeochemical cycle of silica with eutrophication. *Marine Ecology Progress Series* 101:179–192.
Corkidi, L., D. L. Rowland, N. C. Johnson, and E. B. Allen. 2002. Nitrogen fertilization alters the functioning of arbuscular mycorrhizas at two semiarid grasslands. *Plant and Soil* 240:299–310.
Coûteaux, M. M., K. McTiernan, B. Berg, D. Szuberla, and P. Dardennes. 1998. Chemical composition and carbon mineralisation potential of Scots pine needles at different stages of decomposition. *Soil Biology and Biochemistry* 30:583–595.
Coûteaux, M. M., P. Bottner, and B. Berg. 1995. Litter decomposition, climate, and litter quality. *Trends in Ecology and Evolution* 10:63–66.
Crutzen, P. J., and M. O. Andreae. 1990. Biomass burning in the tropics: Impact on atmospheric chemistry and biogeochemical cycles. *Science* 250:1669–1677.
D'Almeida, G. A. 1986. A model for Saharan dust transport. *Journal of Climatology and Applied Meterology* 24:903–916.
Daniel, M. H. B., A. A. Montebello, M. C. Bernardes, J. P. H. B. Ometto, P. B. Camargo, A. V. Krusche, M. V. Ballester, R. L. Victoria, and L. A. Martinelli. 2002. Effects of urban sewage on dissolved oxygen, dissolved organic carbon, and organic carbon, and electrical conductivity of small streams along a gradient of urbanization in the Piracicaba River basin. *Water Air and Soil Pollution* 136:189–206.
Drinkwater, L. E., P. Wagoner, and M. Sarrantonio. 1998. Legume-based cropping systems have reduced carbon and nitrogen losses. *Nature* 396:262–265.
Driscoll, C. T., G. B. Lawrence, A. J. Bulger, T. J. Butler, C. S. Cronan, C. Eagar, K. F. Lambert, G. E. Likens, J. L. Soddard, and K. C. Weathers. 2001. Acidic deposition in

the northeastern United States: Sources and inputs, ecosystem effects, and management strategies. *BioScience* 51:180–198.

Egerton-Warburton, L. M., R. C. Graham, and M. F. Allen. 2001. Reconstruction of the historical changes in mycorrhizal fungal communities under anthropogenic nitrogen deposition. *Proceedings of the Royal Society of London Series B-Biological Sciences* 268:2479–2484.

Elser, J. J., T. H. Chryanowski, R. W. Sterner, and K. H. Mills. 1998. Stoichiometric constraints on food-web dynamics: A whole lake experiment in the Canadian shield. *Ecosystems* 3:293–307.

Evans, R. D., and J. R. Ehleringer. 1993. A break in the nitrogen cycle in aridlands? Evidence from ï^{15}N of soils. *Oecologia* 94:314–317.

Falkowski, P., R. J. Scholes, E. Boyle, J. Canadell, D. Canfield, J. Elser, N. Gruber, K. Hibbard, P. Högberg, S. Linder, F. T. Mackenzie, B. Moore III, T. Pedersen, Y. Rosenthal, S. Seitzinger, V. Smetacek, and W. Steffen. 2000. The global carbon cycle: A test of our knowledge of Earth as a system. *Science* 290:291–296.

Falkowski, P. G., R. T. Barber, and V. Smetacek. 1998. Biogeochemical controls and feedbacks on ocean primary production. *Science* 281:200–206.

Fearnside, P. M. 2002. Greenhouse gas emissions from hydroelectric reservoir (Brazil's Tucuruí Dam) and the energy policy implications. *Water Air and Soil Pollution* 133:69–96.

Fenn, M. E., M. A. Poth, J. S. Baron, B. T. Bormann, D. W. Johnson, A. D. Lemly, S. G. McNulty, D. F. Ryan, and R. Stottlemyer. 1998. Nitrogen excess in North American ecosystems: Predisposing factors, ecosystem responses, and management strategies. *Ecological Applications* 8:706–733.

Filoso, S., L. A. Martinelli, M. R. Williams, L. B. Lara, A. Krusche, M. V. Ballester, R. L. Victoria, and P. B. Camargo. Land use and nitrogen export in the Piracicaba River basin, Southeast Brazil. *Biogeochemistry*. In press.

FAO (Food and Agriculture Organization of the United Nations). 2001. FAOSTAT Agriculture Data. http://apps.fao.org.

Galloway, J. N., and E. B. Cowling. 2002. Reactive nitrogen and the world: 200 years of change. *Ambio* 31:64–71.

Gleick, P. H. 2000. The changing water paradigm: A look at 21st century water development. *Water International* 25:127–138.

Goodale, C. L., M. J. Apps, R. A. Birdsey, C. B. Field, L. S. Heath, R. A. Houghton, J. C. Jenkins, G. H. Kohlmaier, W. Kurz, S. Liu, G.-J. Nabuurs, S. Nilsson, and A. Z. Shvidenko. 2002. Forest carbon sinks in the northern hemisphere. *Ecological Applications* 12:891–899.

Groffman, P. M., R. Brumme, K. Butterbach-Bahl, K. E. Dobbie, A. R. Mosier, D. Ojima, H. Papen, W. J. Parton, K. A. Smith, and C. Wagner-Riddle. 2000. Evaluating annual nitrous oxide fluxes at the ecosystem scale. *Global Biogeochemical Cycles* 14:1061–1070.

Hibbard, K. A., S. Archer, D. S. Schimel, and D. W. Valentine. 2001. Biogeochemical changes accompanying woody plant encroachment in a subtropical savanna. *Ecology* 82:1999–2011.

Hobbie, S. E. 1996. Temperature and plant species control over litter decomposition in Alaskan tundra. *Ecological Monographs* 66:503–522.

———. 2000. Interactions between litter lignin and soil nitrogen availability during leaf litter decomposition in a Hawaiian montane forest. *Ecosystems* 3:484–494.

Hobbie, S. E., and F. S. Chapin. 1998. The response of tundra plant biomass, aboveground production, nitrogen, and CO_2 flux to experimental warming. *Ecology* 79:1526–1544.

Hobbie, S. E., and P. M. Vitousek. 2000. Nutrient regulation of decomposition in Hawaiian forests: Do the same nutrients limit production and decomposition? *Ecology* 81:1867–1877.

Howarth, R. W., H. S. Jensen, R. Marino, and H. Potsma. 1995. Transport to and processing of P in near-shore and oceanic waters. Pp. 323–346 in *Phosphorus in the global environment: Transfers, cycles, and management,* edited by H. Tiessen. SCOPE 54. Chichester, U.K.: John Wiley and Sons.

Howarth, R. W., F. Chan, and R. Marino. 1999. Do top-down and bottom-up controls interact to exclude nitrogen-fixing cyanobacteria from the plankton of estuaries? Explorations with a simulation model. *Biogeochemistry* 46:203–231.

Howarth, R. W., E. W. Boyer, W. J. Pabich, and J. N. Galloway. 2002. Nitrogen use in the United States from 1961–2000 and potential future trends. *Ambio* 31:88–96.

Howarth, R. W., D. Walker, and A. Sharpley. 2002. Sources of nitrogen pollution to coastal waters of the United States. *Estuaries* 25:656–676.

Howarth, R. W., R. Marino, D. P. Swaney, and E. W. Boyer. 2003. Wastewater and watershed influences on primary productivity and oxygen dynamics in the Lower Hudson River Estuary. In *Science and Management of the Hudson River,* edited by J. Levinton. New York: Springer. In press.

Hughes, R. F., J. B. Kauffman, and D. L. Cummings. 2002. Dynamics of aboveground and soil carbon and nitrogen stocks and cycling of available nitrogen along a land-use gradient in Rondônia, Brazil. *Ecosystems* 4:244–259.

Hunt, H. W., E. R. Ingham, D. C. Coleman, E. T. Elliot, and C. P. P. Reid. 1988. Nitrogen limitation of production and decomposition in prairie, mountain meadow, and pine forest. *Ecology* 69:1009–1016.

IPCC (Intergovernmental Panel on Climate Change). 2000. *Special report on emissions scenarios.* Cambridge: Cambridge University Press.

Jackson, R. B., S. R. Carpenter, C. H. Dahm, D. M. McKnight, R. J. Naiman, S. L. Postel, and S. W. Running. 2001. Water in a changing world. *Ecological Applications* 11:1027–1045.

Joabsson, A., and T. R. Christensen. 1999. Vascular plant controls on methane emissions from northern peatforming wetlands. *Trends in Ecology and Evolution* 14:385–388.

Joabsson, A., and T. R. Christensen. 2001. Methane emissions from wetlands and their relationship with vascular plants: An Arctic example. *Global Change Biology* 7:919–932.

Karl, D. M., K. Bjorkman, J. Dore, L. Fujicki, D. Hebel, T. Houlihan, R. Letelier, and L. Tupas. 2001. Ecological nitrogen-to-phosphorus stoichiometry at Station ALOHA. *Deep Sea Research* 48:1529–1566.

Karl, D. M., A. Michaels, B. Bergman, D. Capone, E. Carpenter, R. Letelier, F. Lipschultz, H. Paerl, D. Sigman, and L. Stal. 2002. Dinitrogen fixation in the world's oceans. *Biogeochemistry* 57/58:47–98.

Kauffman, J. B., D. G. Cummings, D. E. Ward, and R. Babbit. 1995. Fire in the Brazilian Amazon I: Biomass, nutrient pools, and losses in slashed primary forests. *Oecologia* 113:415–427.

Klimont, Z., J. Cofala, W. Schöpp, M. Amann, D. Streets, Y. Ichikawa, and S. Fujita. 2001. *IIASA special report: Projections of SO_2, NO_x, NH_3 and VOC emissions in East Asia up to 2030.* Available at www.iiasa.ac.at/~rains.

Lelieveld, J., P. J. Crutzen, V. Ramanathan, M. O. Andreae, C. A. M. Brenninkmeijer, T. Campos, G. R. Cass, R. R. Dickerson, H. Fischer, J. A. de Gouw, A. Hansel, A. Jefferson, D. Kley, A. T. J. de Laat, S. Ial, M. G. Lawrence, J. M. Lobert, O. L. Mayol-Bracero, A. P. Mitra, T. Novakov, S. J. Oltmans, K. A. Prather, T. Reiner, H. Rodhe, H. A. Scheeren, D. Sikka, and J. Williams. 2001. The Indian Ocean experiment: Widespread air pollution from South and Southeast Asia. *Science* 291:1031–1036.

Lilleskov, E. A., T. J. Fahey, T. R. Horton, and G. M. Lovett. 2002. Belowground ectomycorrhizal fungal community change over a nitrogen deposition gradient in Alaska. *Ecology* 83:104–115.

Lima, I. B. T., R. L. Victoria, E. M. L. M. Novo, B. J. Feigl, M. V. R. Ballester, and J. P. Ometto. 2002. Methane, carbon dioxide, and nitrous oxide emissions from two Amazonian reservoirs during high water table. *Verhandlungen der Internationale Vereinigung für Theoretische und Angewandte Limnologie* 28:1–5.

Lloyd, A. H., and C. L. Fastie. 2002. Spatial and temporal variability in the growth and climate response of treeline trees in Alaska. *Climate Change* 52:481–509.

Mackay, N. A., and J. J. Elser. 1998. Nutrient recycling by *Daphnia* reduces N_2 fixation by cyanobacteria. *Limnology and Oceanography* 43:347–354.

Magill, A. H., and J. D. Aber. 1998. Long-term effects of experimental nitrogen additions on foliar litter decay and humus formation in forest ecosystems. *Plant and Soil* 203:301–311.

Malmer, N., and B. Wallén. 1996. Peat formation and mass balance in subarctic ombrotrophic peatlands around Abisko, northern Scandinavia. Pp. 79–92 in *Plant ecology of the subarctic Swedish Lapland*, edited by P. S. Karlsson and T. V. Callaghan. Ecological Bulletins 45. Copenhagen: Blackwell Scientific.

Marino, R. F., F. Chan, R. Howarth, M. Pace, and G. Likens. 2002. Ecological and biogeochemical interactions constrain planktonic nitrogen fixation in estuaries. *Ecosystems* 5:719–725.

Martin, J. H., and S. Fitzwater. 1988. Iron deficiency limits phytoplankton growth in the northeast Pacific subarctic. *Nature* 331:341–343.

Martinelli, L. A., A. Krusche, R. L. Victoria, P. B. Camargo, M. C. Bernardes, E. S. Ferraz, J. M. Moraes, and M. V. Ballester. 1999. Effects of sewage on the chemical composition of Piracicaba River, Brazil. *Water Air and Soil Pollution* 110:67–79.

Matson, P. A., W. J. Parton, A. G. Power, and M. J. Swift. 1997. Agricultural intensification and ecosystem properties. *Science* 277:504–509.

Matson, P. A., K. A. Lohse, and S. J. Hall. 2002. The globalization of nitrogen deposition: Consequences for terrestrial ecosystems. *Ambio* 31:113–119.

McIsaac, G. F., M. B. David, G. Z. Gertner, and D. A. Goolsby. 2001. Eutrophication: Nitrate flux in the Mississippi River. *Nature* 414:166–167.

Melillo, J. M., J. D. Aber, P. A. Steudler, and J. P. Schimel. 1982. Nitrogen and lignin control of hardwood leaf litter decomposition dynamics. *Ecology* 63:621–626.

Michaels, A., D. Olson, J. Sarmiento, J. Ammerman, K. Fanning, R. Jahnke, A. Knap,

F. Lipschult, and J. Prospero. 1996. Inputs, losses, and transformations of nitrogen and phosphorus in the pelagic North Atlantic Ocean. *Biogeochemistry* 35:181–226.

Moore, J. K., S. C. Doney, D. M. Glover, and I. Y. Fung. 2001. Iron cycling and nutrient-limitation patterns in surface waters of the world ocean. *Deep Sea Research II: Tropical Studies in Oceanography* 49:463–507.

Moulin, C., C. E. Lambert, F. Dulac, and U. Dayan. 1997. Control of atmospheric export of dust from North Africa by the North Atlantic Oscillation. *Nature* 387:691–694.

Nadelhoffer, K. J., B. A. Emmett, P. Gundersen, O. J. Kjønaas, C. J. Koopmans, P. Schleppi, A. Tietema, and R. F. Wright. 1999. Nitrogen deposition makes a minor contribution to carbon sequestration in temperate forests. *Nature* 398:145–148.

Naiman, R. J., and M. G. Turner. 2000. A future perspective on North America's freshwater ecosystems. *Ecological Applications* 10:958–970.

NRC (National Research Council). 1993. *Managing wastewater in coastal areas.* Washington, D.C.: National Academy Press.

———. 2000. *Clean coastal waters: Understanding and reducing the effects of nutrient pollution.* Washington, D.C.: National Academy Press.

Nykänen, H., J. Heikkinen, L. Pirinen, K. Tiilikainen, and P. J. Martikainen. Annual CO_2 exchange and CH_4 fluxes on a subarctic palsa mire during climatically different years. 2003. *Global Biogeochemical Cycles* 17:1018, doi:10.1029/2002GB001861.

Ollinger, S. V., J. D. Aber, P. B. Reich, and R. J. Freuder. 2002. Interactive effects of nitrogen deposition, tropospheric ozone, elevated CO_2, and land-use history on the carbon dynamics of northern hardwood forests. *Global Change Biology* 8:545–562.

Papen, H., and K. Butterbach-Bahl. 1999. A 3-year continuous record of nitrogen trace gas fluxes from untreated and limed soils of a N-saturated spruce and beech forest in Germany: N_2O emissions. *Journal of Geophysical Research* 104:18487–18503.

Pimm, S., G. Russell, J. Gittelman, and T. Brooks. 1995. The future of biodiversity. *Science* 269:347–350.

Postel, S. 1999. *Pillar of sand: Can the irrigation miracle last?* New York: W. W. Norton.

Prather, M., D. Ehhalt, F. Dentener, D. Derwent, E. Dlugokencky, E. A. Holland, I. Isaksen, J. Katima, V. Kirchhoff, P. A. Matson, P. Midgley, W. Mingxing, and contributing authors. 2001. Atmospheric chemistry and trace gases: Third Assessment Report (TAR), Working Group I. In *Intergovernmental Panel on Climate Change (IPCC): Climate change 2000.* Cambridge: Cambridge University Press.

Prescott, C. E., B. E. Kishchuk, and G. F. Weetman. 1995. Long-term effects of repeated N fertilization and straw application in a jack pine forest. 3, Nitrogen availability in the forest floor. *Canadian Journal of Forest Research* 25:1984–1990.

Rabalais, N. N., R. E. Turner, D. Justic, Q. Dortch, W. J. Wiseman, and B. K. Sen Gupta. 1996. Nutrient changes in the Mississippi River and system responses on the adjacent continental shelf. *Estuaries* 19:386–407.

Ramakrishnan, P. S. 1992. *Shifting agriculture and sustainable development: An interdisciplinary study from north-eastern India.* Park Ridge, N.J.: Parthenon.

Ramaswamy, V., O. Boucher, F. J. Haigh, D. Hauglustaine, F. J. Haywood, G. Myhre, T. Nakajima, G. Y. Shi, and S. Solomon. 2001. Radiative forcing of climate change: Third Assessment Report (TAR), Working Group I. In *Intergovernmental Panel on Climate Change (IPCC): Climate change 2000.* Cambridge: Cambridge University Press.

Reiners, W. A. 1986. Complementary models for ecosystems. *American Naturalist* 127:59–73.

Sala, O. E., F. S. Chapin, J. J. Armesto, E. Berlow, J. Bloomfield, R. Dirzo, E. Huber-Sanwald, L. F. Huenneke, R. B. Jackson, A. Kinzig, R. Leemans, D. M. Lodge, H. A. Mooney, M. Oesterheld, N. L. Poff, M. T. Sykes, B. H. Walker, M. Walker, and D. H. Wall. 2000. Global biodiversity scenarios for the year 2100. *Science* 287:1770–1774.

Schlesinger, W. H. 1977. Carbon balance in terrestrial detritus. *Annual Review of Ecology and Systematics* 8:51–81.

Schlesinger, W. H., J. F. Reynolds, G. L. Cunningham, L. F. Huenneke, W. M. Jarrell, R. A. Virginia, and W. G. Whitford. 1990. Biological feedbacks in global desertification. *Science* 247:1043–1047.

Schnell, S., and G. M. King. 1994. Mechanistic analysis of ammonium inhibition of atmospheric methane consumption in forest soils. *Applied and Environmental Microbiology* 60:3514–3521.

Schultze, E. D., A. J. Dolman, P. Jarvis, R. Valentini, P. Smith, P. Ciais, J. Grace, S. Linder, and C.Büning, eds. 2001. *The carbon sink: Absorption capacity of the European terrestrial biosphere.* Brussels: European Commission.

Shaver, G. R., W. D. Billings, F. S. Chapin, A. E. Giblin, K. Nadelhoffer, W. C. Oechel, and E. B. Rastetter. 1992. Global change and the carbon balance of arctic ecosystems. *BioScience* 42:433–441.

Sterner, R. W., and J. J. Elser. 2002. *Ecological stoichiometry: The biology of elements from molecules to the biosphere.* Princeton: Princeton University Press.

Steudler, P. A., R. D. Bowden, J. M. Melillo, and J. D. Aber. 1989. Influence of nitrogen fertilization on methane uptake in temperate forest soils. *Nature* 341:314–316.

Sturm, M., C. Racine, and K. Tape. 2001. Increasing shrub abundance in the Arctic. *Nature* 411:546–547.

Svensson, B. H., T. R. Christensen, E. Johansson, and M. Öquist. 1999. Interdecadal variations in CO_2 and CH_4 exchange of a subarctic mire: Stordalen revisited after 20 years. *Oikos* 85:22–30.

Tibbets, J. 2002. Coastal cities: Living on the edge. *Environmental Health Perspectives* 110:A674–A681.

Tilman, D., J. Fargione, B. Wolff, C. D'Antonio, A. Dobson, R. Howarth, D. Schindler, W. H. Schlesinger, D. Simberloff, and D. Swackhamer. 2001. Forecasting agriculturally driven global environmental change. *Science* 292:281–284.

Titus, J. H., and J. Leps. 2000. The response of arbuscular mycorrhizae to fertilization, mowing, and removal of dominant species in a diverse oligotrophic wet meadow. *American Journal of Botany* 87:392–401.

UNEP (United Nations Environmental Programme). 2002. *Global environment outlook 3.* London: Earthscan.

United Nations Population Division. 2002. *World urbanization prospects: The 2001 revision.* New York: United Nations.

Valentini, R., G. Matteucci, A. J. Dolman, E.-D. Schulze, C. Rebmann, E. J. Moors, A. Granier, P. Gross, N. O. Jensen, K. Pilegaard, A. Lindroth, A. Grelle, C. Bernhofer, T. Grunwald, M. Aubinet, R. Cuelemans, A. S. Kowalski, T. Vesala, U. Rannik, P. Berbigier, D. Loustau, J. Guomundsson, H. Thorgeirsson, A. Ibrom, K. Morgenstern,

R. Clement, J. Moncrieff, L. Montagnani, S. Minerbi, and P. G. Jarvis. 2000. Respiration as the main determinant of carbon balance in European forests. *Nature* 404:861–865.

van Breemen, N., E. W. Boyer, C. L. Goodale, N. A. Jaworski, K. Paustian, S. P. Seitzinger, K. Lajtha, B. Mayer, D. van Dam, R. W. Howarth, K. J. Nadelhoffer, M. Eve, and G. Billen. 2002. Where did all the nitrogen go? Fate of nitrogen inputs to large watersheds in the northeastern U.S.A. *Biogeochemistry* 57/58:267–293.

Vitousek, P. M. 1982. Nutrient cycling and nutrient use efficiency. *American Naturalist* 119:553–572.

Vitousek, P. M., L. R. Walker, L. D. Whiteaker, D. Mueller-Dombois, and P. A. Matson. 1987. Biological invasion by *Myrica faya* alters ecosystem development in Hawaii. *Science* 238:802–804.

Vitousek, P. M., H. A. Mooney, J. Lubchenco, and J. M. Melillo. 1997. Human domination of Earth's ecosystems. *Science* 277:494–499.

Vörösmarty, C. J., P. Green, J. Salisbury, and R. B. Lammers. 2000. Global water resources: Vulnerability from climate change and population growth. *Science* 289:284–288.

West, N. E. 1990. Structure and function of microphytic soil crusts in wildland ecosystems of arid to semiarid regions. *Advances in Ecological Research* 20:179–223.

World Health Organization. 2000. *Global water supply and sanitation assessment 2000 report.* Geneva and New York: World Health Organization and United Nations Children's Fund.

WRI, UNEP, and World Bank (World Resources Institute, United Nations Environmental Programme, and World Bank). 1996. *World resources 1996–1997: The urban environment.* Washington, D.C.: World Resources Institute.

Wulff, F., A. Stigebrandt, and L. Rahm. 1990. Nutrient dynamics of the Baltic Sea. *Ambio* 19:126–133.

Zheng, X., C. Fu, X. Xu, X. Yan, Y. Huang, S. Han, F. Hu, and G. Chen. 2002. The Asian nitrogen cycle case study. *Ambio* 31:79–87.

3

Disturbance and Element Interactions

Bruce A. Hungate, Robert J. Naiman, Mike Apps, Jonathan J. Cole, Bedrich Moldan, Kenichi Satake, John W. B. Stewart, Reynaldo Victoria, and Peter M. Vitousek

Disturbances are punctuated, episodic events that result in element redistribution within and between ecosystems, involving element transport between the biosphere, hydrosphere, lithosphere, and atmosphere. Disturbances redistribute and mobilize elements in ratios that frequently differ from ecosystem stoichiometries, mediate element interactions on large spatial and temporal scales, and alter element distributions. Ecosystem responses to disturbances are regulated by positive and negative feedbacks involving abiotic and biotic processes, including the strong effects of individual species on element cycles. Disturbances are events with relatively short time constants that can have large impacts and in many cases are the result of human activities. Disturbances differ in the extent and nature of element redistributions they cause because they differ in the degree to which they disrupt the chemical, physical, and biological characteristics of ecosystems. By redistributing elements, disturbances influence interactions among element cycles.

Disturbances Redistribute and Mobilize Elements in Novel Ratios

Disturbances mobilize and redistribute elements within and among the four major reservoirs of materials on the Earth's surface: lithosphere, hydrosphere, atmosphere, and biosphere. For example, floods transport elements between the lithosphere and the hydrosphere, influencing the biosphere in the process. Similarly, volcanoes transport elements from the lithosphere to the atmosphere, and fires from the biosphere to the atmosphere (Figure 3.1). For many elements, the

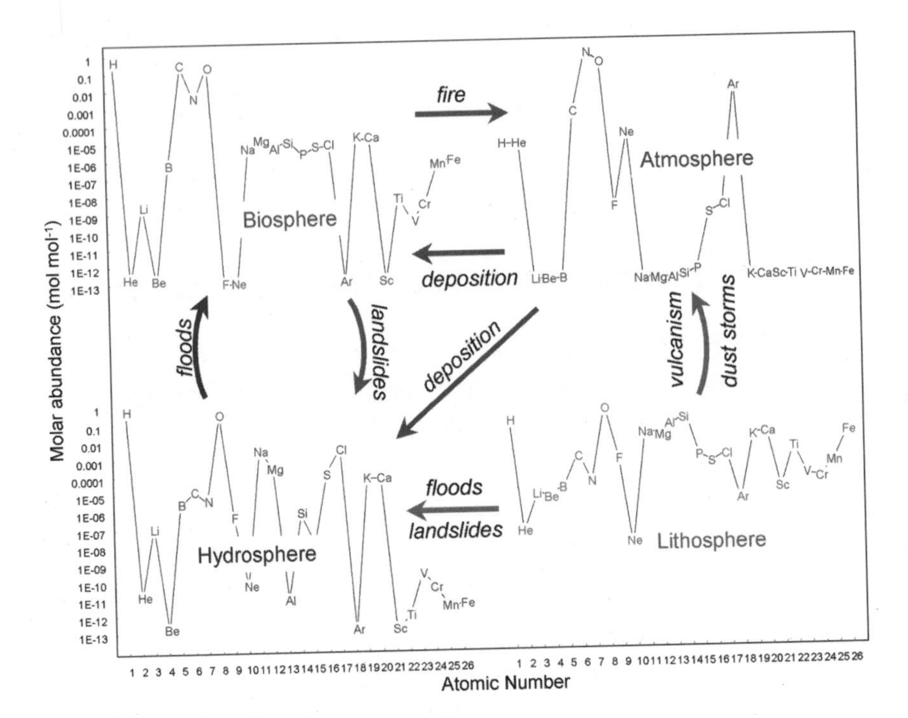

Figure 3.1. Elemental composition of the biosphere, atmosphere, hydrosphere, and lithosphere for the lightest twenty-six elements and examples of disturbances resulting in element transport between these spheres. Disturbance redistributes elements in novel ratios both because of the differences in element profiles between these spheres and because mobility varies among elements and depends on the type and intensity of the disturbance. Data on molar abundances are compiled from Deevey (1970), Schlesinger (1997), Lambers, Chapin, and Pons (1998), Nozaki (2001), Park and Schlesinger (2002).

atmosphere is only a temporary reservoir; the hydrosphere, for example, ultimately receives many elements mobilized by fire, and the biosphere receives many elements released by volcanic activity (Vitousek, Chapter 6, this volume). Because these spheres differ widely in elemental composition, and because some elements are more mobile than others, the ratios of elements released through disturbances can be quite different.

The ratio of elements mobilized by disturbance also generally differs from the stoichiometric requirements of organisms (Sterner and Elser 2002). For example, volatile element losses following fire in both Mediterranean and managed tropical forests are relatively richer in N than in P, K, Na, Ca, Mg, and S compared with the relative abundances of these elements in the preburn vegetation (Table 3.1). Thus, downwind or downslope ecosystems will receive these elements in ratios

Table 3.1. Effects of fire disturbance on element redistribution for a tropical forest under plantation management, for Mediterranean scrub forests, and for pine and eucalyptus forests

		Proportional change in molar element ratios caused by fire disturbance					
		P:N	*Ca:N*	*Mg:N*	*Na:N*	*K:N*	*S:N*
Volatile	Mediterranean	−0.2	0.3	0.0	−0.4	−0.4	—
	Tropical	−0.5	−0.6	−0.6	−0.7	−0.5	−0.2
Ash	Mediterranean	11.1	8.6	8.0	24.0	3.5	—
	Tropical	14.5	16.8	11.9	13.7	16.9	5.2
Leaching	Mediterranean	0.5	2.5	2.2	18.5	4.2	—
	Tropical	3.9	0.9	1.2	12.5	14.6	1.6
Erosion (vs. soil)	Pine	15.8	—	—	—	0.2	—
	Eucalyptus	3.2	—	—	—	0.3	—

Sources: Plantation management: Mackensen et al. 1996; Mediterranean scrub forests: Soto, Basanta, and Diaz-Fierros 1997; and pine and eucalyptus forests: Thomas, Walsh, and Shakesby 1999.

Note: For volatile, ash, and leaching categories, values shown are the proportional change in molar element ratios after fire compared with ratios in the preburn vegetation. For example, volatile losses after fire in the Mediterranean scrub forest have a molar P:N ratio that is 0.8 times the molar P:N in the preburn vegetation, indicating that volatile losses of N are relatively greater than of P. By contrast, leaching losses after fire in the tropical forest have a molar P:N ratio that is 3.9 times that of the preburn vegetation, indicating proportionally greater leaching losses of P. For erosion, values shown are the proportional change in molar element ratios after fire compared with ratios in the preburn surface soil. Note that P:N ratios of eroded sediment after fire were nearly 16 times higher in the pine forest but only 3 times higher under eucalyptus. By contrast, K:N ratios in eroded sediment were similar for the two forests, about 1.25 times higher in both cases. Dashes (—) indicate no change in element ratios caused by fire.

that differ from their own nutrient requirements, possibly altering element interactions, such as nutrient limitation of net primary productivity. By contrast, for the forests considered here, the ash formed and the materials leached following fire are relatively richer in P, K, Na, Ca, Mg, and S than in N compared with requirements of preburn vegetation (Table 3.1). Therefore, the ratios in which elements are redistributed following disturbance depend on the pathway through which they are mobilized. Similarly, traits of the biological community before disturbance can also influence the ratios in which elements are mobilized by disturbance: Compared with the P:N ratio in the surface soils before fire disturbance, the P:N ratio in sediments eroded after fire were nearly sixteen times higher in a pine forest, but only three times higher in a eucalyptus forest (Table 3.1). In this way, disturbances are similar to many human perturbations to global element cycles (Austin et al., Chapter 2, this volume), releasing suites of elements of varying mobility and in ratios that depart from the stoichiometric requirements of organisms.

Disturbances Influence Biogeochemical Interactions at Large Spatial and Temporal Scales

The assumption that element cycles exhibit approximately steady-state behavior—while a useful abstract construction—likely underestimates the biogeochemical importance of disturbance. Disturbance regimes affect net carbon balance over large regions in terrestrial and aquatic ecosystems and long-distance transport of elements between ecosystems, as well as important element interactions within ecosystems. These impacts are extensive and persistent, creating long-lasting and far-reaching ecological legacies (e.g., Foster, Fluet, and Boose 1999; Goodale and Aber 2001).

Disturbances can potentially affect large-scale biosphere-atmosphere element exchange in both terrestrial and aquatic ecosystems. The carbon balance of the Amazon is influenced by seasonal inundation of the flood plain, which converts C sequestered in biomass to CO_2 and thereby alters the C balance of the entire catchment (Richey et al. 2002). Although lakes and rivers tend to be net sources of CO_2 to the atmosphere (Cole et al. 1994; Cole and Caraco 2001), when these systems are flooded they are far more supersaturated with CO_2 than when they are within their usual channels, amplifying the efflux of CO_2 from water to air (Richey et al. 2002). Changes in this hydrological disturbance regime could influence net carbon balance at the catchment scale.

Similarly, pervasive disturbance by fire, windthrow, and insect outbreaks influence carbon balance in terrestrial ecosystems. Rather than a landscape of different ecosystems approximately at steady state, the view of the natural world as a shifting mosaic of patches following trajectories after disturbance is important for large-scale biogeochemistry. For example, models of forest ecosystems that incorporate disturbance predict biomass and element stocks in vegetation approximately consistent with those observed in forest inventories, but excluding disturbance overestimates biomass, likely because of overestimates of tree age (Sykes and Prentice 1996). Assessments of carbon budgets on relatively short time scales emphasize the importance of disturbance regime: in Canada forests switched from a sink of about 225 teragrams (Tg) C per year to a source of 75 Tg C per year over a period of twenty years as a result of a more than twofold increase in the area annually disturbed by fire and insects (Kurz and Apps 1999).

Disturbances redistribute elements not only within ecosystems, but also between them, again in ratios depending on element mobility. For example, scouring floods in streams or rivers remove biomass and nutrient capital but in particular enhance the removal of particle-associated elements compared with base-flow conditions. The single 100-year flood can transport more sediment than centuries of normal-flow conditions (Bormann and Likens 1979). Similarly, disturbance in terrestrial ecosystems can alter nutrient losses, in turn influencing element interactions in streams. In the Amazon conversion of forest to pasture increases losses of particulates and phosphorus but decreases losses of NO_3^-, resulting in a switch from P to N limitation of algal productivity in streams draining the managed watersheds (Neill et al. 2001). Disturbances can also alter ecosystem structure and thereby affect the rate and ratios at which elements are delivered from donating ecosystems upstream or upwind. For example, forest edges, often themselves the result of a disturbance, intercept more $SO_4^=$, NO_3^-, NH_4^+, and Ca^{++} through atmospheric deposition compared with forest interiors, but the relative increases were greatest for Ca^{++} (58 percent) and smallest for $SO_4^=$ (17 percent) (Weathers, Cadenasso, and Pickett 2001). In this way, forest edges alter the magnitude and relative abundances of element redistribution following disturbance.

Element transport following disturbance can occur at the global scale, creating biogeochemical connections between distant ecosystems and mediating important element interactions. For example, the intensity and frequency of dust storms increase in dry and devegetated areas, such that the flux of dust from the terrestrial surface to the ocean was an order of magnitude higher during glacial periods compared with interglacial periods (Basile et al. 1997). Terrestrial dust is the major source of Fe to the oceans, an element in low supply for phytoplankton in many ocean regions. Thus, greater dust input during glacial periods stimulated oceanic productivity, substantially reducing atmospheric CO_2 concentrations. Ultimately, high rates of dust and associated Fe inputs may have resulted in a shift from Fe to P limitation of oceanic productivity during glacial periods (Falkowski, Barber, and Smetacek 1998). Additionally, dust storms in Asia are the major source of phosphorus to highly weathered tropical soils in Hawaii, with most of the inputs having occurred during glacial periods favoring dust transport. While P limits plant production in these soils, dust inputs partly ameliorate this constraint (Chadwick et al. 1999).

A Biogeochemical Legacy of Disturbance: N Limitation of NPP?

Disturbances redistribute elements depending on their frequency, intensity, and duration—the disturbance regime. For example, disturbance can enhance nitrogen losses from terrestrial ecosystems (Likens et al. 1970; Vitousek et al. 1979), necessitating biogeochemical recovery during early succession (Gundersen and

Bashkin 1994). Equally important, the disturbance regime may govern the nature of key element interactions long after the disturbance event itself, creating long-lasting biological and biogeochemical legacies. For example, widespread occurrence of N limitation of primary production in terrestrial ecosystems—perhaps the archetypal example of a critical element interaction—may arise because of widespread disturbance of the terrestrial landscape. A simple model of C-N-P interactions illustrates how disturbance frequency and intensity alters N stocks and the nature of N limitation to net primary productivity.

The model is similar to one developed by Vitousek and Field (1999) to evaluate controls on N fixation in terrestrial ecosystems. It mineralizes N from soil organic matter when the C:N ratio falls below a critical value; N is taken up from the available pool up to some maximum capacity (and N limits production when uptake is less than that capacity); and N is lost from the system when available N remains in the soil after plants have taken up the maximum they can. P is modeled similarly, except that it is less mobile in that only a fraction of available P is lost following plant uptake. Production is taken as the minimum of N-supported and P-supported productivity, with uptake of the more abundant element scaled to that of the less abundant. We model perennial plants (with a C, N, and P residence time of ten years) but do not include wood. N and P enter the system annually at a low rate (5 percent) relative to plants' uptake capacity. In addition, we simulate biological N fixation as occurring when there is available P, N availability is low, and some light reaches the ground surface (biomass is less than 80 percent of maximum biomass).

We evaluated three classes of disturbance arrayed along a gradient in the quantity of plant biomass that is killed or consumed by disturbance. The disturbances roughly correspond to windthrow, herbivory, and fire and capture a range in disturbance intensity: With simulated windthrow almost all of the C, N, and P in plant biomass is converted to litter during each disturbance; herbivory transfers 40 percent of the C (to account for herbivore respiration) and all of the N and P in plants to litter; and fire volatilizes almost all plant C and N and puts almost all of the P into available forms. We examined four disturbance frequencies—disturbance every 50 years, 200 years, and 500 years and no disturbance during a 2,000-year simulation—and considered these effects in the presence and absence of biological N fixation.

In this model, N capital in soil reflects the balance of N inputs and losses over the 2,000-year simulation (Table 3.2). The soil N (and P) pool was reduced under all of the disturbance frequencies and intensities we considered. For windthrow and herbivory, the mechanism responsible for N depletion was loss of available N shortly after disturbance, when mineralization is temporarily greater than plant demand for N. Postdisturbance losses also occurred following fire, but

Table 3.2. N pool sizes in soil, as a function of disturbance frequency and intensity and the presence or absence of N fixation

	Disturbance frequency			
Disturbance intensity	*50 years*	*200 years*	*500 years*	*None*
High (fire)				
No N fixation	316	891	1,156	1,225
N fixation	610	981	1,168	
Medium (herbivory)				
No N fixation	808	1,143	1,198	1,225
N fixation	895	1,143	1,198	
Low (windthrow)				
No N fixation	879	1,153	1,203	1,225
N fixation	955	1,153	1,206	

Note: Values in g N m^{-2}, averaged over the past 500 years of each simulation.

in addition fire itself volatilized N in plant biomass and so N pools were smaller when fire was the agent of disturbance.

In the absence of N fixation, the final N pool was slightly less (overall losses were slightly greater) under herbivory than windthrow, because the high C:N ratio of litter produced by windthrow led to greater N immobilization than occurred with herbivory. With high-frequency fire, the soil N pool contained little more than a quarter as much N as in the no-disturbance case. In the presence of N fixation, fire increased P availability in the postdisturbance interval, enhancing rates of N fixation. Consequently, the N pool in the high-frequency, high-intensity case was about twice as great in the presence of N fixation than in its absence.

In addition, we used the model to explore the effects of disturbance frequency and intensity, and the presence or absence of N fixation, on the existence of N limitation of primary production. We determined the fraction of time during the last 500 years of each simulation during which either N mineralization or available P supplied less N or P than the maximum needed to sustain productivity. Without disturbance, sufficient N accumulates from the low annual rates of input to keep the simulated N supply adequate for plant growth. The loss of N during and following disturbance reduces N supply during the regrowth phase and so induces N limitation to plant growth. With low-frequency and medium- or low-intensity disturbance, enough N accumulated during the postdisturbance interval that plant growth was not limited by N supply most of the time (Table 3.3); light-N interactions suppressed N fixation throughout the disturbance event itself and the subsequent ecosystem recovery.

Table 3.3. Percentage of years in which N or P limits plant production during the last 500 years of each simulation

	Disturbance frequency			
Disturbance intensity	*50 years*	*200 years*	*500 years*	*None*
High (fire)				
N limit				
N fixation	4.0	0.4	0.8	0.0
No N fixation	100.0	97.0	62.4	0.0
P limit				
N fixation	78.6	91.0	52.8	0.0
No N fixation	0.0	0.0	0.0	0.0
Medium (herbivory)				
N limit				
N fixation	3.2	55.8	33.6	0.0
No N fixation	88.0	92.8	33.6	0.0
P limit				
N fixation	66.8	33.8	0.0	0.0
No N fixation	0.0	0.0	0.0	0.0
Low (windthrow)				
N limit				
N fixation	12.0	92.0	27.4	0.0
No N fixation	88.0	92.4	30.6	0.0
P limit				
N fixation	54.0	0.0	0.0	0.0
No N fixation	0.0	0.0	0.0	0.0

In contrast, in the absence of N fixation and with high-intensity (fire) and high-frequency disturbance regimes, N supply always limited growth—even in the immediate postdisturbance interval when N availability was high in most other cases. In the other situations N supply is nonlimiting only immediately after disturbance; the greater frequency of N limitation under the intermediate- to low-frequency disturbance regimes reflects the number of disturbances during 500 years in these treatments. Where N fixation can occur, both frequent and intense disturbance created conditions that favored fixation—and P limitation was more important than N limitation under these conditions. Of course, understanding controls of N fixation (Vitousek and Field 1999) and of N losses in addition to disturbance-mediated pathways (Hedin, Arnesto, and Johnson 1995; Vitousek et al. 1998) is as important as understanding N losses caused by disturbance.

Consequences of Element Redistribution

Element redistribution during disturbances leads to positive and negative feedback responses within ecosystems, further altering element distribution (positive feedbacks) or returning the system to predisturbance conditions (negative feedbacks), together shaping the trajectory of ecosystem responses. Ecosystem responses to element redistribution are modulated by the postdisturbance biotic assemblage (and their characteristic stoichiometric requirements) and by abiotic conditions, including those modified during disturbance (Figure 3.2). Some responses promote recovery to predisturbance assemblages, such as those accumulating elements that were disproportionately lost or mobilized during disturbance (e.g., photosynthesis, N_2 fixation, and P uptake following fire). These act as negative feedbacks, tending toward restoring the predisturbance stoichiometry. Other responses push the ecosystem along a new trajectory, potentially amplifying losses of elements released during disturbance or promoting accumulation of elements in relative proportions markedly different from those in the predisturbance assemblage (e.g., colonizers or invaders with different biological stoichiometries; Mack, D'Antonio, and Ley 2001). Such responses act as positive feedbacks, causing further element redistribution within and between ecosystems. Both positive and negative feedbacks can occur simultaneously in a single ecosystem's response to a disturbance event. For example, fire can increase soil NO_3^- (Johnson et al. 2001), promoting further N losses through leaching and denitrification, while simultaneously enhancing the growth of N_2-fixing species (Hungate et al. 1999).

Human-induced global changes are likely to amplify ecosystem responses supporting positive feedbacks to element redistribution. For example, disturbances in forests subject to chronic acid deposition or in rivers experiencing nutrient loading are likely to reorganize along some new trajectory of element distribution following disturbance (Lawrence, David, and Shortle 1995). Similarly, disturbance often facilitates invasion by exotic species, potentially altering element distribution as well as other critical ecological processes, including the frequency and intensity of subsequent disturbances (D'Antonio and Vitousek 1992).

Element redistribution during a disturbance event shapes the postdisturbance biological assemblage. Because different organisms require resources in different ratios, the change in the relative availabilities of elements following disturbance has the potential to influence species composition of the recolonizing community. For example, there are well-documented effects of the ratios of incoming materials (e.g., N, P, Si) for phytoplankton species compositions in aquatic systems (Tilman 1977, 1982). The nature of the postdisturbance biological assemblage, in turn, influences the trajectory along which elements reassemble and recombine in ecosystems following disturbance. For example, life-history characteristics (e.g.,

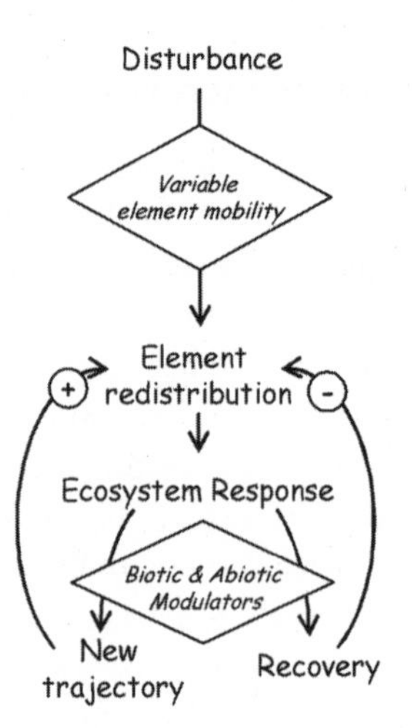

Figure 3.2. Variable element mobility results in element redistribution following disturbance. Ecosystem responses, modulated by biota and abiotic conditions, involve positive and negative feedbacks (shown by arrows marked + and -, respectively), reinforcing or ameliorating the nature of postdisturbance element redistribution.

dispersal ability, size, growth rate) influence which organisms recolonize after disturbance, characteristics that may be unrelated to ratios of element availability in the environment but may nevertheless be associated with characteristic biological stoichiometries. In this way, element interactions that occur through biological element uptake following disturbance will reflect both the effects of disturbance on the relative availabilities of different elements and the characteristic stoichiometries of recolonizing organisms (Sterner and Elser 2002).

Together, changes in the relative availabilities of elements and changes in the microenvironment in response to disturbance can result in a cascade of element interactions, some with important biogeochemical consequences. For example, inundation events in aquatic ecosystems (e.g., beaver dams, formation of oxbow lakes, water impoundments) cause settling of particles and associated elements (Si, P), leading to benthic anoxia and associated changes in element cycling. Beaver dams, for example, have been shown to alter methane emissions (Naiman, Manning, and Johnson 1991). In an experimental flooding of a small Canadian wetland forest, Kelly et al. (1997) observed an enormous increase in the rates of CH_4 loss to the atmosphere and the formation of methyl mercury. Similarly, some flooding events in rivers in southern Brazil cause large-scale resuspension of organic matter in the water column, increasing biological oxygen demand, reducing O_2, and in turn causing massive fish kills and large releases of CO_2 and CH_4 to the atmosphere (Fearnside 1997; St. Louis et al. 2000).

Repeated disturbances can influence abiotically mediated element interactions that shape the rate and nature of ecosystem response to subsequent disturbance. For example, fire disturbance in some grassland ecosystems can create a positive feedback cycle involving disturbance regime, charcoal formation, and soil fertility.

First, burning itself increases the probability of future burns, increasing disturbance frequency (e.g., D'Antonio and Vitousek 1992). Burning mobilizes P, causing some to be lost, but causes greater losses of C and N in ratios and forms depending on fire intensity. Hot fires release C and N through volatile pathways, whereas cooler fires form charcoal, which is relatively inert in soils and thus unchanged after several thousand years (Lertzman et al. 2002; Cochrane and Schulze 1999a, b). Such soils have been described in some detail in the Yucatan, Mexico, and the Cerrado and Amazon regions of Brazil. In the Yucatan soils, charcoal accumulation reduces site fertility by restricting N supply and altering moisture retention (Tiessen and DeJong, pers. comm.). In this situation fire disturbance either releases or ties up C and N in noncycling forms, likely influencing productivity and composition of the recovering vegetation and associated element interactions.

Biologically mediated element interactions can also strongly shape ecosystem response to disturbance. For example, development of riparian vegetation in the Pacific Northwest depends on biological mechanisms responsible for nutrient accretion following flood or sandbar disturbance (Naiman, Bilby, and Bisson 2000). Specifically, the accumulation of nitrogen in such cases can occur not only via the establishment of N_2-fixing alder, but also through exogenous inputs from salmon carcasses. Where nutrient inputs through salmon carcasses occur, establishment of N_2-fixing alder declines, suggesting a trade-off, and perhaps a complementarity, between these two sources of nutrients (Helfield and Naiman 2002). Likely because of differences in stoichiometry and quality of alder and salmon tissue, however, the nature of the riparian community depends on which source of nutrient input dominates (Bartz and Naiman, submitted). Where salmon inputs dominate, overstory stem density is significantly greater, primarily due to a fivefold increase in the number of large-diameter willows (*Salix* spp.). Understory stem density is lower, with all common genera but *Viburnum* (typical of N-rich sites) contributing to the difference. Ground layer species richness is also lower, as are nonvascular and dwarf shrub species covers. Spruce and willow, in particular, respond strongly to the increased availability of marine-derived nutrients, increasing their growth by as much as 200–300 percent with the consequences being expressed throughout the plant-soil community (Naiman et al. 2002). In this case the response of the riparian system to disturbance strongly depends on biologically mediated element interactions.

Limitations of Our Assessment

Our treatment focuses on terrestrial ecosystems, with some examples from freshwaters, but largely excludes marine environments. Particularly in the open ocean, the concept of short-term natural disturbance is not considered in the same light as in terrestrial ecosystems for a number of reasons. While a passing hurricane, for example, undoubtedly disrupts element cycles in the ocean water beneath it, the relatively rapid exchange of water means that the effects will be diluted more quickly than measurements might occur. Longer-term disturbances or oscillations, such as the El Niño/Southern Oscillation or the Intertropical Convergence Zone, which alter upwelling and stratification, have profound effects on the open ocean (Karl 1999). The much shorter life cycles of planktonic organisms compared with trees also implies that reorganization after disturbance can be much more rapid in open-water systems. On the other hand, disturbances on land clearly impact oceanic systems, for example, with fish assemblages developing around collections of terrestrially derived organic matter debris (Maser and Sedell 1994), now in some cases replaced with artificial "fish aggregation devices" to stimulate development of fisheries.

What We Have Learned

Elements are required by organisms and mobilized by geochemical processes in characteristic ratios (stoichiometries). Although studying the cycling of individual elements can be valuable, stoichiometry provides a more complete perspective, and one that becomes increasingly useful when considering how disturbance redistributes elements within and between ecosystems. Within individual ecosystems these stoichiometries vary over time, shifting the balance and ratios of element stocks in biotic, chemical, and physical compartments in ways that can be described with complex systems tools (e.g., Holling 2001).

Natural and anthropogenic disturbances affect both biological and geological components of element cycling, disproportionately transforming some elements into biologically available forms and mobilizing others to ecosystems downslope, downstream, and downwind. The changes occur both immediately (at the time of disturbance) and over time as the disturbed ecosystem tends toward a new steady state. If a threshold in environmental space has been crossed, and a change of ecosystem trajectory results from the disturbance, the new steady state may be very different than the original one. Thus the concept of an average ecosystem with a fixed stoichiometry may be misleading, particularly in a changing global environment. Disturbances are pervasive; attempting to evaluate element cycling and interactions in the absence of disturbance gives misleading views of the states of ecosystems and how their dynamics are regulated.

Challenges for the Future

There is a critical need to link biological and geochemical stoichiometries underlying element cycles and their interactions to understand their expression as biotic communities and processes. How do organisms' responses to disturbance influence rates of primary mineral weathering and biosphere-atmosphere fluxes? How do changes in weathering of particular minerals influence organisms? For example, biological influences on Si mobilization through soils are rarely considered, but Si transport can shape the functioning of downstream aquatic ecosystems, and silicate weathering is driven in large part by organisms, representing a persistent sink for CO_2.

Although the role of disturbance in shaping ecosystems is widely acknowledged, regional and global models incorporating the effects of disturbance on element interactions are not. Dynamic vegetation models would improve by operating on multiple time scales required to capture phenomena such as the exposure of mineral soil caused by windthrow versus the subsequent weathering of the exposed material, and by considering element transport to other ecosystems where they may trigger an entirely new sequence of elemental reactions. We suggest that a new, integrating framework for the models could (1) reflect the changes in the regional mosaic pattern over time, (2) incorporate understanding of life cycle trajectories (i.e., changes in the internal functioning of the disturbed ecosystems over their entire life cycle) and thereby capture the range of states in the mosaic at a particular point in time, and (3) facilitate the exchanges and fate of material between the patches in the mosaic. Such an approach could help generate hypotheses about the influences of element interactions on processes at the landscape and global scales.

Current knowledge is limited in part by which elements have appropriate isotopes that can be determined easily—and by cultural differences between biologists and geochemists on which elements are of interest. For example, isotopes of Ca and Si may provide fundamental information that will help explain how transfers between terrestrial and aquatic ecosystems are regulated. More widespread use of a number of existing and new technologies will help evaluate interactions between disturbances and the biological and geological components of element cycling (Ollinger et al., Chapter 4, this volume).

Finally, what is the role of natural disturbance in a human-dominated world? Human activity not only causes new disturbances and disturbances that mimic and/or modify the effects of natural disturbance, but it also alters the frequency, intensity, and duration of "natural" disturbance to the point that the dichotomy becomes artificial. Can we link the perspectives that have guided our analyses of natural and anthropogenic disturbance into a more unified framework? We face challenges in incorporating these changes in our projections of future feedbacks through the interlinking of biogeochemical cycles.

Acknowledgments

We thank all the participants of the SCOPE Element Interactions Workshop for useful suggestions during the formulation of this chapter. Terry Chapin, Chris

Field, Jane Marks, and Jerry Melillo provided valuable comments on the manuscript.

Literature Cited

Bartz, K. K., and R. J. Naiman. Impacts of salmon-borne nutrients on riparian soils and vegetation in southeast Alaska. *Ecology*, submitted.

Basile, I., F. E. Grousset, M. Revel, J. R. Petit, P. E. Biscaye, and N. I. Barkov. 1997. Patagonian origin of glacial dust deposited in East Antarctica (Vostok and Dome C) during glacial stages 2, 4, and 6. *Earth and Planetary Science Letters* 146:573–589.

Bormann, F. H., and G. E. Likens. 1979. *Pattern and process in a forested ecosystem.* New York: Springer-Verlag.

Chadwick, O. A., L. A. Derry, P. M. Vitousek, B. J. Huebert, and L. O. Hedin. 1999. Changing sources of nutrients during four million years of ecosystem development. *Nature* 397:491–497.

Cochrane, M. A., and M. D. Schulze. 1999a. Fire as a recurrent event in tropical forests of the eastern Amazon: Effects on forest structure, biomass, and species composition. *Biotropica* 31:2–16.

———. 1999b. Forest fires in the Brazilian Amazon. *Conservation Biology* 12:948–950.

Cole, J. J., and N. F. Caraco. 2001. Carbon in catchments: Connecting terrestrial carbon losses with aquatic metabolism. *Marine and Freshwater Research* 52:101–110.

Cole, J. J., N. F. Caraco, G. W. Kling, and T. Kratz. 1994. Carbon dioxide supersaturation in the surface waters of lakes. *Science* 265:1568–1570.

D'Antonio, C. M., and P. M. Vitousek. 1992. Biological invasions by exotic grasses, the grass-fire cycle, and global change. *Annual Review of Ecology and Systematics* 23:63–87.

Deevey, E. S., Jr. 1970. Mineral cycles. *Scientific American* 223:149–158.

Falkowski, P. G., R. T. Barber, and V. Smetacek. 1998. Biogeochemical controls and feedbacks on ocean primary production. *Science* 281:200–206.

Fearnside, P. M. 1997. Greenhouse-gas emissions from Amazonian hydroelectric reservoirs: The examples of Brazil's Tucurui Dam as compared to fossil fuel alternatives. *Environmental Conservation* 24:64–75.

Foster, D. R., M. Fluet, and E. R. Boose. 1999. Human or natural disturbance: Landscape-scale dynamics of the tropical forests of Puerto Rico. *Ecological Applications* 9:555–572.

Goodale, C. L., and J. D. Aber. 2001. The long-term effects of land-use history on nitrogen cycling in northern hardwood forests. *Ecological Applications* 11:253–267.

Gundersen, P., and V. N. Bashkin. 1994. Nitrogen cycling. In *Biogeochemistry of small catchments: A tool for environmental research,* edited by B. Moldan and J. Cerny. SCOPE 51. Chichester, U.K.: John Wiley and Sons.

Hedin, L. O., J. J. Arnesto, and A. H. Johnson. 1995. Patterns of nutrient loss from unpolluted, old-growth temperate forests: Evaluation of biogeochemical theory. *Ecology* 76:493–509.

Helfield, J. M., and R. J. Naiman. 2002. Salmon and alder as nitrogen sources to riparian forests in a boreal Alaskan watershed. *Oecologia* 133:573–582.

Holling, C. S. 2001. Understanding the complexity of economic, ecological, and social systems. *Ecosystems* 4:390–405.

Hungate, B. A., P. Dijkstra, D. W. Johnson, C. R. Hinkle, and B. G. Drake. 1999. Elevated CO_2 increases nitrogen fixation and decreases soil nitrogen mineralization in Florida scrub oak. *Global Change Biology* 5:781–790.

Johnson, D. W., B. A. Hungate, P. Dijkstra, G. Hymus, and B. G. Drake. 2001. Effects of elevated CO_2 on soils in a Florida scrub oak ecosystem. *Journal of Environmental Quality* 30:501–507.

Karl, D. M. 1999. A sea of change: Biogeochemical variability in the North Pacific subtropical gyre. *Ecosystems* 2:181–214.

Kelly, C. A., J. W. M. Rudd, R. A. Bodaly, N. T. Roulet, V. L. St. Louis, A. Heyes, T. R. Moore, S. Schiff, R. Aravena, K. J. Scott, B. Dyck, R. Harris, B. Warner, and G. Edwards. 1997. Increases in fluxes of greenhouse gases and methyl mercury following flooding of an experimental reservoir. *Environmental Science and Technology* 31:1334–1344.

Kurz, W. A., and M. J. Apps. 1999. A 70-year retrospective analysis of carbon fluxes in the Canadian forest sector. *Ecological Applications* 9:526–547.

Lambers, H., F. S. Chapin III, and T. L. Pons. 1998. *Plant physiological ecology.* New York: Springer-Verlag.

Lawrence, G. B., M. B. David, and W. C. Shortle. 1995. A new mechanism for calcium loss in forest-floor soils. *Nature* 378:162–164.

Lertzman K., D. Gavin, D. Hallet, L. Brubaker, D. Lepofsky, and R. Mathewes. 2002. Long-term fire regime estimated from soil charcoal in coastal temperate rainforests. *Conservation Ecology* 6:5–17.

Likens, G. E., F. H. Bormann, N. M. Johnson, D. W. Fisher, and R. S. Pierce. 1970. Effects of forest cutting and herbicide treatment on nutrient budgets in the Hubbard Brook watershed-ecosystem. *Ecological Monographs* 40:23–47.

Mack, M. C., C. M. D'Antonio, and R. E. Ley. 2001. Alteration of ecosystem N dynamics by exotic plants: A case study of C4 grasses in Hawaii. *Ecological Applications* 11:1323–1335.

Mackensen, J., D. Hölscher, R. Klinge, and H. Fölster. 1996. Nutrient transfer to the atmosphere by burning of debris in eastern Amazonia. *Forest Ecology and Management* 86:121–128.

Maser, C. J., and R. Sedell. 1994. *From the forest to the sea: The ecology of wood in streams, rivers, estuaries, and oceans.* Delray Beach, Fla.: St. Lucie Press.

Naiman, R. J., T. Manning, and C. A. Johnson. 1991. Beaver population fluctuations and tropospheric methane emissions in boreal wetlands. *Biogeochemistry* 12:1–15.

Naiman, R. J., R. E. Bilby, and P. A. Bisson. 2000. Riparian ecology and management in the Pacific coastal rain forest. *BioScience* 50:996–1011.

Naiman, R. J., R. E. Bilby, D. E. Schindler, and J. M. Helfield. 2002. Pacific salmon, nutrients, and the dynamics of freshwater and riparian ecosystems. *Ecosystems* 5:399–417.

Neill, C., L. A. Deegan, S. M. Thomas, and C. C. Cerri. 2001. Deforestation for pasture alters nitrogen and phosphorus in soil solution and streamwater of small Amazonian watersheds. *Ecological Applications* 11:1817–1828.

Nozaki, Y. 2001. Elemental distribution overview. In *Encyclopedia of ocean sciences,* Vol.

2, edited by J. H. Steele, S. A. Thorpe, and K. K. Turekian. San Diego: Academic Press.
Park, H., and W. H. Schlesinger. 2002. Global biogeochemical cycle of boron. *Global Biogeochemical Cycles* 16(4), doi:1072, 10.1029/2001GB001766.
Richey, J. E., J. M. Melack, A. K. Aufdenkampe, V. M. Ballester, and L. L. Hess. 2002. Outgassing from Amazonian rivers and wetlands as a large tropical source of atmospheric CO_2. *Nature* 416:617–620.
Schlesinger, W. H. 1997. *Biogeochemistry: An analysis of global change.* San Diego: Academic Press.
Soto, B., R. Basanta, and F. Diaz-Fierros. 1997. Effects of burning on nutrient balance in an area of gorse (*Ulex europaeus* L.) scrub. *The Science of the Total Environment* 204:271–281.
St. Louis, V. L., C. A. Kelly, E. Duchemin, J. W. M. Rudd, and D. M. Rosenberg. 2000. Reservoir surfaces as sources of greenhouse gases to the atmosphere: A global estimate. *BioScience* 50: 766–775.
Sterner, R. W., and J. J. Elser. 2002. *Ecological stoichiometry: The biology of elements from molecules to the biosphere.* Princeton: Princeton University Press.
Sykes, M. T., and I. C. Prentice. 1996. Carbon storage and climate change in Swedish forests: A comparison of static and dynamic modelling approaches. In *Forest ecosystems, forest management, and the global carbon cycle,* edited by M. P. Apps and D. T. Price. NATO ASI Series I, Vol. 40. Berlin: Springer-Verlag.
Thomas, A. D., R. P. D. Walsh, and R. A. Shakesby. 1999. Nutrient losses in eroded sediment after fire in eucalyptus and pine forests in the wet Mediterranean environment of northern Portugal. *Catena* 36:283–302.
Tilman, D. 1977. Resource competition between planktonic algae: An experimental and theoretical approach. *Ecology* 58:338–348.
Tilman, D. 1982. *Resource competition and community structure.* Monographs in Population Biology. Princeton: Princeton University Press.
Vitousek, P. M., and C. B. Field. 1999. Ecosystem constraints to symbiotic nitrogen fixers: A simple model and its implications. *Biogeochemistry* 46:179–202.
Vitousek, P. M., L. O. Hedin, P. A. Matson, J. H. Fownes, and J. Neff. 1998. Within-system element cycles, input-output budgets, and nutrient limitation. Pp. 432–451 in *Successes, limitations, and frontiers in ecosystem science,* edited by M. Pace and P. Groffman. New York: Springer.
Vitousek, P. M., J. R. Gozk, C. C. Grier, J. M. Melillo, W. A. Reiners, and R. L. Todd. 1979. Nitrate losses from disturbed ecosystems. *Science* 204:469–474.
Weathers, K. C., M. L. Cadenasso, and S. T. A. Pickett. 2001. Forest edges as nutrient and pollutant concentrators: Potential synergisms between fragmentation, forest canopies, and the atmosphere. *Conservation Biology* 15:1506–1514.

4

New Frontiers in the Study of Element Interactions

Scott Ollinger, Osvaldo Sala, Göran I. Ågren, Björn Berg, Eric Davidson, Christopher B. Field, Manuel T. Lerdau, Jason Neff, Mary Scholes, and Robert Sterner

In 1922 the Russian geologist Vladimir Vernadsky coined the term "biogeochemistry" in recognition of the intimate connections that exist between living organisms and the physical components of the Earth. This insight came after many failed attempts to understand the chemical makeup of the Earth's atmosphere and crust through strictly geochemical processes, ignoring the influence of element interactions mediated by living organisms. In his landmark book *Biosfera* (The Biosphere), which was completed in 1926 at Charles University in Prague, Vernadsky wrote:

> A characteristic role is also played by the respiration and feeding of organisms, through which the organisms actively select the materials necessary for life. . . . Living matter builds bodies of organisms out of atmospheric gases such as oxygen, carbon dioxide and water, together with compounds of nitrogen and sulfur, converting these gases into liquid and solid combustibles that collect the cosmic energy of the sun. After death, it restores these elements to the atmosphere by means of life's processes. . . . Such a close correspondence between terrestrial gasses and life strongly suggests that the breathing of organisms has primary importance in the gaseous system of the biosphere; in other words, it must be a planetary phenomenon (Vernadsky 1926).

Since the publication of *Biosfera*, the discipline Vernadsky named has seen impressive growth in its collective knowledge of element cycling, achieved largely through numerous studies of the fluxes, budgets, and chemical transformations of individual elements. Although this line of investigation has been enormously fruitful, a growing number of important questions in ecosystem science involve poorly understood interactions among two or more elements.

The importance of multiple element interactions stems from the fact that pairs and even groups of elements often participate in a series of common biological and geochemical reactions. Because carbon represents a common currency among all life forms, many multiple-element interactions are mediated through the relative ratios of carbon to other nutrients (e.g., Sterner and Elser 2002). For instance, if the availability of a given element limits the rate of carbon fixation by primary producers, other elements that are cycled through biomass production and turnover can be indirectly affected. As an example, research in the Hawaiian Islands has revealed element interactions during ecosystem development that involve a change from nitrogen limitation to phosphorus limitation over geological time scales (Vitousek et al. 2002). Although supplies of N relative to plant demand increase steadily over this progression, the absolute rate of N cycling slows as ecosystems become increasingly P-limited. This stems from the maintenance of relatively conservative N:P ratios in vegetation (Crews et al. 1995) and the dependence of N cycling on organic matter production and turnover.

Despite these insights, growth in scientists' understanding of multiple-element interactions has been restricted by several factors, including the diversity of forms in which a single element can occur (carbon, in particular) and the fact that element ratios in organisms are not always conservative. At present, the degree to which element interactions across a range of ecological systems are controlled by these sources of complexity versus general patterns of organism stoichiometry is simply not known.

Our lack of understanding is highlighted best by the fact that few ecosystem models explicitly include element interactions, a shortcoming that limits their applicability for many scientific and policy issues. For example, the notion of nitrogen saturation in temperate forest ecosystems involves a condition where excess N loading via atmospheric deposition leads to a shift from nutrient limitation by N to limitation by some other element (Aber et al. 1989). A noteworthy ingredient of this shift is its decoupling from the normal mechanisms of ecosystem development demonstrated in Hawaii. An important feature of N saturation is the increasing production and leaching of nitrate from soils, a process that can also cause depletion of positively charged ions such as calcium. Despite growing evidence that this process has begun to affect forests in heavily impacted regions (e.g., Peterjohn, Adams, and Gilliam 1996; Aber et al. 2003), few models have attempted to explicitly capture multiple-element chemical processes in soils (e.g., Gbondo-Tugbawa et al. 2001) and none include the plant-soil feedbacks necessary to simulate the shift in element limitations.

As the biogeochemistry research community is confronted with limitations of single element models, the need for new analytical and conceptual tools will become increasingly apparent. In all fields of science, major breakthroughs often follow technological or conceptual advancements, so a challenge that exists just before those breakthroughs involves recognizing the pending advancements that will be of greatest benefit. The purpose of this chapter is to discuss important limitations in several areas of element interaction research and to highlight several forthcoming conceptual and/or methodological approaches that may help overcome these hurdles. Given the large number of issues that involve multiple element interactions and the diverse array of new approaches becoming available, it would be impossible to conduct an exhaustive review that has relevance for all ecological systems. Instead, our approach is to discuss several persistent challenges and highlight a few recent developments that have promising potential for addressing them. It is our hope that this discussion will inspire new ways of addressing issues beyond the specific examples we offer here.

Current Limitations in Element Interactions Research

Major obstacles for advancing our understanding of element interactions are often manifested as inconsistencies between how we formalize our knowledge into models and the data available to describe the systems being modeled. This discrepancy can be due either to a lack of relevant data, as a result of technical difficulties in making certain measurements, or to a deeper mismatch between parameters and processes included in models and measurable features of the real world. In this section, we discuss examples of several common limitations that influence the applicability of currently available biogeochemistry models, both conceptual and methodological.

Discrepancies between Model Formulations and Observable Ecosystem Properties

Carbon Dynamics in Terrestrial Soils

A common difficulty in matching model constructs with observations of nature involves the issue of how to capture complex or highly variable properties in a functionally efficient manner. A good example of this challenge involves models of carbon turnover in terrestrial soils, where organic carbon is lumped into a small number of discrete states that are meant to represent a broad continuum of turnover times. Although it is widely acknowledged that what we simplistically refer to as soil carbon contains a complex mixture of chemical compounds with

Box 4.1. Carbon quality and turnover time of soil organic matter

Carbon in soil organic matter is distributed over many different compounds. These compounds have a huge range of degradability traits (qualities) and, consequently, turnover times. In models these different turnover times are represented in different ways. One alternative is to consider the full distribution of turnover times. This approach leads to models that are rather mathematically complex. A simpler approach is to split the distribution of carbon into a small number of discrete classes, as represented in the figure by the vertical lines. The problem is that this partitioning is rather arbitrary and does not correspond to observed fractions. Partitioning the carbon according to some empirical procedure, such as light and heavy fractions, is likely to cut the carbon density distribution in still some other way that is not directly coupled to turnover times. This approach is exemplified with the broken line, where the area under the line would be the light fraction and the area above the line the heavy fraction.

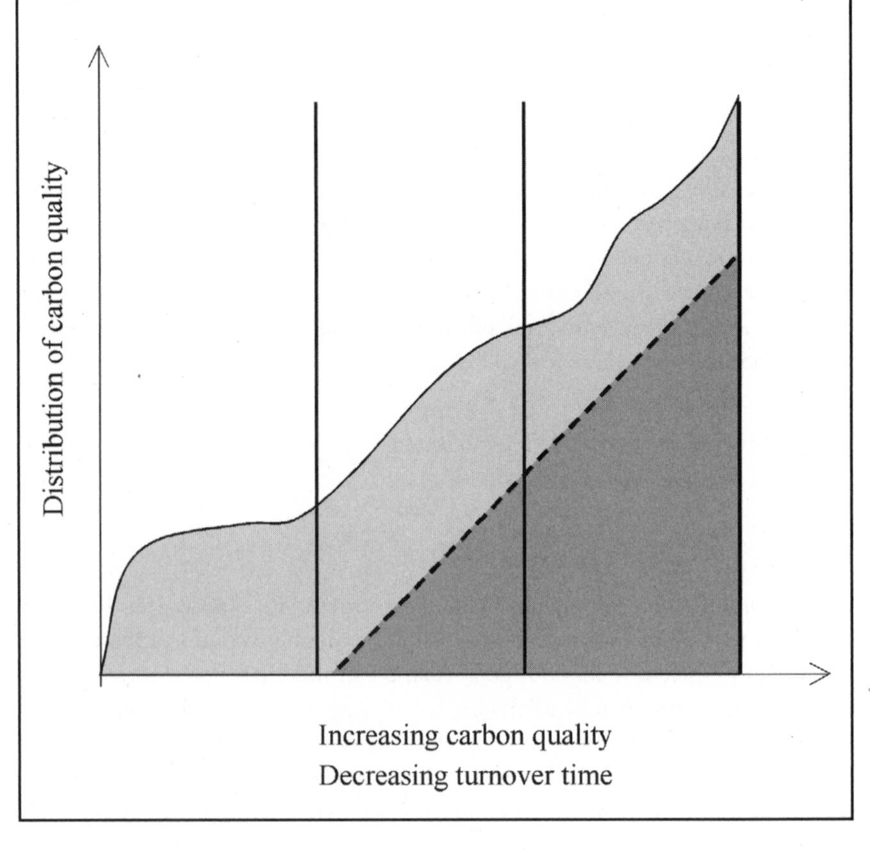

different levels of accessibility for decomposers, the question of how to represent that complexity has been far from obvious.

The method of aggregating soil carbon compounds into distinct groups (such as fast, medium, and slow turnover pools) is attractive because it offers a certain degree of functional utility. Because soil material does not actually exist in such discrete forms, however, it presents the problem of how to parameterize model variables that lack real-world counterparts and cannot be explicitly measured. In addition, there is the challenge of deciding how many compartments such a model should include. Although finer divisions may come closer to mimicking reality, such an approach becomes unwieldy as the number of compartments grows. The use of a few highly aggregated compartments may be functionally desirable, but this approach presents increasing difficulties of linking measurements to corresponding theories within the model (Ågren and Bosatta 1998).

Maintaining the integrity of theories and mechanisms within a model is most necessary where models are expected to predict phenomena that have not been explicitly prescribed or that fall beyond the range of known observations. The most salient examples of this need involve processes in which thresholds play an important role. Net nitrification in forest soils, for example, seems to occur only when the soil C:N ratios fall below approximately 24–25 (MacDonald et al. 2002; Ollinger et al. 2002). In a well-constructed model such a result should come from underlying interactions between soil C and N dynamics without being prescribed a priori.

Element Ratios and Threshold Effects

The difficulty of simulating systems that include important threshold responses may stem from the overriding prevalence of single element nutrient models and models that erroneously assume a linear response to driving environmental variables (i.e., because available data are not sufficient to define a full response surface). An additional challenge arises in systems where biogeochemical cycles are influenced by ecological processes that span multiple levels of organization, particularly those that are usually not included in ecosystem-level studies (e.g., population or community-level feedbacks).

An experimental demonstration of this type of effect was presented by Sommer (1992), who showed how an algae-Daphnia system switched between algae-dominated and grazer-dominated when the dilution rate exceeded a narrowly defined value. At low rates of P turnover, low algal production and high algal C:P ratios severely inhibited grazer growth rates and maintained an algae-dominated community. As P availability rose above a certain critical threshold, however, algal quality improved to the point where consumption rates rose markedly and the experimental ecosystems supported large numbers of grazers. In this system the threshold response was driven by differences in the C:P ratios of algae compared with the requirements of grazers. For models to simulate this dynamic in real ecosystems, they must capture interactions between the biogeochemical processes regulating P cycling, variability in algal P concentrations, and factors controlling algae-grazer competitive relations.

Lack of Data due to Difficulties Making Basic Measurements

Sampling Individual Species within Microbial Communities

In terrestrial plant communities, the ability to sample and analyze tissues from individual organisms has allowed us to gain insights into the functioning and flexibility of species within the community. This type of knowledge has led to important theories about how dynamics of species and ecological communities scale to the level of the entire ecosystem. In microbial communities, which consist of the most numerous and biogeochemically significant organisms on Earth, we have little insight into these relationships because we lack basic methods for sampling and analyzing individual organisms or species.

Microorganisms are arguably the most important detritivores globally and are temporary sources and sinks of plant-essential nutrients. Microbial populations, biomass, and nutrient content are notoriously difficult to measure in soils. Although soil microbial C, N, and P are commonly estimated following fumigation of a soil sample with chloroform, this process solubilizes most cell walls and releases cytoplasmic contents. Because not all microbial cell constituents are recovered, a conversion factor must be used to estimate their nutrient contents (Horwath and Paul 1994). Despite the large amount of effort that has been put toward estimating these conversion factors and understanding how they vary among soils, large uncertainties remain. Moreover, these methods ignore differences *within* microbial communities that may be related to rates of nutrient cycling.

In aquatic systems as well, size-based fractionations might yield samples heavily dominated by individual microbial populations, but separation of populations, or even functional groups, is problematic. Nutrient studies in lake ecosystems are typically based on analysis of particulate matter, but the material sampled can include both living and nonliving components, and the living fraction might contain algae, bacteria, and protozoa in varying abundances. What we lack is an easily applied method to determine the elements held within each individual component of the system.

Determining the Flexibility of Microbial Stoichiometry

In addition to measurements of nutrient stocks in microbial biomass, rates of flows of nutrients through the microbial biomass pool are needed for improved understanding of ecosystem function and multielemental interactions. A common simplification in conceptual and numerical models is that C, N, and P stoichiometry is fixed in the microbial biomass of a given ecosystem, but this is unlikely to be true. If microbial stoichiometry were fixed, then separate functions for each element would not be needed. We know, however, that the processes involved in element linkages can be flexible, which makes the description of these multielemental interactions more complicated. In the case of microbial biomass, for example, each species may have some plasticity in its C:N and C:P ratios because N and P, when available in excess, are taken up and temporarily stored within the biomass. These ratios may also change as species composition changes over relatively short generation time periods in response to varying resources.

In addition to differences among species within a microbial community, differences among organic compounds within the cells of a single species may also affect multiple element interactions. For example, C and N stored in cell walls occur in different ratios and cycle at different rates than do C and N in extracellular enzymes and cellular metabolites used for osmotic regulation. We currently have very little understanding of how much of the N immobilized by soil microorganisms is used for cell growth (requiring proportionally all the C and other elements needed for cell biomass) and how much of the N taken up by microorganisms is subsequently exuded as extracellular enzymes and cellular metabolites (requiring only the N:element ratios of those particular compounds). New approaches are needed to identify the microbial players in a given ecological community and to quantify the importance of various organic compounds with regard to pools and fluxes of elements.

Limitations in Conceptual Understanding

Single versus Multiple Resource Limitation

According to Liebig's law of the minimum, a biological process, such as crop productivity, is limited by the single resource that is in shortest supply relative to demand (von Liebig 1847). If that limitation is alleviated, as in fertilizer amendment to an agricultural field, then that process increases until some other resource becomes limiting. Although Liebig's law is not explicitly included in most biogeochemical models, the concept of "a limiting nutrient" for a given ecosystem is persistent in our thinking. For example, vegetation growing on highly weathered soil substrates is commonly considered P limited, while vegetation on younger substrates is commonly considered N limited (Crews et al. 1995). These generalizations are based on an understanding of soil chemical properties as they vary during soil weathering and are supported by nutrient amendment experiments and patterns of foliar concentrations and nutrient use efficiencies (Vitousek 1984; Vitousek and Farrington 1997; Harrington, Fownes, and Vitousek 2001). Although these simplifications have been instructive, they do not preclude the possibility—indeed, the likelihood—of important multielemental interactions when a wider variety of processes and time scales are considered.

When nutrient manipulation experiments demonstrate a synergistic response to more than one element, the response is usually described by an interaction term resulting from conventional analysis of variance. Although the ANOVA may be necessary for demonstrating statistical significance, the ecological significance is likely to lie in the interpretation of how and why the response of X_1 to X_2 is different when X_3 is present in greater or lesser quantity. This ecological interpretation usually requires expression within a modeling framework and may also require disaggregation of the response variable. For example, if X_1 is plant growth and X_2 and X_3 are N and P concentrations in the soil, it may be necessary to divide the plant biomass growth response into responses of individual species within the community or to analyze separately the responses of roots, stems, and foliage. This simple example ignores much more complex processes, such as herbivory, enzyme production, or symbioses that may be equally important. Although parsimonious explanations and simple models are usually desirable, multielemental interactions clearly require some degree of added complexity in conceptual and numerical models.

Limitations of the Optimization Paradigm

Another common paradigm used in ecosystem models is the notion of optimization, which assumes that organisms optimize the use and allocation of resources in a way that maximizes their functional efficiency. This assumption holds great conceptual value and has been effectively applied to simulating resource limitations in a variety of habitats (e.g., Armstrong 1999; de Mazancourt, Loreau, and Abbadie 1999). There are, however, at least two important pitfalls that must be considered for effective use in element cycling models. These pitfalls involve the question of what processes are being optimized and at what level of organization optimization should occur. Terrestrial ecosystem models, for example, often carry the assumption that optimal resource allocation should lead to a maximization of biological production. Evolutionary pressures, however, favor adaptations that maximize survival and reproduction rather than growth (Arendt 1997). Although growth rate and reproductive success are often related, resource allocation patterns that appear to be suboptimal can be favored when they provide an advantage in terms of competitive interactions with neighboring individuals.

Further, because selection operates on individuals rather than on ecological communities, we should not assume that optimization will necessarily produce maximal functioning over entire ecosystems. As an example, most efforts to predict leaf area index in plant canopies assume that optimization should lead to maximum photosynthetic capacity of the entire canopy. Anten and Hirose (2001), however, suggested that this approach is conceptually flawed because evolutionary mechanisms act not on whole plant canopies, but on the individuals within them. Using a model that contrasted individual and canopy-level photosynthesis, the authors showed that foliar allocation patterns that optimize photosynthesis over the canopy are not evolutionarily stable because at this leaf area index (LAI), any individual within the stand can increase its growth rate by increasing its individual leaf area.

Ecological Interactions and Species Effects

The ways in which individual species influence biogeochemical cycles has been an active area of investigation and debate for many years. Because functional differences between species often involve differences in the proportions and forms in which various elements are used, differences in species composition within ecological communities lead to basic differences in element interactions. For example, Binkley and Valentine (1991) showed differences in soil properties and element cycles among fifty-year-old forest plantations in Connecticut, U.S.A., that corresponded to differences in carbon quality and cation use among the species planted. Understanding differences in element use between species is particularly important for predicting temporal dynamics because species composition in ecological communities is rarely static. Changes over time that result from patterns of disturbance and succession complicate our understanding of element interactions because our understanding of the mechanisms driving those changes is incomplete. In addition, the temporal dynamics of element interactions due to species change is unlikely to coincide directly with the species change itself, because species can have important legacy effects on the environments in which they exist, long after they have been replaced by other species.

Nutrient Availability and Historical Human Land Use

The previous examples involved situations in which population-, community-, and ecosystem-level processes interact in ways that challenge traditional biogeochemistry models. Other examples of how factors that affect element interactions can cross traditional disciplinary boundaries involve processes mediated by

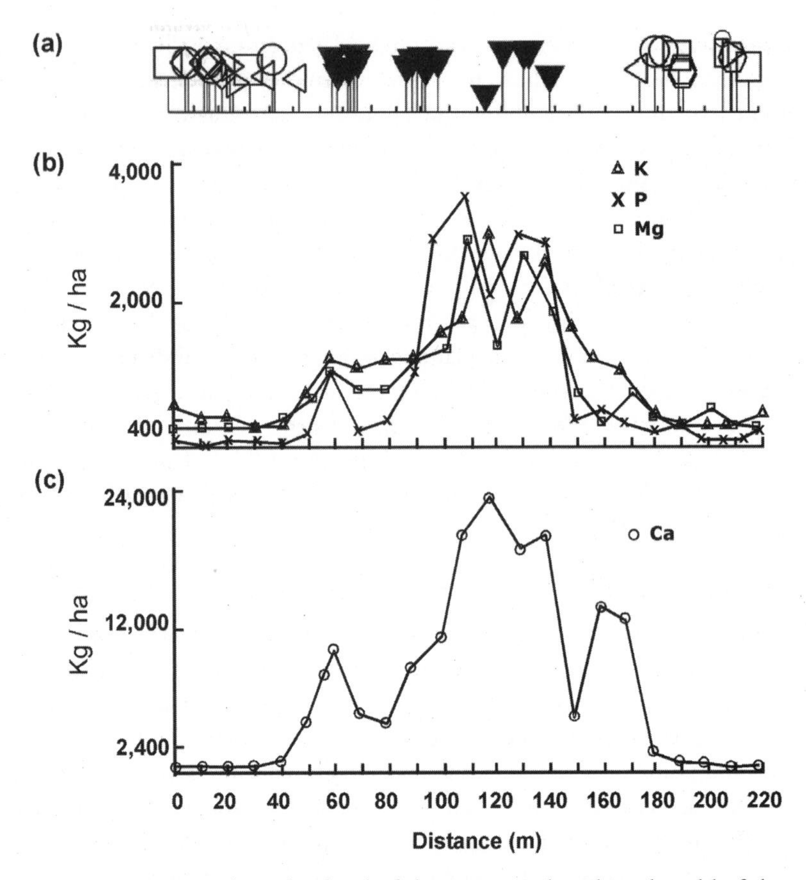

Figure 4.1. A transect through a fine-leafed savanna patch within a broad-leafed savanna, reflecting the long-term legacy of the African Iron Age Tswana people. (a) Distribution of woody plants: solid symbols are fine-leafed trees; (b and c) soil nutrient chemistry (Scholes and Walker 1993).

human social or economic activity. An instructive demonstration of this comes from the nutrient-poor savannas of southern Africa as described by Scholes and Walker (1993). These ecosystems are extensive on the sandy, highly leached and acidic soils that occupy the majority of the region and are typically dominated by broad-leaved vegetation such as *Burkea africana.* Discrete patches of nutrient-rich fine-leaved savanna, however, are occasionally found within these broad-leaved vegetation communities. The distribution of these patches bears no obvious relationship to physical features of the landscape but rather has been tied to archaeological remnants of past human societies.

Settlements of the African Iron Age Tswana people occurred in the region more than 700 years ago but typically lasted for just a few decades in any one location. These settlements led to the accumulation of large amounts of nutrients in soils as a result of firewood collection and the corralling of animals at night. The accumulation of nutrients, in turn, led to a complete switch in the vegetation type and ecosystem dynamics that in some cases have persisted for many centuries (Figure 4.1). It is difficult to understand how relatively short-term nutrient inputs have produced such long-term changes, but one possible explanation is that the added nutrients altered the fundamental plant-soil feedback mechanisms whereby additional phosphorus from decomposing animal tissues made it possible for *Acacias* to become established. Because *Acacias* are di-nitrogen fixers, their presence would further the process of nutrient enrichment. Once established, these sites also attract a variety of native herbivores, which may contribute to the ongoing maintenance of their atypical biogeochemical status. This combination of historical human land use and shifting ecological feedback mechanisms has been characterized quantitatively (Scholes and Walker 1993), but it represents an enormous and, as yet, unmet challenge for process modeling.

New Approaches and Recent Advancements in Element Interactions Research

Although issues discussed in the preceding section represent important and in some cases daunting challenges to element interactions research, it can be both instructive and encouraging to consider areas in which methodological and conceptual breakthroughs have led to rapid and substantial gains in our understanding. As an early example, the biogeochemical theories proposed by Vladimir Vernadsky represented a significant conceptual advancement, yet they could also be viewed as a product of preceding methodological advancements that provided element analyses of atmospheric gases, rocks, and soils.

Many years later, development of methods for retrieving and analyzing gases trapped in ice cores provided a new view of feedbacks between element interactions and climate dynamics that spanned geological time scales (e.g., Mayewski and White 2002). Another example of concordant technological development and scientific understanding can be seen in the technologies associated with trace metal analysis and the role of iron limitation in the open ocean (Ducklow, Oliver, and Smith, Chapter 16, this volume). Development of methods for clean water sampling led to the important discovery that di-nitrogen fixation can be limited by nanomolar levels of iron (e.g., Kolber et al. 1994). In this section we highlight several more recent advancements that may have significant potential to foster rapid changes in element cycling research.

Resolving the Chemical Identity of Element Observations

Identification and quantification of materials at a molecular level is a crucial step in the study of element interactions. Most of the major advances in our understanding of processes governing these interactions are grounded in detailed knowledge of chemical compounds. As the preceding sections described, barriers to our understanding of these interactions often arise from an inability to identify and/or measure the distribution of elements across various chemical compounds. For example, we have known for more than twenty years that litter decomposition and associated nutrient transformations are related to the structure and concentration of lignin and nitrogen in plant tissues (e.g., Berg and Staaf 1980; Melillo, Aber, and Muratore 1982; Berg and McClaugherty 2003). Nevertheless, there can still be large unexplained variation in the relationships between lignin, N, and litter decomposition patterns. Recent advances in analytical techniques such as near infrared spectroscopy (NIRS) and pyrolysis-gas chromatography/mass spectrometry-combustion interface-isotope ratio-mass spectrometry (py-GC/MS-C-IRMS) allow a more comprehensive understanding of decomposition at the molecular level and, in particular, the relationship between litter carbon quality and decomposition rate (Joffre et al. 2001; Gleixner et al. 2002).

Solid and Liquid Phase Analyses

For soils and water samples, techniques such as ^{13}C, ^{15}N, and ^{32}P nuclear magnetic resonance (NMR) have been used with varying success across a range of ecosystems. In aquatic ecosystems NMR techniques have been used to better understand the nature of carbon/nitrogen bonding in dissolved organic matter (McKnight et al. 1997). In terrestrial ecosystems at least two studies have combined ^{13}C, ^{15}N, and ^{32}P NMR techniques to evaluate the degree to which N and P mineralization are coupled to C (Gressel et al. 1996). The combined analysis of the structural basis for C, N, and P bonding and change through time focuses on (and challenges) evidence behind long-standing arguments for coupled C-N and decoupled C-P during decomposition/mineralization (McGill and Cole 1981). New techniques involving py-GC/MS-C-IRMS and NMR offer significant insight into decomposition patterns, the preservation of organic matter through time, and the relationship between N availability and decomposition rates (Gleixner, Bol, and Balesdent 1999; Preston 2001; Gleixner et al. 2002; Neff et al. 2002). Despite these advances, there is a clear need to better integrate ecological and molecular techniques and to improve communication between investigators working at multiple scales. Further, to date, there has not been as much advancement in the application of advanced

Box 4.2. Abiotic nitrate immobilization in soils: The ferrous wheel hypothesis

An example of multiple element interactions is the "ferrous wheel hypothesis" for abiotic nitrate immobilization in soils, offered by Davidson, Chorover, and Dail (2003). Carbon compounds derived from photosynthate reduce Fe(III) in soil minerals, producing reactive Fe(II) species that reduce nitrate to nitrite, and nitrite subsequently reacts with receptive moieties in dissolved organic matter (DOM) to produce dissolved organic-N (DON). This process could be an important mechanism by which inorganic-N from atmospheric deposition is converted to organic-N within soils without the large C requirement necessitated by biological assimilation. Iron is abundant as Fe(III) in most well-drained mineral soils, and interactions of nitrate and iron are known to be important in aquatic ecosystems (Senn and Hemond 2002), but little is known about iron speciation and reactivity in soil microsites.

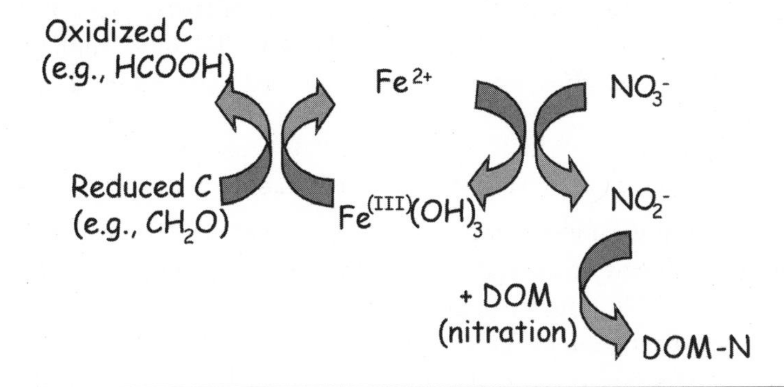

analytical techniques to element interactions at a molecular level. In the future, compound-specific techniques that link structural identification of biomarkers with not only $^{12,13}C$ but also $^{14,15}N$ could significantly advance our understanding of elemental interactions at a molecular level in both aquatic and terrestrial environments.

Gas Phase Analyses

Perhaps the best-known "new advance" in atmospheric measurement techniques involves the development and widespread application of robust open-path infrared gas analyzers. These instruments exponentially increased the number of commercially available eddy-covariance systems and allowed many researchers to measure high-frequency CO_2 and H_2O fluxes at the canopy scale. The wealth of new data these measurements have provided have given us a much better sense of both the variability in CO_2 and H_2O fluxes in natural and agricultural ecosystems and of the mechanisms underlying these variations.

Other techniques are emerging that have already proven helpful for making unique measurements and for making measurements at shorter time scales than were previously possible. Several of these methods involve laser spectroscopy of air samples and the quantification of particular compounds (and in some cases, their isotopic ratios). Tunable diode lasers (TDL) are the best known of these techniques and have been used for canopy measurements of methane, nitrous oxide, and carbon dioxide (Fowler et al. 1995). Recently TDLs have been used for long-term (longer than one year) studies of nitric acid and nitrogen dioxide fluxes at the canopy scale, results from which have been essential for balancing the reactive nitrogen budget in northeastern forests (C. Volpe Horii, unpublished results). Currently, TDLs are beginning to be used for the determination of isotopic ratios during eddy covariance studies for particular compounds of interest (e.g., CO_2; Mcmanus et al. 2002). A next generation of lasers, quantum cascade lasers, promises to bring the analytical capabilities of TDL down to the size of a briefcase, which would free researchers from the need to have a steady supply of liquid nitrogen.

Methods and Models of Atmosphere Biosphere Exchange

Improved conceptual understanding has sometimes been coupled with the simultaneous development of instrumentation and modeling. Emissions of volatile organic carbon (VOC) from vegetation are a case in point. As portable, battery-powered instrumentation for VOC measurements became available, many more plant taxa were sampled across the surface of the globe (Lerdau, Guether, and Monson 1997; Valentini et al. 1997). This increased sampling led to much-improved regional inventories, and these, in turn, allowed better-constrained models of element interactions in the troposphere (Guenther et al. 1995, 1996, 1999). In the early 1990s real-time analyzers for specific compounds allowed identification of the biophysical controls over production and emission. This mechanistic understanding led to the development of mechanistic flux models that were applicable across broad spatial and temporal scales (Harley, Monson, and Lerdau 1999). Such effective models have proven invaluable for determining the redox potential of the atmosphere and the interactions among VOCs, reactive forms of nitrogen, and ozone (Guenther et al. 1999).

It is unlikely that such rapid progress could have been made in the absence of these novel instruments. An additional benefit of this rapid accumulation of knowledge has been an improved understanding and quantification of VOC fluxes with respect to the global carbon cycle (Lerdau, Chapter 9, this volume).

Advancements in Element Analysis for Individuals and Species

In recent years considerable progress has been made in characterization of population sizes, and in some cases detailed size-based or pigment-based characterizations can be made with flow cytometry. Though very helpful in dealing with biomass, flow cytometry does not measure element content. It would be a major breakthrough if flow cytometry could be coupled to a method to estimate element abundance (e.g., x-rays or near infrared spectroscopy [NIRS]). At the molecular level, probes for specific genetic material can now be used to assay particular microbial function groups, such as nitrifying bacteria in soils. In the future continued development of such methods for gaining information about microbial populations (broadly defined, including heterotrophic bacteria, small algae, and others) will greatly improve our knowledge of element cycle linkage in terrestrial and aquatic ecosystems, particularly when biogeochemical measurements of elements can be coupled to population censuses.

A promising example of how this may be achieved involves the combined use of new molecular techniques and stable isotope analyses. This strategy was suggested by Nannipieri, Badalucco, and Landi (1994) as an approach for measuring specific microbial pools in soil, and more recently Radajewski et al. (2000) reported on a specific application. After treating soil with a known ^{13}C source, they were able to extract and separate ^{13}C-labeled DNA and then characterize it (both taxonomically and functionally) by gene probing and sequence analysis. This technique may offer a useful new means of estimating the size of active microbial C pools: by multiplying the ratio of ^{13}C-DNA to total DNA by the soil microbial C pool. In addition, the taxonomic and functional characterization provides a more precise understanding of how microbial communities are affected by various types of added C-substrate.

Molecular Techniques for Nutrient Limitations

One of the most difficult issues associated with element interaction is determining when and how nutrient limitation of organisms occurs. Although it is relatively easy to measure the concentration of an element in a soil sample, it is much more difficult to determine whether that particular element limits a particular biotic activity. A significant methodological advance in the study of element interactions will, we hope, soon come with the development and application of bioreporter probes. Bioreporter organisms are genetically modified cells, so far either bacteria or algae, that have had a fluorescence-producing gene coupled to a regulatory gene that is sensitive to physiological limitation of a single element or resource. Bioreporters have been developed, for example, to measure Fe-limitation (Durham et al. 2002). Bioreporters for other elements are under development.

A similar approach to measurement of element availability is in immunoassays for individual proteins. In Fe-limited algae, the respiratory enzyme flavodoxin substitutes for the Fe-containing ferridoxin when cellular iron is deficient. Immunoassays have been developed that can interrogate the cells from natural populations to determine which of these two enzymes they carry (McKay et al. 1999). Methods development allowing for routine application of bioreporters, immunoassays, and other physiological studies of nutrient availability could mean a major breakthrough, because this approach uses the response of an organism in the habitat itself to measure the availability of nutrients. We look forward to putting a mixture of bioreporters into water samples or soil samples and letting the emitted fluorescence of given wavelengths "report" the ratios of available nutrients in those substrates.

Thus far, this chapter has concentrated largely on the difficulties of identifying microorganisms and their role in element interactions. This emphasis occurs partly because plant species and their element content have generally been easier to identify, at least for aboveground plant parts. Identifying plant roots in the soil profile, however, has been extremely difficult. Ecologists recognize a large variability in rooting depth among plant species (Canadell et al. 1996; Jackson et al. 1996) and, consequently, hypothesize that changes in plant-species composition would have major consequences for the cycling in ecosystems. Testing of this hypothesis requires unequivocal identification of roots in the soil profile. Recently, Jackson et al. (1999) were able to identify roots of tree species at different depths in the soil profile by comparing the root DNA sequences against a reference database developed for aboveground plant parts. Studies like this will become more common in the near future because reference databases are growing at a rapid rate, saving researchers' time and resources.

Scaling Element Interactions from Sites to Regions

A common theme in all fields of science is the question of how to take information collected at one scale and apply it to processes that manifest themselves at another. In ecosystem science this challenge is particularly relevant because processes that are critical to the discipline occur at spatial scales ranging from microscopic to global and at temporal scales ranging from seconds to millennia. In terrestrial systems, methods by which leaf-, plant-, and stand-level observations can be related to landscapes, regions, and continents have been the subject of much discussion (e.g., Ehleringer and Field 1993; Reich, Turner, and Bolstad 1999). Although a variety of approaches have been proposed, there is widespread agreement that remote sensing holds a central and irreplaceable role. Remote platforms provide the only means of viewing large portions of the Earth's surface at regular intervals, and the selective absorption and reflectance of light by plant tissues allows optical sensors to gather tremendous amounts of ecologically relevant information.

Recent Advancements in Remote Sensing

Satellite and aircraft remote-sensing data provide some of the best options for spatial scaling of ecological processes (Ustin, Smith, and Adams 1993). For biogeochemical analysis, however, remote-sensing data generally provide only a fraction of the information scientists require. Many important ecosystem properties cannot be directly detected but must be estimated from algorithms that relate raw reflectance to more useful landscape or vegetation properties. These algorithms are often challenging to derive and can be subject to substantial uncertainties. To date, a large number of remote-sensing applications in ecosystem science have addressed patterns of productivity and carbon uptake, but fewer have explicitly addressed issues of element cycling or, more rarely, multiple element interactions.

One common approach to conducting spatially explicit analyses of ecosystem production has been to use spectral vegetation indices, such as the normalized difference vegetation index (NDVI), that relate to vegetation properties such as cover type or LAI (leaf area per unit ground area). These data can then be used as input to ecosystem models capable of simulating a variety of other processes (e.g., Ollinger, Aber, and Federer 1998). The use of LAI as an estimator of productivity has been most effectively demonstrated across large moisture gradients where substantial variation in LAI can be observed (e.g., Gower, Vogt, and Grier 1992; Fassnacht and Gower 1997). The approach, however, is more limited at finer spatial scales and within regions where moisture regimes and LAI are less variable, but where variation in production nevertheless occurs owing to variation in soil nutrient availability and/or the production efficiency of foliage (Coops and Waring 2001). Additional limitations of LAI-based methods stem from factors that cause both vegetation indices and LAI to exhibit nonlinear, asymptotic relationships with rates of production (Gower, Reich, and Son 1993; Reich, Turner, and Bolstad 1999).

In recent years a variety of new remote-sensing techniques have been developed that may offer improved access to ecosystem structure and function. Synthetic aperture radar and laser altimetry (e.g., light detection and ranging, or LIDAR) provide information on the size and spatial distribution of canopy elements that can provide the basis for biomass estimates. The utility of these techniques is increasingly established, based on data from aircraft and short space flights, but satellites with appropriate sensors and global coverage are not yet deployed. Multiangle remote sensing provides another avenue for approaching canopy structure (Diner et al. 1999). Multiangle data are now available from space-based platforms such as the Multiangle Imaging Spectro Radiometer (MISR). Hyperspatial data, with a spatial resolution of one meter or better, are now available from satellite sensors, and investigators are beginning to explore the potential of new algorithms based on the size, spacing, and dynamics of the crowns of individual trees (Coops and Culvenor 2000).

Another intriguing innovation in remote sensing is the development of aircraft- and satellite-based imaging spectrometers, also known as high-spectral-resolution, or hyperspectral, sensors. Whereas traditional remote-sensing instruments divide the reflectance spectra into a small number of discrete and discontinuous spectral bands, typically averaged over a wide range of wavelengths, hyperspectral sensors measure continuous spectra using a larger number of much narrower bands. This capacity has proven especially useful for biogeochemistry research because spectral information contained in these detailed images can be related to concentrations of specific elements in plant foliage (e.g., Wessman et al. 1988; Martin and Aber 1997; Asner 1998). Spectral detection of biochemical constituents in plant tissue grew initially from work with benchtop spectrometers, which highlighted reflectance features associated with C-H and N-H bonds. Air- and space-borne instruments provide essentially the same type of spectral data but present the added challenge of accounting for signal variation due to atmospheric absorption and variation in canopy structure. In one recent application remote estimates of canopy N concentrations in a temperate forest landscape were combined with extensive field measurements of forest C and N cycling to yield mapped estimates of aboveground net primary production (NPP) and soil C:N ratios (Colorplate 1; Ollinger et al. 2002; Smith et al. 2002). Spatial estimates of leaf N and soil C:N could, in turn, be related to losses of nutrients and dissolved organic carbon from watersheds.

Although hyperspectral instruments have been widely explored with aircraft platforms, the number of sensors in operation is still quite small and the limited opportunities for obtaining data have prevented its widespread use. In addition, the challenges of collecting, storing, and interpreting hyperspectral data are still substantial and will require additional methodological development before the full range of its capabilities and/or limitations is appreciated. Recent development of new satellite sensors (e.g., NASA's Hyperion instrument aboard the Earth Observer 1 [EO-1] platform) make further exploration of this potential promising.

A Challenge for the Future: Resolving Fundamental Patterns of Element Interactions

The world is currently experiencing unprecedented fluxes of biologically and geochemically active elements, but we lack the ability to forecast the consequences of

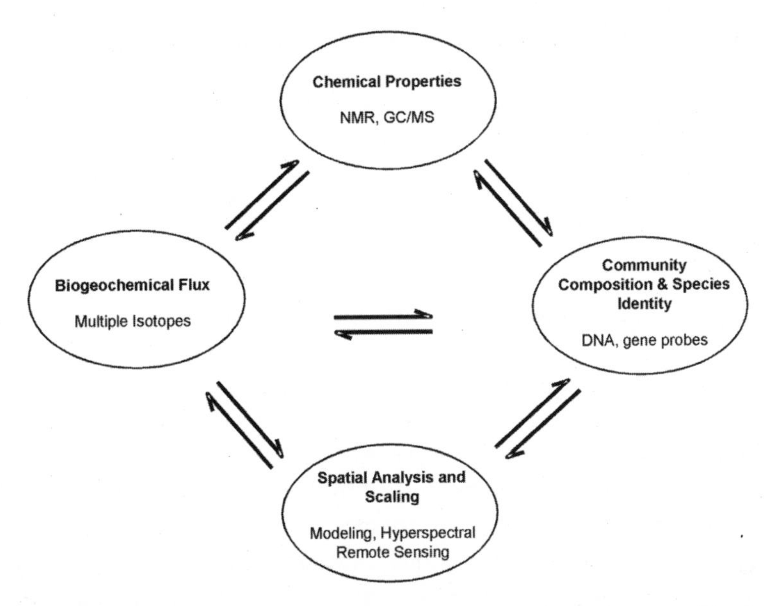

Figure 4.2. Potential for future element interactions research involving combined application of new methodological approaches

these fluxes (Austin et al., Chapter 2, this volume), in part, because our basic understanding of how element interactions are played out is insufficient. We have seen cases where organism stoichiometry provides a useful conceptual framework, but there are also numerous sources of variability that reduce the applicability of simple element ratios. What we presently lack is a basic understanding of the degree to which multiple element interactions follow globally generalizable patterns versus local sources of complexity.

Fundamental biogeochemical mechanisms such as nutrient turnover, atmospheric deposition, and biological N_2 fixation are common across ecosystems, but their relative importance varies tremendously across elements and environmental conditions. Experimentation in element interactions poses unique challenges because, in addition to the inherent complexity of biogeochemical interactions, their drivers are distributed heterogeneously across the globe. For example, whereas patterns of atmospheric nitrogen deposition are driven by human industrial and agricultural activities, patterns of geologically derived elements, such as phosphorus, typically follow the geological origin of the underlying parent material. The biota, which largely controls cycling of elements, also has independent patterns resulting from biogeographic mechanisms. Many large-scale experiments, those of both a manipulative nature and an observational nature, in a range of different ecosystems have added to our understanding of biogeochemistry. We need to move beyond these experiments, however, in trying to explore elemental interactions in a more comprehensive way, making use of a combination of the techniques discussed in the preceding sections (Figure 4.2).

A Suggested Experimental Approach for Basic Element Interactions Research

Single-factor manipulation experiments, statistically analyzed with analysis of variance techniques (e.g., response of element *X* to an addition of element *Y*), capture the behavior of a system at only two points and allow only partial understanding of element interactions. Compared with a world in which tremendous heterogeneity and nuanced responses are the norm rather than the exception, it is worth considering whether experimental studies should focus more on multifactorial/multitreatment experiments that more closely resemble the range of conditions present in natural systems. Such experiments would allow scientists to move beyond simple tests of significance toward derivation of full response surfaces, giving better indications of possible thresholds and a continuum of ecosystem responses. Experiments also need to be planned with a longer time frame (minimum of five years) than traditional funding cycles allow because element manipulations can involve long lag times before interactive responses become evident.

We fully recognize, however, that detailed multifactor and treatment experiments are not possible, or even desirable, everywhere. We suggest that parallel series of experiments could be linked to help address the central questions about element interactions. This series of experimental studies might have two tiers of tightly coordinated manipulative experiments and a dense network of monitoring sites. The first tier would be made up of a few multifactorial experiments that would explore major element interactions and their causal mechanisms. At this level the detailed chemical and biological identity techniques already described could be used to explore mechanistic responses to varying treatments in great detail. The first-tier experiments would be complemented by a larger number of less-intensive manipulative experiments distributed along broad gradients of the major biogeochemical drivers. These experiments would provide information on the response surface of the major driving variables and provide validation of the mechanisms suggested by the first-tier experiments.

Finally, a dense network of element cycling monitoring sites established across broad environmental gradients would complement manipulative experiments by providing information across a much wider range of conditions. This network would provide information on a range of biogeochemical parameters and their

Table 4.1. Elements of a proposed experimental framework for examining basic fundamental patterns of element interactions

Site level	*Elements of experimental design*	*Criteria for site or measurement selection*	*Duration and process level detail of measurements*	*Measurements targeted at key uncertainties and utilizing new technologies*
Tier 1	Observational (process based) measurements Multifactorial and/or multilevel manipulative experiments	Representative of mechanisms (and not necessarily climatic, edaphic, or organism/community type) Accessibility	Seconds to years High detail	Molecular and microbial identity/composition •Element pool sizes and fluxes •Ecosystem responses to perturbations (mediated by elemental availability) •Organism-ecosystem interactions
Tier 2	Experiments or observations arrayed along broad gradients of major biogeochemical drivers Broadly distributed experiments or observations that utilize similar measurements/manipulations	Representative of the important biogeochemical drivers (e.g., soil or vegetation type, climatic zones)	Months to decades Intermediate detail	•Ecosystem response to multiple stressors (e.g., N, CO_2, land use) •Elemental effects of change in biological community composition in time and space
Tier 3	Monitoring networks and environmental gradients for large-scale (regional to global) observations of element cycling	Broad geographic distribution	Years to centuries Low detail	•Satellite remote sensing of biosphere properties involving two or more elements

distribution and, we believe, would add tremendous value to regional to global-scale biogeochemical studies. Ecologists have often taken advantage of the natural variation that occurs along environmental gradients, and several efforts are already underway that could be complemented by the addition of element cycling measurements. The International Geosphere-Biosphere Programme (IGBP) has expanded this concept to the global scale with the idea of "transects." They have proposed a global system of about twelve transects, each of which would be thousands of kilometers long, spanning major gradients believed to be susceptible to human-induced changes and climate change. As an example, the Kalahari Transect, mostly in Botswana, is one of three proposed transects spanning the precipitation gradient between the humid tropics and the subtropical deserts. Koch et al. (1995) proposed the following set of benefits that could be gained from these transect efforts:

- The consequences of future changes could be inferred from patterns currently exhibited along the existing gradient.
- Threshold effects may be identified along a continuous environmental gradient.
- A mechanism for extrapolation would be gained that links the site-based work of ecologists to the regional scale of climatologists and policy makers.

Finally, many limitations in trying to quantify elemental interactions exist because of the constraints of time, available labor, and financial resources. Limitations in the availability of financial and human resources preclude replicating key experiments across the globe in a way that would match the existing combinations of drivers. Not only are resources scarce and insufficient, but they also have a skewed distribution, with a large fraction concentrated in developed countries. Most developing countries have smaller gross national products and devote a smaller fraction to research and development than developed countries.

Data Management and Synthesis

Besides collecting new data, we recognize that the scientific endeavor of understanding multiple element interactions needs to improve the management and archiving of existing data. Remarkable progress has occurred as a result of the systematic analysis and synthesis of data that were collected all over the world with different purposes. For example, compilations of existing data sets have given us important new insight into a wide range of topics, including global patterns of N cycling and ^{15}N distributions (Handley et al. 1999), linkages between vegetation N status and C uptake (Reich, Walters, and Ellsworth 1997; Green, Erickson, and Kruger 2003), plant response to altered CO_2 concentrations (Medlyn et al. 2001), and regional to global controls on C fixation (Sala et al. 1988). In many cases, it is not justified to collect new global data sets when reevaluation of existing data is feasible. For multiple scientific endeavors, there are examples of national-level organizations developed to facilitate the organization, storage, and distribution of data. For biodiversity, national services such as CONABIO (Comisión Nacional de Biodiversidad) in Mexico and others now play important roles in managing scientific data. Similar examples exist in many countries for weather data, remote-sensing information, and genetic research. International efforts are being developed under the auspices of the United Nations. Despite these signs of progress, a great deal of biogeochemical data remains broadly distributed and poorly characterized. The lack of centralized data storage and organization limits advances in modeling and synthesis and should be a central focus of new efforts for improved information management at a national or international level.

Conclusions

As with many fields of science, progress in the study of element interactions has often involved an interplay between technological advancements, which provide new information on poorly understood processes, and conceptual breakthroughs, which give us new perspectives on the systems we study and place new emphasis on the need for additional measurements. In this chapter we have highlighted several recent examples of this interaction and have suggested areas in which present knowledge is constrained by data availability, conceptual understanding, or a mismatch between the two.

The status of current ecosystem models illustrates some of the limitations in our present state of knowledge. Despite the well-known role of multiple element interactions in a number of biogeochemical processes, few ecosystem models explicitly include interactions beyond those that involve N and C. To a large extent, this reflects a lack of basic understanding regarding the processes controlling multiple element interactions and the degree to which they can be generalized at various spatial or temporal scales. In other cases model limitations arise either from limitations of the underlying paradigms on which they are based (e.g., whole-system optimization or single-element resource limitation) or from a failure to capture processes that act across multiple levels of organization (e.g., population, community, and ecosystem). Finally, models are frequently limited by a shortage of data available for parameterization or validation. This shortage stems both from the inherent difficulties of measuring certain biogeochemical variables and from discrepancies between variables used in models and those that can be readily measured in the field.

Current gaps in data availability require additional methodological development in a number of areas, including determination of chemical identity, species identity, and approaches for scaling existing information through space and time. Methods for determining chemical identity are important because the chemical properties of organic compounds often influence element interactions in ways that cannot be inferred from element ratios alone. Similarly, because the forms and ratios in which elements are used by organisms can vary among species, methods for determining species identity are important in systems where sampling individuals has historically been problematic (e.g., microbial communities, belowground plant tissues). Examples of recent technological innovations that offer promise in these areas include the use of NMR techniques for analyzing the structure and composition of complex organic compounds, development of bioreporters that produce observable responses to nutrient limitation by specific elements, and combined application of flow cytometry, stable isotopes, and DNA analysis for determining the influence of microbial species or functional groups.

The need to develop methods for scaling information on element interactions stems from the fact that important mechanisms occur at spatial scales ranging from microscopic to global and at temporal scales ranging from seconds to millennia. For example, relating microbial processes that affect nutrient dynamics in soils to patterns of atmosphere-biosphere exchange that affect element cycling globally requires well-validated methods by which information can be translated from one scale to another. Recent improvements in remote-sensing techniques such as laser altimetry and imaging spectroscopy offer great promise in this arena. For these approaches to explicitly address patterns of multiple element interactions, however, particularly with respect to elements that cannot be observed directly (e.g., those occurring in underlying soil substrate), they require further development and integration with other methods.

Lying beneath all of these individual challenges is the fact that the discipline of biogeochemistry still has a limited theoretical basis for understanding multiple element interactions. Although the importance of ecological stoichiometry as a potential controlling mechanism has been appreciated since Redfield's seminal paper on the nutrient ratios of plankton in marine ecosystems (Redfield 1958), we are also well aware of the many factors confounding a simple extension of organism-based ratios to global element cycling patterns. We suggest that one strategy for overcoming this hurdle would be to invest in a series of coordinated experiments intended to examine the response of individual elements to changes in the pools or fluxes of other elements. The structure of these experiments would include both intensive multifactorial manipulations designed to elucidate fundamental mechanisms and spatially distributed studies that focus on specific interactions but across broad environmental gradients. The intended outcome of these experiments, in combination, would be a series of multielement response surfaces and a better understanding of the degree to which element interactions are controlled by globally generalizable patterns versus local sources of complexity.

Literature Cited

Aber, J. D., K. Nadelhoffer, P. Steudler, and J. M. Melillo. 1989. Nitrogen saturation in northern forest ecosystems. *BioScience* 48:921–934.

Aber, J. D., C. L. Goodale, S. V. Ollinger, M.-L. Smith, A. H. Magill, M. E. Martin, J. L. Stoddard, and NERC Participants. 2003. Is nitrogen altering the nitrogen status of northeastern forests? *BioScience* 54:375–389.

Ågren, G. I., and E. Bosatta. 1998. *Theoretical ecosystem ecology: Understanding element cycles.* Cambridge: Cambridge University Press.

Anten, N. P. R., and T. Hirose. 2001. Limitations on photosynthesis of competing individuals in stands and the consequences for canopy structure. *Oecologia* 129:636–636.

Arendt, J. D. 1997. Adaptive intrinsic growth rates: Integration across taxa. *Quarterly Review of Biology* 72:149–177.

Armstrong, R. A. 1999. An optimization-based model of iron-light-ammonium colimitation of nitrate uptake and phytoplankton growth. *Limnology and Oceanography* 44:1436–1446.

Asner, G. P. 1998. Biophysical and biochemical sources of variability in canopy reflectance. *Remote Sensing of Environment* 64:234–253.

Berg, B., and H. Staaf. 1980. Decomposition rate and chemical changes of Scots pine needle litter. II. Influence of chemical composition. Pp. 373–390 in *Structure and function of northern coniferous forests: An ecosystem study,* edited by T. Persson. Ecological Bulletins (Stockholm) 32.

Berg, B., and C. McClaugherty. 2003. *Plant litter: Decomposition, humus formation, carbon sequestration.* Heidelberg: Springer-Verlag.

Binkley, D., and D. Valentine. 1991. 50-year biogeochemical effects of green ash, white-pine, and Norway spruce in a replicated experiment. *Forest Ecology and Management* 40:13–25.

Canadell, J., R. B. Jackson, J. R. Ehleringer, H. A. Mooney, O. E. Sala, and E. D. Schulze. 1996. Maximum rooting depth of vegetation types at the global scale. *Oecologia* 108:583–606.

Coops, N. C., and D. Culvenor. 2000. Utilizing local variance of simulated high-spatial resolution imagery to predict spatial pattern of forest stands. *Remote Sensing of Environment* 71:248–260.

Coops, N. C., and R. H. Waring. 2001. The use of multiscale remote sensing imagery to derive regional estimates of forest growth capacity using 3-PGS. *Remote Sensing of Environment* 75:324–334.

Crews, T. E., K. Kitayama, J. H. Fownes, D. A. Herbert, D. Mueller-Dombios, and P. M. Vitousek. 1995. Changes in soil phosphorus fractions and ecosystem dynamics across a long chronosequence in Hawaii. *Ecology* 76:1407–1424.

de Mazancourt, C., M. Loreau, and L. Abbadie. 1999. Grazing optimization and nutrient cycling: Potential impact of large herbivores in a savanna system. *Ecological Applications* 9:784–797.

Davidson, E. A., J. Chorover, and D. B. Dail. 2003. A mechanism of abiotic immobilization of nitrate in forest ecosystems: The ferrous wheel hypothesis. *Global Change Biology* 9:228–236.

Diner, D. J., G. P. Asner, R. Davies, Y. Knyzikhin, J. Muller, A. W. Nolin, B. Pinty, C. B. Schaaf, and J. Stroeve. 1999. New directions in earth observing: Scientific applications of multiangle remote sensing. *Bulletin of the American Meteorological Society* 80:2209–2228.

Durham, K. A., D. Porta, M. R. Twiss, R. M. L. McKay, and G. S. Bullerjahn. 2002. Construction and initial characterization of a luminescent Synechococcus sp. PCC 7942 Fe-dependent bioreporter. *Fems Microbiology Letters* 209:215–221.

Ehleringer, J. R., and C. B. Field, eds. 1993. *Scaling physiological processes: Leaf to globe.* San Diego: Academic Press.

Fassnacht, K. S., and S. T. Gower. 1997. Interrelationships among the edaphic and stand characteristics, leaf area index, and aboveground net primary production of upland forest ecosystems in north central Wisconsin. *Canadian Journal of Forest Research* 27:1058–1067.

Fowler, D., K. J. Hargreaves, U. Skiba, R. Milne, M. S. Zahniser, J. B. Moncrieff, I. J. Beaverland and M. W. Gallagher. 1995. Measurements of CH_4 and N_2O fluxes at the landscape scale using micrometeorological methods. *Philosophical Transactions of the Royal Society of London Series A: Mathematical, Physical and Engineering Sciences* 351:339–355.

Gbondo-Tugbawa, S. S., C. T. Driscoll, J. D. Aber, and G. E. Likens. 2001. Evaluation of an integrated biogeochemical model (PnET-BGC) at a northern hardwood forest ecosystem. *Water Resources Research* 37:1057–1070.

Gleixner, G., R. Bol, and J. Balesdent. 1999. Molecular insight into soil carbon turnover. *Rapid Communications in Mass Spectrophotometry* 13:1278–1283.

Gleixner, G., N. Poirier, R. Bol, and J. Balesdent. 2002. Molecular dynamics of organic matter in a cultivated soil. *Organic Geochemistry* 33:357–366.

Gower, S. T., K. A. Vogt, and C. C. Grier. 1992. Carbon dynamics of Rocky Mountain Douglas fir: Influence of water and nutrient availability. *Ecological Monographs* 62:43–65.

Gower, S. T., P. B. Reich, and Y. Son. 1993. Canopy dynamics and aboveground production of five tree species with different leaf longevities. *Tree Physiology* 12:327–345.

Green, D. S., J. E. Erickson, and E. L. Kruger. 2003. Foliar morphology and canopy nitrogen as predictors of light-use efficiency in terrestrial vegetation. *Agricultural and Forest Meteorology* 115:163–171.

Gressel, N., J. G. Mccoll, C. M. Preston, R. H. Newman, and R. F. Powers. 1996. Linkages between phosphorus transformations and carbon decomposition in a forest soil. *Biogeochemistry* 33:97–123.

Guenther, A., C. N. Hewitt, D. Erickson, R. Fall, C. Geron, T. Graedel, P. Harley, L. Klinger, M. Lerdau, W. A. Mckay, T. Oierce, B. Scholes, R. Steinbrecher, R. Tallamraju, J. Taylor, and P. A. Zimmerman. 1995. Global models of natural volatile organic compound emissions. *Journal of Geophysical Research* 100:8873–8892.

Guenther, A., L. Otter, P. Zimmerman, J. Greenberg, R. Scholes, and M. Scholes. 1996. Biogenic hydrocarbon emissions from southern African savannas. *Journal of Geophysical Research* 101:25859–25865.

Guenther, A., B. Baugh, G. Brasseur, P. Greenberg, L. Klinger, D. Serca, and Vierling.

1999. Isoprene emission estimates and uncertainties for the Central African EXPRESSO study domain. *Journal of Geophysical Research* 104:30625–30639.

Handley, L. L., A. T. Austin, D. Robinson, C. M. Scrimegour, J. A. Raven, T. H. E. Heaton, S. Schmidt, and G. R. Stewart. 1999. The 15N natural abundance (d15N) of ecosystem samples reflects measures of water availability. *Australian Journal of Plant Physiology* 26:185–199.

Harley, P., R. Monson, and M. T. Lerdau. 1999. Ecological and evolutionary aspects of isoprene emission from plants. *Oecologia* 118: 109–123.

Harrington, R. A., J. H. Fownes, and P. M Vitousek. 2001. Production and resource use efficiencies in N- and P-limited tropical forests: A comparison of responses to long-term fertilization. *Ecosystems* 4:646–657.

Horwath, W. R., and E. A. Paul. 1994. Microbial biomass. Pp. 753–773 in *Methods of soil analysis,* Part 2, *Microbial and biochemical properties,* edited by R. W. Weaver. SSSA Book Series No. 5. Madison, Wisc.: Soil Science Society of America.

Jackson, R. B., J. Canadell, J. R. Ehleringer, H. A. Mooney, O. E. Sala, and E. D. Schulze. 1996. A global analysis of root distributions for terrestrial biomes. *Oecologia* 108:389–411.

Jackson, R. B., L. A. Moore, W. A. Hoffmann, W. T. Pockman, and C. R. Linder. 1999. Ecosystem rooting depth determined with caves and DNA. *Proceedings National Academy of Sciences* 96:11387–11392.

Joffre, R., G. I. Ågren, D. Gillon, and E. Bosatta. 2001. Organic matter quality in ecological studies: Theory meets experiment. *Oikos* 93:451–458.

Koch, G. W., R. J. Scholes, W. L. Steffen, P. M. Vitousek, and B. H. Walker, eds. 1995. *The IGBP terrestrial transects: Science plan.* IGBP Global Change Report No. 36. Stockholm, Sweden: International Geosphere-Biosphere Programme.

Kolber, Z. S., R. T. Barber, K. H. Coale, S. E. Fitzwater, R. M. Greene, K. S. Johnson, S. Lindley, and P. G. Falkowski, 1994. Iron limitation of phytoplankton photosynthesis in the equatorial Pacific Ocean. *Nature* 371:145–149.

Lerdau, M. T., A. Guether, and R. Monson. 1997. Plant production and emission of volatile organic compounds. *BioScience* 47:373–383.

MacDonald, J. A., N. B. Dise, E. Matzner, M. Armburster, P. Gundersen, and M. Forsius. 2002. Nitrogen input together with ecosystem nitrogen enrichment predict nitrate leaching from European forests. *Global Change Biology* 8:999–1017.

Martin, M. E., and J. D. Aber. 1997. Estimation of forest canopy lignin and nitrogen concentration and ecosystem processes by high spectral resolution remote sensing. *Ecological Applications* 7:441–443.

Mayewski, P. A., and F. White. 2002. *The ice chronicles: The quest to understand global climate change.* Hanover, N.H.: University Press of New England.

McGill, W. B., and C. V. Cole. 1981. Comparative aspects of cycling of organic C, N, S, and P through soil organic matter. *Geoderma* 26:267–286.

McKay, R. M. L., J. La Roche, A. F. Yakunin, D. G. Durnford, and R. J. Geider. 1999. Accumulation of ferredoxin and flavodoxin in a marine diatom in response to Fe. *Journal of Phycology* 35:510–519.

McKnight, D. M., R. Harnish, R. L. Wershaw, J. S. Baron, and S. Schiff. 1997. Chemical characteristics of particulate, colloidal, and dissolved organic material in Loch Vale watershed, Rocky Mountain National Park. *Biogeochemistry* 36:99–124.

Mcmanus, J. B., M. S. Zahniser, D. D. Nelson, L. R. Williams, and C. E. Kolb. 2002.

Infrared laser spectrometer with balanced absorption for measurement of isotopic ratios of carbon gases. *Spectrochemica ACTA Part A: Molecular and Biomolecular Spectroscopy* 58:2465–2479.

Medlyn, B. E., C. V. M. Barton, M. S. J. Broadmeadow, R. Cuelemans, P. DeAngelis, M. Forstreuter, M. Freeman, S. B. Jackson, S. Kellomäki, E. Laitat, A. Rey, B. D. Sigurdsson, J. Strassemeyer, K. Wang, P. S. Curtis, and P. G. Jarvis. 2001. Stomatal conductance of forest species after long-term exposure to elevated CO_2 concentration: A synthesis. *New Phytologist* 149:247–264.

Melillo, J. M., J. D. Aber, and J. M. Muratore. 1982. Nitrogen and lignin control of hardwood leaf litter decomposition dynamics. *Ecology* 63:621–626.

Nannipieri, P., L. Badalucco, and L. Landi. 1994. Holistic approaches to study of populations, nutrient pools, and fluxes: Limits and future research needs. Pp. 231–238 in *Beyond the biomass: Compositional and functional analysis of soil microbial communities,* edited by K. Ritz, J. Dighton, and K. E. Giller. New York: John Wiley and Sons.

Neff, J. C., A. R. Townsend, G. Gleixner, S. Lehman, J. Turnbull, and W. Bowman. 2002. Variable effects of nitrogen additions on the stability and turnover of soil carbon. *Nature* 419:915–917.

Ollinger, S. V., M. L. Smith, M. E. Martin, R. A. Hallett, C. L. Goodale, and J. D. Aber. 2002. Regional variation in foliar chemistry and soil nitrogen status among forests of diverse history and composition. *Ecology* 83:339–355.

Ollinger, S. V., J. D. Aber, and C. A. Federer. 1998. Estimating regional forest productivity and water balances using an ecosystem model linked to a GIS. *Landscape Ecology* 13:323–34.

Peterjohn, W. T., M. B. Adams, and F. S. Gilliam. 1996. Symptoms of nitrogen saturation in two central Appalachian hardwood forest ecosystems. *Biogeochemistry* 35:507–522.

Preston, C. M. 2001. Carbon-13 solid-state NMR of soil organic matter: Using the technique effectively. *Canadian Journal of Soil Science* 81:255–270.

Redfield, A. C. 1958. The biological control of chemical factors in the environment. *American Scientist* 46:205–221.

Radajewski, S., P. Ineson, N. R. Parekh, and C. J. Murrell. 2000. Stable-isotope probing as a tool in microbial ecology. *Nature* 403:646–649.

Reich, P. B., M. B. Walters, and D. S. Ellsworth. 1997. From tropics to tundra: Global convergence in plant functioning. *Proceedings of the National Academy of Sciences of the United States of America* 94:13730–13734.

Reich, P. B., D. P. Turner, and P. Bolstad. 1999. An approach to spatially-distributed modeling of net primary production (NPP) at the landscape scale and its application in validation of EOS NPP products. *Remote Sensing of Environment* 70:69–81.

Sala, O. E., W. J. Parton, W. K. Lauenroth, and L. A. Joyce. 1988. Primary production of the central grassland region of the United States. *Ecology* 69:40–45.

Scholes, R. J., and B. H. Walker. 1993. *An African savanna: Synthesis of the Nylsvley study.* Cambridge: Cambridge University Press.

Senn, D. B., and H. F. Hemond. 2002. Nitrate controls on iron and arsenic in an urban lake. *Nature* 296:2373–2376.

Smith, M. L., S. V. Ollinger, M. E. Martin, J. D. Aber, and C. L. Goodale. 2002. Direct prediction of aboveground forest productivity by remote sensing of canopy nitrogen. *Ecological Applications* 12:1286–1302.

Sommer, U. 1992. Phosphorous-limited daphnia: Intraspecific facilitation instead of competition. *Limnology and Oceanography* 37:966–973.
Sterner, R. W., and J. J. Elser. 2002. *Ecological stoichiometry: The biology of elements from molecules to the biosphere.* Princeton: Princeton University Press.
Ustin, S. L., M. O. Smith, and J. B. Adams. 1993. Remote sensing of ecological processes: A strategy for developing and testing ecological models using spectral mixture analysis. Pp. 339–357 in *Scaling physiological processes: Leaf to globe,* edited by J. R. Ehleringer and C. B. Field. San Diego: Academic Press.
Valentini, R., S. Greco, G. Seufert, N. Bertin, P. Ciccioli, A. Cecinato, E. Brancaleoni, and M. Frattoni. 1997. Fluxes of biogenic VOC from Mediterranean vegetation by trap enrichment relaxed eddy accumulation. *Atmospheric Environment* 31:229–238.
Vernadsky, V. I. 1926. *The biosphere.* Leningrad: Nauchtechizdat. (Re-released as Vernadsky, V. I. 1998. *The biosphere: Complete annotated edition.* New York: Springer.)
Vitousek, P. M. 1984. Litterfall, nutrient cycling, and nutrient limitations in tropical forests. *Ecology* 65:285–298.
Vitousek, P. M., and H. Farrington. 1997. Nitrogen limitation and soil development: experimental test of a biogeochemical theory. *Biogeochemistry* 37:63–75.
Vitousek, P. M., K. Cassman, C. Cleveland, T. Crews, C. Field, N. Grimm, R. Howarth, R. Marino, L. Martinelli, E. Rastetter, and J. Sprent. 2002. Towards an ecological understanding of biological nitrogen fixation. *Biogeochemistry* 57/58:1–45.
von Liebig, J. 1847. *Chemistry in its applications to agriculture and physiology.* London: Taylor and Walton.
Wessman, C. A., J. D. Aber, D. L. Peterson, and J. M. Melillo. 1988. Remote sensing of canopy chemistry and nitrogen cycling in temperate forest ecosystems. *Nature* 335:154–156.

5

Potential for Deliberate Management of Element Interactions to Address Major Environmental Issues

Filip Moldan, Sybil Seitzinger, Valerie T. Eviner, James N. Galloway, Xingguo Han, Michael Keller, Paolo Nannipieri, Walker O. Smith Jr., and Holm Tiessen

Human activities often decouple elements from their natural stoichiometry through the selective release and mobilization of elements such as C, N, S, and P from their long-term stores (Austin et al., Chapter 2, this volume). Many environmental problems arise from these biogeochemical changes, including rising atmospheric CO_2 and the degradation of the quality of land, air, and water. A mechanistic understanding of elemental interactions provides an opportunity for deliberate manipulations of elements in order to mitigate environmental problems. In this chapter we demonstrate that recoupling of elements can be a successful strategy for minimizing or alleviating the harmful effects of human activities and for restoring degraded ecosystems.

There are two fundamentally different pathways by which elemental interactions can be manipulated to achieve environmental goals. The first is to limit the source of the excess elemental inputs (limiting primary forcing). Examples include decreasing the emissions of pollutants or greenhouse gases or minimizing the applications of fertilizers or other chemicals. A second approach is to minimize the effects of excess elements by introducing another change to counteract the undesirable effects of the primary forcing. Examples of enhanced secondary forcing would be liming of acidified lakes and soils in order to mitigate acidification or enhancing carbon sequestration by fertilizing terrestrial systems with nitrogen and oceans with iron.

Minimizing the direct source of an environmental problem is the most desirable and effective approach. Even though this approach is often relatively straightforward, an understanding of element interactions may be critical for successful mitigation of a problem. For example, reductions in tropospheric ozone (O_3) concentrations by minimizing emissions of volatile organic carbon (VOC) were less efficient than expected, particularly in urban environments, because ozone production was frequently limited by concentrations of nitrogen oxides (NO_x) rather than by VOC (Chameides et al. 1988; NRC 1991).

Mitigation of environmental problems by enhancing secondary forcing is a much less reliable option. Manipulation of the environment, even with the aim to restore it, involves inevitable risks of side-effects that could be difficult to foresee. Manipulating elemental interactions might cure one environmental problem but also create or intensify another. Thus, a broad view needs to be taken in order to evaluate the overall effects of a manipulation on multiple aspects of the environment. To illustrate the differences between the two approaches, we will examine the case of lake acidification. Reductions in the deposition of acidic compounds will likely allow for the recovery of lakes and downstream ecosystems (Galloway, Chapter 14, this volume). Decreases in acid deposition will also improve air quality and acidified soils in the lake's catchment. On the other hand, liming of lakes might be equally effective in restoring lake pH. The extinct organisms may not be brought back by liming, however, and the relative abundance of species established in the lake after the liming might be different from pre-acidification because of the different pathway by which the lake chemistry was achieved. Furthermore, by not treating the source of the problem, the lakes will need to be limed indefinitely and other efforts will be needed to address the negative effects of acid deposition on, for example, air quality, soils, corrosion, and human health. From the environmental point of view, it is clearly more efficient and effective to limit the primary forcing by reducing emissions of pollutants that cause the acidification rather than employing ecosystem manipulations that will address only a subset of the problem and may contribute to other environmental problems.

The popularity of the enhanced secondary forcing approach is more political than ecological. And although it has been argued that limiting the source of the problem would have negative economic impacts, the suite of ecosystem manipulations that are necessary to secondarily address the problems of increased pollution emissions are likely much more expensive and less effective. Minimizing the primary forcing, although desirable, has its limits set by the expanding needs of the world's population (Figure 5.1). Enhanced secondary forcing might be necessary if parts of the environment are to be protected. It could also help to preserve the environment over critical periods of time between when a problem is recog-

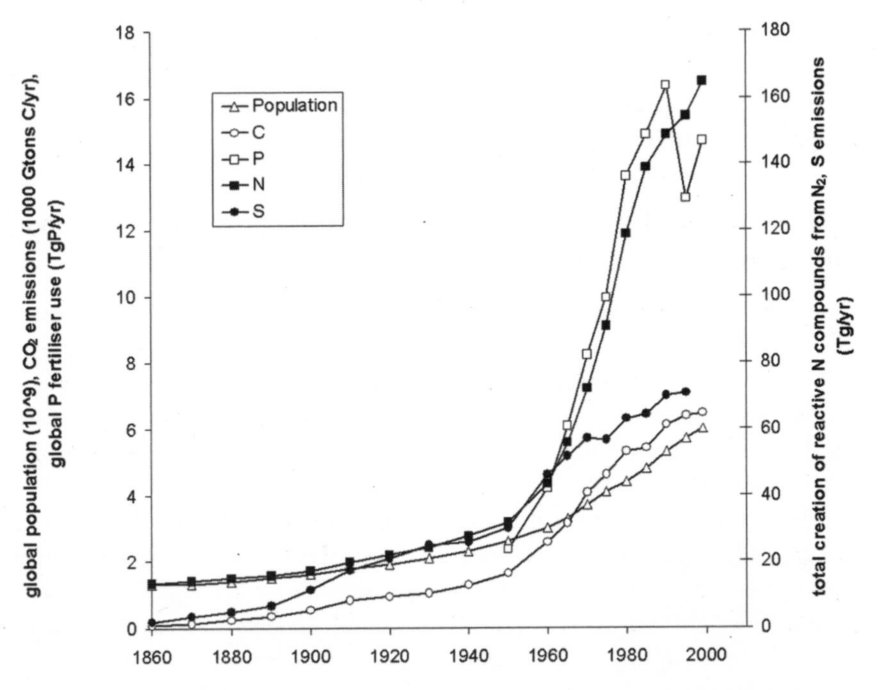

Figure 5.1. Global population (billions), total CO_2 emissions from fossil fuels (1,000 Gtons C/yr), global P fertilizer consumption (Tg P/yr), and total creation of reactive N compounds from N_2 by fossil-fuel combustion, cultivation-induced biological N_2 fixation, and the Haber-Bosch process (Tg/yr). Data on CO_2 emissions from fossil fuel combustion, P fertilizer consumption, and SO_2 emissions from fossil fuel combustion and sulfide ore-smelting were provided by May Ver (University of Hawaii) and reported upon in Mackenzie, Ver, and Lerman (2002). Specific sources of the data were: CO_2, Marland et al. (2002); P, FAO (1950–1999); S and N, Dignon and Hameed (1989), Dignon (1992), Brown, Renner, and Flavin (1997), and Galloway et al. (2003).

nized and when the cause of the problem is under control. Understanding element interactions enables us both to limit the primary forcing and to minimize damage through secondary forcing mechanisms. Which of the two pathways (or what combination of the two) is the best strategy will vary for different environmental problems.

In this chapter we address the potential for deliberate element manipulations both to limit the primary forcing and to enhance the secondary forcing in a broad sense, with some examples on scales from local to global and from different environments and different parts of the globe. The addressed environmental problems are divided into four areas:

1. air, water, and soil quality degradation due to emissions of nitrogen and sulfur
2. increased greenhouse gas emissions
3. nutrient enrichment of coastal ecosystems
4. land degradation

Air, Water, and Soil Quality Degradation Due to Emissions of Nitrogen and Sulfur

Energy production results in the emission of NO_x and SO_2 to the atmosphere. Food production results in the emission of S (H_2S, organic S) and N (NH_3, N_2O, NO, organic N) species to the atmosphere. Large regions of all continents (except Antarctica) have experienced significant increases in these emissions over the past few decades, and there are projections for significant increases in the future, especially in the developing world.

Agroecosystems are defined as cropland and animal production systems (i.e., animal feeding operations, or AFOs, and pastures). Both fertilized fields and AFOs are large sources of NH_3, N_2O, and NO emissions to the atmosphere, as well as N and P losses to aquatic ecosystems. The N and P budgets for an agroecosystem can be described by its inputs, its internal recycling, and its outputs. These outputs are products and losses to water and, for N, to air (Figure 5.2). These losses have negative impacts on both the farm economy and the environment and should be reduced as much as possible in a financially sustainable way.

These emissions to air and water have consequences for the health of people and aquatic and terrestrial ecosystems. SO_2 is converted to sulfate aerosol in the atmosphere, where it reduces radiation and negatively impacts human health. Once deposited to the Earth's surface, it has the potential to acidify both soils and freshwater ecosystems (Galloway, Chapter 14, this volume). N_2O is a greenhouse gas in the troposphere and contributes to stratospheric ozone depletion. The other N species also have environmental impacts. NO_x contributes to increased levels of O_3 in the troposphere (Finlayson-Pitts and Pitts 2000), which can result in decreased productivity of terrestrial ecosystems (Ollinger et al., Chapter 4, this volume) and have significant human health effects (Follet and Follet 2001; Wolfe and Patz 2002). NH_3 emitted to the atmosphere can react with sulfuric acid creating an ammonium sulfate aerosol. NH_3 plays an important role in the direct and indirect effects of aerosols on radiative forcing and thus global climate change (Penner, Robinson, and Woods 2001; Seinfeld and Pandis 1998).

In addition to these direct effects, NO_x emissions contribute to a wide variety of other environmental impacts as they are converted to other chemical species and cycle through environmental reservoirs. Referred to as the "nitrogen cascade" (Galloway et al. in press; Galloway, Chapter 14, this volume) and the "phospho-

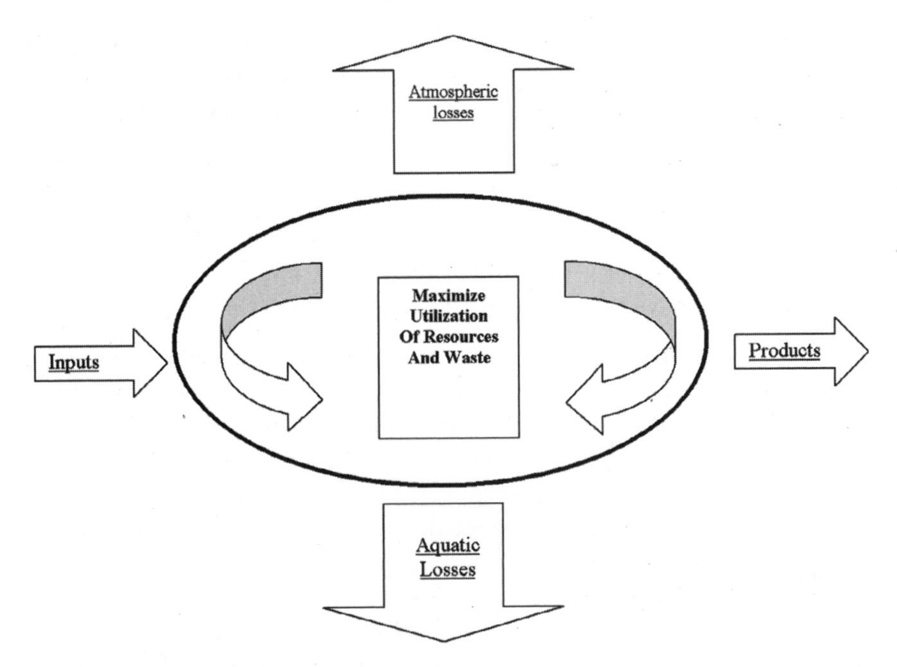

Figure 5.2. Schematic diagram of animal feeding operations (AFO), animals plus associated cropland. Adapted from NRC (2003).

rus cascade" (Newbold et al. 1983), the sequential transfer of reactive N and P through environmental systems results in environmental changes. N and P move through or are temporarily stored within each system, where they contribute to terrestrial N saturation (Aber et al. 1998), biodiversity losses (Eviner and Chapin, Chapter 8, this volume), and freshwater and coastal eutrophication (Rabalais 2002).

Element Interactions That Can Be Used to Address the Issue

Emissions of NO_x and SO_2 from Fossil Fuel Combustion

Understanding the interactions of SO_2 and NO_x with other chemical species has provided the basis for the development of the technology to reduce SO_2 and NO_x emissions to the atmosphere by treatments before, during, or after fossil fuel combustion. SO_2 is removed by reaction with $CaCO_3$ producing $CaSO_4$. NO_x is converted to N_2 by catalytic reaction with NH_3. These technologies have been extensively used in Europe and to a lesser degree in North America to cause substantial reductions in SO_2 and NO_x emissions. Although the application of the technology in developing regions is less common, in some regions SO_2 emissions have been extensive owing to change in fuel type (e.g., China). A major barrier to applying the existing technology in developing regions is cost. To address this issue, current research using our knowledge of elemental interactions is investigating how to make reductions in NO_x emissions more effective and cheaper (e.g., Bradley and Jones 2002).

Emissions from Agroecosystems

Element interactions can be manipulated through a number of pathways in agroecosystems to reduce the losses of reactive N and P to the environment, to increase the recycling of N and P within the agroecosystem, and thus to increase the efficiency of N and P use in agroecosystems. Careful selection of crops and proper composition and timing of fertilizer application and irrigation can maximize crop uptake of available N and P (Matson, Naylor, and Ortiz-Monasterio 1998). More frequent, smaller inputs of fertilizer that are timed with plant demand are more likely to be taken up by plants and microbes and less likely to be lost to runoff or to the atmosphere. N retention of fertilizer additions can be greater from organic fertilizers, which have gradual N release, with the C:N ratio of the organic fertilizer influencing the rate of release. As with inorganic fertilizers, careful attention must be given to the amount and timing of manure application to minimize nutrient losses. Similarly, slow-release inorganic fertilizers, such as S-coated urea, can minimize leaching losses. Nitrification inhibitors can be used with N fertilizers to minimize N losses by NO_3^- leaching. Plant compounds such as phenolics have also been shown to inhibit nitrification, and nitrification can be minimized by growing these species as cover crops or adding their residues to the soil. Minimizing the spatial and seasonal prevalence of bare soil on croplands by cover cropping reduces leaching and erosion losses. In order to maximize the effectiveness of cover crops, plant species should be chosen based on desirable traits such as fast growth and extensive N and P uptake at the time of maximum leaching loss or at a time or depth different from the crop. The C:N ratio of cover crops can determine how quickly the cover crop residue is mineralized or to what extent these nutrients are immobilized by the microbial community.

In animal production systems, diet management ensures that the proper mix of N and P is present for uptake by animals, thus minimizing losses through excretion. Once animal waste is excreted, biogeochemical interactions may be used to maximize the loss of reactive N as N_2 to the atmosphere. For example, N emissions from AFOs could be reduced by using current knowledge of elemental ratios to create N_2 production systems that first maximize nitrification (aerobic systems) and then maximize denitrification to produce N_2 (anaerobic systems).

This approach would be especially appropriate where economic constraints limit waste recycling directly to agroecosystems (NRC 2003).

In a perfect management system all N and P added would be stoichiometrically matched to carbon and other elements in the production of crops or animals. The essence of the practices described is to use the minimum N and P necessary to support a given unit of productivity and to potentially decrease the losses of N and P to the water and atmosphere. The current management of conventional agroecosystems, however, makes some amount of loss inevitable.

Increased Greenhouse Gas Emissions

A serious environmental problem that has received significant attention is the accumulation of carbon dioxide and other radiatively active gases in the atmosphere, which in turn increases the surface temperatures of the world. Limitation of greenhouse gas emissions is the most direct means to reduce atmospheric increases. But because anthropogenic emissions of CO_2, CH_4, and N_2O are directly tied to critical economic activities of food and energy production, strong source controls are politically contentious and difficult to implement. Future increases predicted for the global population and increasing "industrialization" confound attempts to reduce emissions. Even though strong source controls are controversial, it should be noted that there is substantial potential for limiting emissions by conserving energy and by using environment-friendly technologies without significantly affecting lifestyle. An alternative or a complement to limiting the primary forcing by emission controls is the enhancement of sinks, particularly for CO_2, by forcing natural ecosystem processes to enhance the long-term retention of carbon. Understanding of the interaction of elemental cycles and the interactions of mainly C, N, and P will be critical to the management of greenhouse gases. Developing viable long-term solutions is a formidable challenge, as will be discussed.

Element Interactions That Might Be Used to Address the Issue

Terrestrial Ecosystems

In terrestrial ecosystems, carbon storage will depend upon the balance of net primary productivity (NPP) and heterotrophic respiration (Rh) (Randerson et al. 2002). Thus, CO_2 sequestration may be increased either by augmenting NPP or reducing the rate of Rh. A number of element interactions potentially come into play in controlling this balance. For a given area NPP may be limited by availability of energy (radiation), water, and/or of one or several nutrient elements. In

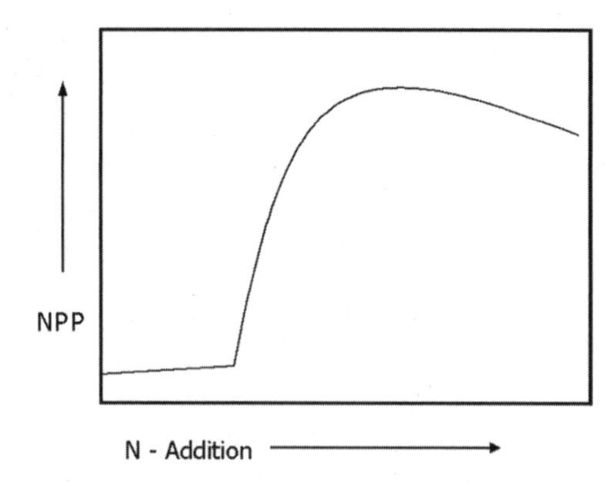

Figure 5.3. Conceptual relationship between net primary production (NPP) and N addition. Adapted from Aber et al. (1989, 1998).

general, we cannot control radiation. Irrigation is used extensively to increase NPP for crops. Fertilization with nutrient elements (N, P, K, etc.), either intentionally or inadvertently (in the case of deposition of atmospheric pollutants), can also increase the NPP of managed and unmanaged terrestrial ecosystems.

Nitrogen is the main limiting nutrient for primary production in many temperate terrestrial ecosystems (Ågren et al., Chapter 7, this volume). In Figure 5.3 we illustrate a theoretical response of productivity to a given level of nitrogen addition. Systems on the left-hand side of the graph are severely nitrogen limited, and these respond to nitrogen addition first by immobilization of added N. The degree to which the added N is partitioned between plants, decomposers, and nonliving organic matter will depend upon the form and timing of the addition, as well as plant species composition (Holland and Carroll, Chapter 15, this volume). NPP will increase with N addition until it reaches a level where it is no longer limiting. Beyond that level, the system will become nitrogen saturated (Aber et al. 1989, 1998) and productivity will decline.

Enhanced sequestration of C in forests associated with increased N inputs has received considerable attention (Schindler and Bayley 1993; Nadelhoffer et al. 1999). The relationship between N inputs and C sequestration, however, is complex. We suggest a framework based on interactions between N and C that may explain responses of decomposition to N additions. A number of studies have found that high availability of soil inorganic N stimulates decomposition of recalcitrant litter and old soil C. In contrast, additions of high-N litter tend to lower utilization of recalcitrant C compounds (Billes, Gandaisriollet, and Bottner 1986; van Ginkel, Gorissen, and Van Veen 1996). While initial decomposition rates of high-N litter are rapid, total mass loss is often lower for high-quality litter than for low-quality litter, with high-N litter making a greater contribution to soil

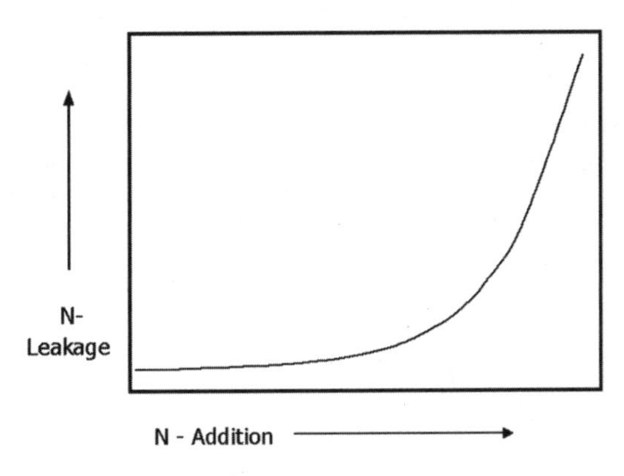

Figure 5.4. Conceptual relationship between N leakage and N addition. Adapted from Aber et al. (1989, 1998).

organic C storage (Berg 2000; Eviner 2001). The effects of N additions on decomposition will also depend on labile C availability to microbes. Labile C can stimulate decomposition of recalcitrant litter and soil organic C under low nutrient conditions, since microbes use labile C as an energy source to metabolize recalcitrant C in search of N. In the presence of high amounts of N, microbes utilize the available C and N and do not decompose more recalcitrant substrates (van Ginkel, Gorissen, and Van Veen 1996).

Based on these experimental results, we hypothesize that when microbes have access to easily available N and C, they will not utilize recalcitrant sources. This hypothesis should apply whether they are presented with these in the form of labile litter, or labile C and inorganic N in soil solution. The presence of either labile C or inorganic N, without the other, will stimulate microbial utilization of more recalcitrant substrates, both as litter and soil organic C. Available C and N in high-quality litter is quickly utilized, but the remaining C in this litter becomes relatively stable. This scenario may be complicated by shifts in the microbial community, which differentially mediate interactions between these elements. In addition, protozoal grazing of rhizosphere bacteria may release N and alter interactions between N and C. Thus, the effect of nitrogen additions on decomposition can vary depending on the amount of nitrogen added, its fate in plant material versus inorganic soil pools, and its influence on labile C through influencing exudation of a species or changing species composition.

There are also possible trade-offs from adding N to sequester C. Additions of N to increase NPP will also increase N leakage from the ecosystem (Aber et al. 1989). Depending upon environmental conditions, N losses may occur in forms such as NO_3^- (lost to groundwater) or N_2O (emitted to the atmosphere). Figure 5.4 illustrates the proportion of N leakage versus N applied. In highly N-limited systems, N leakage will be minimized and the dominant form of N loss will be dissolved organic nitrogen (DON) (Perakis and Hedin 2002). At some level of N addition, leakage of more environmentally sensitive forms such as NO_3^- and N_2O will occur. The leakage will increase, probably nonlinearly, through and beyond the level of N saturation.

Aquatic Ecosystems

Because of their vast volume, world oceans are frequently considered a long-term sink for the increasing CO_2 in the atmosphere. One proposal has been to fertilize the ocean with a limiting nutrient (e.g., iron, or Fe) and thereby stimulate NPP. Assuming vertical flux of carbon to a depth below 3,000 meters, the C from the atmosphere could hypothetically be retained for periods of about 1,000 years. Various regions of the ocean (e.g., the Southern Ocean, the North Pacific, and the South Pacific) are iron-limited (Ducklow, Oliver, and Smith, Chapter 16, this volume), and trace additions result in increased production and biomass accumulation. Because of elemental interactions, however, there are problems associated with enhancing the secondary forcing by ocean fertilization. Removal of iron limitation will result in nitrogen, silica (Si), or phosphorus limitation, thus restricting enhancement of NPP in some regions. Furthermore, secondary limitations may induce significant planktonic assemblage changes, which will then influence export and food web transformations. Addition of significant amounts of Fe to stimulate productivity in the Southern Ocean may result in the generation of anoxic waters at 300–800 meters (Sarmiento and Orr 1991), which would disrupt the extant food web of the region. Most (99.9 percent) of the surface production is remineralized in the upper 3,000 meters (Suess 1981), and hence natural carbon sequestration is a relatively inefficient process. Finally, even if all available surface nutrients were utilized to produce biomass, ocean sequestration of carbon would only decrease *the rate of increase* in atmospheric CO_2 by 10 percent (Sarmiento et al. 1998). Given that surface nutrients cannot be completely removed due to elemental interactions (e.g., Si limitation can preclude complete removal of N and P), the quantitative effect of forced oceanic carbon sequestration is small relative to emission increases.

Freshwater systems store a proportionately large amount of carbon in sediments by virtue of the large bottom area-to-volume ratio and sediments that are highly enriched in particulate organic carbon (POC). Therefore, even though lakes and reservoirs have relatively small areas, they can contribute significantly to carbon sequestration (Dean and Gorham 1998). Carbon turnover in these systems is faster than in the ocean (because of the finite lifetime of most reservoirs and shallow lakes), and so this C storage would be more short term (hundreds of years).

Nutrient Enrichment of Coastal Ecosystems

Nutrient inputs to coastal ecosystems around the world have markedly increased, as discussed in this volume by Austin et al. (Chapter 2). The environmental consequences of nutrient enrichment of coastal ecosystems are now well documented (e.g., Smetacek et al. 1991; Vitousek et al. 1997; NRC 2000; Cloern 2001; Rabalais 2002) and include increased phytoplankton production, increased turbidity with subsequent loss of submerged aquatic vegetation, oxygen deficiency, decreased biodiversity, and changes in species composition and food web structure. The specific effects on a coastal ecosystem depend on a number of factors including both the quantity and quality (forms and ratios) of nutrient inputs (Seitzinger et al. 2002a). The widespread alteration in elemental inputs due to human activity is likely to continue in the future (Kroeze and Seitzinger 1998; Austin et al., Chapter 2, this volume).

Major sources of N and P to coastal ecosystems include agroecosystems, human sewage, and, for N, atmospheric deposition. A portion of the N and P from these sources in watersheds is sequentially transferred to groundwater, riparian zones, wetlands, rivers, and coastal marine ecosystems. At each step along the way N and P can be temporarily stored or permanently removed within each system. This sequential transfer has been referred to as a P cascade (Newbold et al. 1983) and more recently applied to N as an "N cascade" (Galloway et al. 2003; Galloway, Chapter 14, this volume).

Element Interactions That Might Be Used to Address the Issue

Once nutrients are lost from terrestrial systems, there is the potential for nutrient removal at a number of locations along the nutrient cascade before they reach coastal systems. Knowledge of element interactions can be used to optimize nutrient removal along this path. For example, the selection of plant species to store nutrients and influence nutrient recycling rates can be important in transition zones between agricultural lands and aquatic systems, such as riparian strips. Although riparian strips can reduce total P export by 98 percent (Martinelli, Chapter 10, this volume), bioavailable P load may increase in some cases because of leaching from standing vegetation. This possibility highlights the importance of mechanisms other than plant uptake in providing effective control of nutrients. Harvesting aboveground vegetation in buffer zones can remove nitrogen and phosphorus and prevent possible nutrient loss to aquatic systems through leaching or mineralization from residues. This step can be especially important in buffer zones that may become N saturated over time. This solution may not be sustainable, however, since biomass harvest will remove a significant amount of other nutrients over time, impacting the long-term ability of these transition zones to take up nitrogen because of limitation of plant growth by other elements.

Wetlands are another transitional ecosystem that can influence export of nutrients downstream to coastal ecosystems. Wetlands can be important zones of denitrification, and thus remove soluble inorganic N from the water. Denitrification, however, may result in release of N_2O, a potent greenhouse gas. Plant species in wetlands can have significant effects on the conditions necessary for denitrification, including oxygen levels and the presence of organic C. High organic C availability per unit of NO_3^- is key to driving denitrification to N_2, so planting species with high labile C inputs could minimize the trade-off between water and air quality. This approach, however, can also favor dissimilatory nitrate reduction to ammonium, which would preserve N in aquatic systems.

Wetlands may decrease inorganic N inputs to aquatic systems, but they often increase organic N inputs. While these organic inputs are commonly only 20 percent or less bioavailable, they still contribute to coastal eutrophication (Stepanauskas, Leonardson, and Tranvik 1999; Wiegner 2002). Although wetlands may play an important role in minimizing N inputs to aquatic systems, there are trade-offs in management for N versus P. An oxidized soil environment is most likely to cause sorption and precipitation of P on Fe oxides. Under the reducing conditions that cause denitrification and reduce N transfers, Fe^{3+} will be reduced to Fe^{2+} and release bound phosphate to percolating water. Although denitrification occurs at a higher pH than the conversion of Fe^{3+} to Fe^{2+}, it is impossible to manage the reducing potential of riparian zone soils closely enough to control both N and P transfers or retention.

Once N and P enter streams and rivers, a number of factors can affect their transfer to coastal ecosystems. Hydrology is one. For example, at the watershed scale, approximately 30–70 percent of the N inputs to the stream/river network can be denitrified (Seitzinger et al. 2002c). Channeling rivers tends to decrease the residence time of water, leading to a decrease in the interactions of solution and particulate N with the sediments, thus decreasing denitrification rates. Transfer of N and P to terrestrial systems (such as floodplains) also is minimized by channelization, due to faster water flux and less contact with floodplains. Reservoirs in rivers increase water residence time within that reach and thus can increase removal of both N (denitrification, particulate burial) and P (particulate burial). At the watershed scale the effect of reservoirs on nutrient export to coastal systems is highly dependent upon where in the watershed the reservoirs are located and how much of the total N and P pass through the reservoirs (Seitzinger et al. 2002c). Reservoirs are also a sink for dissolved Si.

Nutrients that are not retained within terrestrial ecosystems or removed along the nutrient cascade in riparian zones, wetlands, rivers, and reservoirs enter coastal ecosystems. There they contribute to a wide range of environmental problems, as already discussed. Although we are beginning to understand factors that can reduce N, P, and/or Si within ecosystem types, considerable additional knowledge of element interactions is required to effectively optimize nutrient retention within watersheds to reduce effects of nutrients in coastal systems. For example, within coastal ecosystems advancement in knowledge is needed on which nutrients (N, P, Si) and what nutrient forms (inorganic, organic) are limiting biological productivity over space and time and what the quantitative relationships between nutrient inputs are (quantities, forms, and ratios between them) and the effects on coastal ecosystems. This information must be coupled with advancements in knowledge of the relative magnitude of N, P, and Si (by form) removal within ecosystem types within watersheds and factors controlling their removal. This information then needs to be integrated at the watershed scale. Approaches are required that are applicable across a wide range of geographical regions because coastal eutrophication is a problem worldwide.

Land Degradation

Land use represents another form of primary forcing on the environment. Element loss, induced by decoupling biological elemental cycles from their natural stoichiometry through processes such as harvest, biomass destruction, and cultivation may lead to land degradation. Land use can also cause soil erosion, which is another major factor causing land degradation. In natural ecosystems mineralized N is rapidly incorporated into plant or microbial biomass, while only a small portion is lost by leaching or gaseous loss. P is largely conserved within the organic phase. Continued harvest of plant material removes these nutrients, and continued cultivation of an ecosystem requires the mobilization of N and P for crop export. Without this uncoupling of elements from the natural cycle, no production is possible unless nutrients are externally supplied. Many conventional farms in industrialized nations meet these nutrient demands through additions of inorganic fertilizers. Traditional shifting cultivation achieves this decoupling of nutrient elements by slash and burn. This practice results in substantial nutrient losses because most biomass C and N are released in the fire. P and cations are largely preserved in the ashes, but the vegetation-free soil will subsequently lose some of these elements through leaching or erosion. So although the decoupling of elements is essential to agriculture, the inefficient way in which this is accomplished in many systems causes large losses of organic matter and nutrients.

The effect of grazers is one of partial biomass destruction. Long-term overgrazing in grassland ecosystems not only changes the species composition and productivity through selective foraging and direct destruction, but also decreases soil fertility and changes the hydrological cycles (Dormar and Willms 1998; Wu and Tiessen 2002). In degraded grassland ecosystems of the world, soils are often prone to desertification because of diminishing organic matter input, accelerated evaporation, and destruction of surface soil structure. This desertification is usually accompanied by nutrient loss through surface soil erosion and nutrient leaching.

Finally, land degradation by direct pollution of the soil has become a critical issue in urban areas, mine sites, and other areas contaminated by toxic wastes.

Element Interactions That Might Be Used to Address the Issue

Avoiding the land degradation due to differential nutrient loss requires that elements be re-linked by slowing losses or facilitating the production of biomass and the accretion of organic matter. The concern over soil degradation and the shortage of productive land has prompted a search for alternatives that maintain N and P in organic matter and restore the balance of nutrients. This goal is being accomplished in a number of ways. In subhumid and semiarid environments the introduction of managed short fallows with a high proportion of leguminous trees or the planting of leguminous cover crops restore C and N budgets in the ecosystem (Jama et al. 2000). In humid tropical environments the use of chipped woody vegetation as a soil mulch supplies C and ties up soil N before it can be leached. In many farm operations, the use of plant residues and inorganic fertilizers successfully supplies nutrients, organic matter, and soil cover, restoring element linkages and preventing losses (Palm, Myers, and Nandwa 1997; da Silveira, Tiessen, and Tonneau 2001). Use of leguminous trees in the tropics (MacDicken 1994) in agroforestry and traditional farming systems is an effective way to maximize soil C input and to increase N input through N fixers, while simultaneously alleviating Al-toxicity and overcoming P deficiency (Haynes and Mokolobate 2001). This is an example of how local farmers empirically apply the principle of elemental interactions (C-N-P and Al) to stabilize yield, minimize risk, and maintain soil fertility in their traditional agricultural practices. These cultivation methods have evolved over centuries and can easily be seen in Mesoamerica, the Andean region, the Amazon Basin, and Asia (Altieri 1995; Pretty 1995).

In grazed ecosystems the maintenance of vegetation cover and ecosystem production is critical for protecting the system from the decoupling of nutrient interactions. Therefore, in degraded grassland ecosystems, remediation depends on limiting grazing impact by protecting land and destocking, thereby keeping the cycles of C, N, P, and S in biologically dominated processes rather than in chemically or physically controlled processes. Protection of these systems does not necessarily mean that grazing by animals is completely excluded. To some degree, free-ranging mammalian grazers can increase the availability of diet-enhancing nutrients (e.g., N and Na), thereby increasing their own carrying capacity (McNaughton, Banyikwa, and McNaughton 1997).

Legislation often requires remediation of polluted soils. Remediation is often accomplished by plants, which not only accumulate heavy metals but also promote the degradation of organic pollutants through the microbial community in their rhizosphere. Appropriate selection of plant species is key in successful bioremediation because of plants' differential ability to accumulate toxins (Banuelos 1996; Predieri et al. 2001), exude compounds that form stable complexes with contaminants (Balabane et al. 1999; Dousset et al. 2001), and enhance degradation through the exudation of labile C substrates (Anderson, Kruger, and Coats 1994; Liste and Alexander 1999, 2000). Nutrient imbalance of soil can markedly affect plant growth and also decomposition of recalcitrant pollutants. Similar to the substrate interactions influencing decomposition of plant residues and soil organic matter, the presence of easily degradable organic carbon can affect the degradation of recalcitrant organic pollutants. N inputs also influence this degradation, with high inputs of N fertilizers required in petroleum-polluted soils.

Conclusions

An understanding of elemental interactions is critical to effectively manage a variety of environmental problems. Managing ecosystems requires an understanding of the impact of elemental interactions on the pool sizes, rates, and timing of element cycling. Scientific theory often focuses on generalizations and broad-scale patterns, but management occurs at a local scale. In order for management to be effective, it must call on scientific theories that incorporate the specific interactions at that site and can explain the exceptions to the rule, not just the general trends. This is where a solid understanding of interactions between various elements at different concentrations is invaluable. For example, N additions can enhance microbial utilization of soil organic matter, but also can enhance the stabilization of organic matter. Understanding the role of interactions between N, C, and the microbial community is critical to manage soil C at a local scale.

Effective management of many elements also requires some guidelines for the desired balance of elements, the acceptable ranges of variation around this ratio, and the thresholds beyond which biogeochemical processes significantly change. For example, fertilization of crops or restoration sites by organic additions requires a C:N ratio that provides a balance of nutrient release and retention. Similarly, increasing the ratio of labile C to NO_3^- can enhance N_2 production during denitrification or, alternatively, can lead to dissimilatory nitrate reduction to ammonia and a retention of N in wetland and aquatic systems. Understanding the threshold at which this happens could be critical in managing N losses.

In many cases primary forcing has an impact at geographically distant areas. Air pollution can be transported over long distances. Emissions of greenhouse gases or depletion of stratospheric ozone are examples of environmental problems with global implications in which the geographical location of the sources of primary forcing plays a less important role. On a smaller scale the functioning of coastal ecosystems will depend on how effectively we can manage N, P, and Si export from terrestrial ecosystems. An example of the economic context at different scales is shown by slash-and-burn subsistence agriculture with shortened fallow periods that lead to land degradation. By definition, subsistence agriculture contributes only to the farm families' welfare and has no wider economic benefit. The land degradation resulting from inadequate management affects local and regional land quality. The greenhouse gas emissions produced by the slash and burn and subsequent soil organic matter mineralization are ultimately of global concern. Remedial action must occur in this context, and in the absence of on-farm resources, the regional or even global society will have to subsidize the remedial investment in sustainable land quality. This need is now recognized by major agricultural research and development agencies (Sanchez et al. 1997).

The difficulty of managing N losses in aquatic systems highlights another important management consideration. While many studies focus on the ecosystem effects of a small number of drivers at one time, systems must be managed simultaneously for multiple functions in response to multiple environmental changes. A knowledge of elemental interactions is key to evaluating the trade-offs of any management practice. For example, in order to minimize N input to aquatic systems, we might encourage denitrification. This step can have consequences for air quality, however, if N_2O is released. Similarly, in order to promote C storage, plant species with high litter quality may be utilized, while species with low litter quality are preferable for N immobilization. Because the cycling of elements is controlled by different factors, there will be trade-offs in managing for multiple elements. Ecologists need to highlight the overall impact of any element manipulation, because there are multiple ways to manage any element and the preferred method may be based on the trade-offs for other elements.

Ecology is often seen as a science that is distinct from, for example, physics and chemistry owing to its lack of general laws. Because ecosystems are a result of complex interactions, understanding the biogeochemical consequences of interactions of a large number of elements and environmental conditions, as well as the exceptions to the general patterns, greatly strengthens both the science of ecology and its application.

Overall Points

- Many environmental problems are a result of temporal and spatial decoupling of elements from the range of their natural balance.
- There are two pathways in which elemental interactions could be manipulated to achieve environmental goals. These are (1) limiting primary forcing and (2) enhancing secondary forcing.
- There are three main mechanisms for deliberate management of element interactions: (1) altering the inputs, (2) countering imbalances by selective additions of other elements, and (3) altering the species composition (to manage demand for individual nutrients, to recouple nutrients based on biotic stoichiometry, to overcome toxicity).
- Because elements interact, management practices that influence one elemental cycle will often influence others in different ways. Thus, any given management practice can have trade-offs or multiple benefits, and it is necessary to focus on the overall biogeochemical effects of a management practice to evaluate preferable methods.
- Presenting managers with the mechanistic framework that encompasses the range of biogeochemical interactions allows them to simultaneously manage multiple elements under variable local conditions.
- Element interactions have been effectively used to address environmental problems in a number of cases. In other cases current knowledge suggests that elemental interactions could be used more efficiently.

Specific Examples

Emissions of S and N

- While some of the environmental consequences of S and N emissions are clear, economic considerations often guide policy. Future research should consider the effectiveness and costs of mitigation efforts and compare these with the economic costs and environmental benefits of decreasing emissions and the costs of unmitigated S and N emissions. Such cost-benefit analyses are critical to guide policy decisions.
- Natural ecosystems have tight nutrient cycles linked to organic carbon. Agricultural use disrupts these tight cycles to make elements available for export and harvest but also makes the systems leaky. Recoupling of the nutrients to carbon is needed to reduce environmental impact.
- Where economic constraints limit waste recycling directly to agroecosystems, manipulation of elemental ratios, particularly through the use of coupled aero-

bic and anaerobic disposal systems, provides opportunities to limit waste impacts in general and specifically on denitrification.

Greenhouse Gas Emissions

- Stimulation of productivity by fertilization of both marine and terrestrial ecosystems has been proposed to decrease the net emissions of CO_2. In marine ecosystems the predicted maximum decrease in the rate of increase of atmospheric CO_2 that could be accomplished by Fe fertilizer is only 10 percent. In terrestrial ecosystems the relative effects of nutrient addition on productivity versus decomposition need to be considered.
- Total radiative impact associated with any management practice must be assessed by looking at the weighted production of all greenhouse gases (e.g., CO_2 sinks against N_2O and CH_4 sources).

Coastal Ecosystems

- Coastal eutrophication depends on controls of N, P, and Si in watersheds and in airsheds. Export of nutrients from terrestrial to aquatic systems can be minimized by optimizing crop utilization of applied nutrients, effectively managing manure from animal production systems, utilizing advanced wastewater treatment, and adopting technologies to reduce NO_x emissions from fossil fuel combustion.
- Reductions in nutrient export to aquatic systems can also be achieved by recoupling inorganic nutrients by integrating them with organic matter or by using elemental interactions to stimulate alternative loss pathways such as denitrification. Potential tools for these mitigation efforts are cover cropping, buffer strips, and wetlands. The quantity and stoichiometry of nutrient retention or export is a function of land use and land cover. Careful choice of the configuration of land use units in the landscape is key to managing nutrient losses.

Land Degradation

- Degradation of organic pollutants by microbial communities and heavy metal uptake by plants can be limited by nutrient availability.

Literature Cited

Aber, J. D., W. H. McDowell, K. J. Nadelhoffer, A. Magill, G. Berntson, M. Kamakea, S. G. McNulty, W. Currie, L. Rustad, and I. Fernandez. 1998. Nitrogen saturation in temperate forest ecosystems: Hypotheses revisited. *BioScience* 48:921–934.

Aber, J. D., K. J. Nadelhoffer, P. Steudler, and J. M. Melillo. 1989. Nitrogen saturation in northern forest ecosystems. *BioScience* 39:378–386.

Altieri, M. A. 1995. *Agroecology: The science of sustainable agriculture.* Boulder, Colo.: Westview Press.

Anderson, T. A., E. L. Kruger, and J. R. Coats. 1994. Enhanced degradation of a mixture of three herbicides in the rhizosphere of a herbicide-tolerant plant. *Chemosphere* 28:1551–1557.

Balabane, M., D. Faivre, F. van Oort, and H. Dahmani-Muller. 1999. Mutual effects of soil organic matter dynamics and heavy metals fate in a metallophyte grassland. *Environmental Pollution* 105:45–54.

Banuelos, G. S. 1996. Managing high levels of boron and selenium with trace element accumulator crops. *Journal of Environmental Science and Health.* Part A: *Environmental Science and Engineering and Toxic and Hazardous Substance Control* 31:1179–1196.

Berg, B. 2000. Litter decomposition and organic matter turnover in northern forest soils. *Forest Ecology and Management* 133:13–22.

Billes, G., N. Gandais-Riollet, and P. Bottner. 1986. Effect of Gramineae culture on the decomposition of C-14 and N-15-labeled litter in soil under controlled conditions. *Acta Oecologica-Oecologia Plantarum* 7:273–286.

Bradley, M. J., and B. M. Jones. 2002. Reducing global NOx emissions: Promoting the development of advanced energy and transportation technologies. *Ambio* 31:141–149.

Brown, L. R., M. Renner, and C. Flavin. 1997. *Vital signs: The environmental trends that are shaping our future.* New York: Worldwatch Institute.

Chameides, W. L., R. W. Lindsay, J. Richardson, and C. S. Kiang. 1988. The role of biogenic hydrocarbons in urban photochemical smog: Atlanta as a case study. *Science* 241:1473–1475.

Cloern, J. E. 2001. Our evolving conceptual model of the coastal eutrophication problem. *Marine Ecology Progress Series* 210:223–253.

da Silveira, L. M., H. Tiessen, and J. P. Tonneau. 2001. Organic matter management in family agriculture of semiarid Paraiba, Brazil. *Nutrient Cycling in Agroecosystems* 61:215–223.

Dean, W. E., and E. Gorham. 1998. Magnitude and significance of carbon burial in lakes, reservoirs, and peatlands. *Geology* 26:535–538.

Dignon, J. 1992. NOx and SOx emissions from fossil fuels: A global distribution. *Atmospheric Environment* 26:1157–1163.

Dignon, J., and S. Hameed. 1989. Global emissions of nitrogen and sulfur oxides from 1860 to 1980. *Journal Air Pollution Control Association* 39:180–186.

Dormar, J. F., and W. D. Willms. 1998. Effect of forty-four years of grazing on fescue grassland soils. *Journal of Range Management* 51:122–126.

Dousset, S., J. L. Morel, A. Jacobson, and G. Bitton. 2001. Copper binding capacity of root exudates of cultivated plants and associated weeds. *Biology and Fertility of Soils* 34:230–234.

Eviner, V. T. 2001. Linking plant community composition and ecosystem dynamics: Interactions of plant traits determine the ecosystem effects of plant species and plant species mixtures. Ph.D. diss., University of California, Berkeley.

FAO (Food and Agriculture Organization of the United Nations). 1950–1999. *Annual fertilizer review.* Rome.

Finlayson-Pitts, B. J., and J. N. Pitts. 2000. *Chemistry of the upper and lower atmosphere: Theory, experiments, and applications.* London: Academic Press.
Follett, J. R., and R. F. Follett. 2001. Utilization and metabolism of nitrogen by humans. Pp. 65–92 in *Nitrogen in the environment: Sources, problems, and management,* edited by R. F. Follett and J. L. Hatfield. Amsterdam: Elsevier Science.
Galloway, J. N., J. D. Aber, J. W. Erisman, S. P. Seitzinger, R. W. Howarth, E. B. Cowling, and B. J. Cosby. 2003. The nitrogen cascade. *BioScience* 53:341–356.
Haynes, R. J., and M. S. Mokolobate. 2001. Amelioration of Al toxicity and P deficiency in acid soils by addition of organic residues: A critical review of the phenomenon and the mechanisms involved. *Nutrient Cycling in Aroecosystems* 59:47–63.
Jama, B., C. A. Palm, R. J. Buresh, A. Niang, C. Gachengo, G. Nziguheba, and B. Amadalo. 2000. Tithonia diversifolia as a green manure for soil fertility improvement in western Kenya: A review. *Agroforestry Systems* 49:201–221.
Kroeze, C., and S. P. Seitzinger. 1998. Nitrogen inputs to rivers, estuaries, and continental shelves and related nitrous oxide emissions in 1990 and 2050: A global model. *Nutrient Cycling in Agroecosystems* 52:195–212.
Liste, H. H., and M. Alexander. 1999. Rapid screening of plants promoting phenanthrene degradation. *Journal of Environmental Quality* 28:1376–1377.
———. 2000. Plant-promoted pyrene degradation in soil. *Chemosphere.* 40:7–10.
MacDicken, K. G. 1994. *Selection and management of nitrogen-fixing trees.* Morrilton, Ark.: Winrock International Institute for Agricultural Development.
Mackenzie, F. T., L. M. Ver, and A. Lerman. 2002. Century-scale nitrogen and phosphorus controls of the carbon cycle. *Chemical Geology* 190:13–32.
Marland, G., T. A. Boden, and R. J. Andres. 2002. Global, regional, and national CO_2 emissions. In *Trends: A compendium of data on global change.* Oak Ridge, Tenn.: Carbon Dioxide Information Analysis Center, Oak Ridge National Laboratory, U.S. Department of Energy.
Matson, P., R. Naylor, and P. Ortiz-Monasterio. 1998. Integration of environmental, agronomic, and economic aspects of fertilizer management. *Science* 280:112–115.
McNaughton, S. J., F. F. Banyikwa, and M. M. McNaughton. 1997. Promotion of the cycling of diet-enhancing nutrients by African grazers. *Science* 278:1798–1800.
Nadelhoffer, K. J., B. A. Emmet, P. Gundersen, O. J. Kjønaas, C. J. Koopmans, P. Schleppi, A. Tietema, and R. F. Wright. 1999. Nitrogen deposition makes a minor contribution to carbon sequestration in temperate forests. *Nature* 398:145–148.
NRC (National Research Council). 1991. Rethinking the ozone problem in urban and regional air pollution. Washington, D.C.: National Academy Press.
———. 2000. Clean coastal waters: Understanding and reducing the effects of nutrient pollution. Washington, D.C.: National Academy Press.
———. 2003. Air emissions from animal feeding operations: Current knowledge, future needs. Final Report, Ad Hoc Committee on Air Emissions from Animal Feeding Operations, Committee on Animal Nutrition, Washington, D.C.
Newbold, J. D., J. W. Elwood, R. V. O'Neill, and A. L. Sheldon. l983. Phosphorus dynamics in a woodland stream ecosystem: A study of nutrient spiralling. *Ecology* 64:1249–1265.
Palm, C. A., R. J. K. Myers, and S. M. Nandwa. 1997. Combined use of organic and inorganic nutrient sources for soil fertility maintenance and replenishment. Pp. 193–

217 in *Replenishing soil fertility in Africa*, edited by R. J. Buresh, P. Sanchez, and F. Calhoun. SSSA Special Publication No. 51. Madison, Wisc.: Science Society of America.

Penner, M., C. Robinson, and M. Woods. 2001. The response of good and poor aspen clones to thinning. *Forestry Chronicle* 77:874–884.

Perakis, S. S., and L. O. Hedin. 2002. Nitrogen loss from unpolluted South American forests mainly via dissolved organic compounds. *Nature* 415:416–419.

Predieri, S., J. Figaj, L. Rachwal, E. Gatti, and F. Rapparini. 2001. Selection of woody species with enhanced uptake capacity: the case-study of Niedwiady resort pollution by pesticides stored in bunkers. *Minerva Biotecnologica* 13:111–116.

Pretty, J. 1995. *Regenerating agriculture.* Washington, D.C.: World Resources Institute.

Rabalais, N. 2002. Nitrogen in aquatic ecosystems. *Ambio* 31:102–112.

Randerson, J. T., F. S. Chapin, J. Harden, J. C. Neff, and M. Harmon. 2002. Net ecosystem production: A comprehensive measure of net carbon accumulation by ecosystems. *Ecological Applications* 12:937–947.

Sanchez, P. A., K. D. Shepherd, M. J. Soule, F. M. Place, R. J. Buresh, A. N. Izac, A. U. Mokwunye, F. R. Kwesiga, C. G. Ndiritu, and P. L. Woomer. 1997. Soil fertility replenishment in Africa: An investment in natural resource capital. Pp. 1–46 in *Replenishing soil fertility in Africa*, edited by R. J. Buresh, P. A. Sanchez, and F. Calhoun. SSSA Special Publication No. 51. Madison, Wisc.: Science Society of America.

Sarmiento, J. L., T. M. C. Hughes, R. J. Stouffer, and S. Manabe. 1998. Simulated response of the ocean carbon cycle to anthropogenic climate warming. *Nature* 393:245–249.

Sarmiento, J. L., and J. C. Orr. 1991. Three-dimensional simulations of the impact of Southern Ocean nutrient depletion on atmospheric CO_2 and ocean chemistry. *Limnology and Oceanography* 36:1928–1950.

Schindler, D. W., and S. E. Bayley. 1993. The biosphere as an increasing sink for atmospheric carbon: Estimates from increased nitrogen deposition. *Global Biogeochemical Cycles* 7:717–733.

Seinfeld, J. H., and S. N. Pandis. 1998. *Atmospheric chemistry and physics: From air pollution to global change.* New York: John Wiley and Sons.

Seitzinger, S. P., C. Kroeze, A. F. Bouwman, N. Caraco, F. Dentener, and R. V. Styles. 2002a. Global patterns of dissolved inorganic and particulate nitrogen inputs to coastal systems: Recent conditions and future projections. *Estuaries* 25:640–655.

Seitzinger, S. P., R. W. Sanders, and R. V. Styles. 2002b. Bioavailability of DON from natural and anthropogenic sources to estuarine plankton. *Limnology and Oceanography* 47:353–366.

Seitzinger, S. P., R. V. Styles, E. W. Boyer, R. B. Alexander, G. Billen, R. Howarth, B. Mayer, and N. van Breemen. 2002c. Nitrogen retention in rivers: Model development and application to watersheds in the northeastern U.S.A. *Biogeochemistry* 57:199–237.

Smetacek, V., U. Bathmann, E.-M. Nöthig, and R. Scharek. 1991. Coastal eutrophication: Causes and consequences. Pp. 251–279 in *Ocean margin processes in global change*, edited by R.F.C. Mantoura. New York: John Wiley and Sons.

Stepanauskas, R., L. Leonardson, and L. Tranvik. 1999. Bioavailability of wetland-derived DON to freshwater and marine bacterioplankton. *Limnology and Oceanography* 44:1477–1485.

Suess, E. 1981. Phosphate regeneration from sediments of the Peru continental margin by dissolution of fish debris. *Geochimica et Cosmochimica Acta* 45:577–588.

van Ginkel, J. H., A. Gorissen, and J. A. Van Veen. 1996. Long-term decomposition of grass roots as affected by elevated atmospheric carbon dioxide. *Journal of Environmental Quality* 25:1122–1128.

Vitousek, P. M., J. D. Aber, R. W. Howarth, G. E. Likens, P. A. Matson, D. W. Schindler, W. H. Schlesinger, and D. Tilman. 1997. Human alteration of the global nitrogen cycle: Sources and consequences. *Ecological Applications* 7:737–750.

Wiegner, T. N. 2002. Microbial and photochemical degradation of allochthonous dissolved organic matter within river systems and watersheds. Ph.D. diss., Rutgers, The State University of New Jersey.

Wolfe, A., and J. A. Patz. 2002. Nitrogen and human health: Direct and indirect impacts. *Ambio* 31:120–125.

Wu, R., and H. Tiessen. 2002. Effect of land use on soil degradation in alpine grassland soil, China. *Soil Science Society of America Journal* 66:1648–1655.

PART II
Theory

6

Stoichiometry and Flexibility in the Hawaiian Model System

Peter M. Vitousek

Most analyses of ecosystem-level nutrient cycling include multiple elements but consider them one at a time, separately evaluating controls of N, P, and cation cycling. At one level this is a reasonable approach, because element cycles differ substantially in their sources and controlling mechanisms. Organisms, however, require a full suite of essential elements simultaneously, and the supply of each essential element can thereby affect the cycling of all the others. One way to incorporate these effects is to evaluate element interactions, the reciprocal influences of element cycles (Bolin and Cook 1983). The concept of stoichiometry, however, provides a more fundamental and useful approach. In this chapter I discuss how stoichiometry—and its complement, flexibility—can be used to analyze nutrient cycling and limitation.

The basis of the stoichiometric approach is straightforward (Sterner and Elser 2002). Chemical reactions occur at characteristic ratios of reactants and yield characteristic ratios of products, all of which can be defined in terms of their elemental composition. Moreover, many biochemical reactions are catalyzed by enzymes that themselves have defined elemental compositions and take place within organisms that have more or less defined elemental compositions. Chemical reactions and organisms both require all of their reactants and catalysts—and in the case of organisms, their structures—if they are to proceed and/or grow. Although the elemental composition of products, reactants, catalysts, and structures is an incomplete description of reactions (or organisms), elements are the most conservative component of these reactions. Unlike energy, unlike organic or inorganic compounds, elements are neither created nor consumed—and it is possible to calculate a mass balance for any element in any reaction (other than fission, fusion, or radioactive decay) and for any organism.

Stoichiometric approaches have long been embedded in ecology—for example, they underlie the use of critical C:N ratios in decomposition and nutrient release. Stoichiometry was applied explicitly on a very broad scale by Redfield (1958), who described relationships among C, N, P, and S in marine algae and bacteria, and consequently among the cycles of C, N, P, S, and O in the ocean. One legacy of his pioneering analysis is that C:N:P ratios in marine phytoplankton are termed the Redfield ratios. Reiners (1986) later described "the stoichiometry of life" as one of the fundamental bases of ecosystem ecology, suggesting that all living organisms could be separated into "protoplasmic life"—which might follow the Redfield ratios—and structural components that can be enriched in particular elements (C for terrestrial plants, Ca and P for terrestrial vertebrates).

More recently, Sterner and Elser (2002) built upon their own and others' research to broaden the scope of the stoichiometric approach substantially, developing and evaluating its implications on levels of organization from organelles to ecosystems. Among many contributions, they analyzed the variability in element ratios within as well as among groups of organisms, demonstrating that the marine phytoplankton discussed by Redfield (1958) have the least variable ratios, whereas terrestrial plants are the most variable. Much of the variation within groups of organisms is caused by differences in the quantity and biochemistry of structural tissues, as Reiners (1986) suggested, and some is due to storage or "luxury consumption" (uptake in excess of immediate requirements) of elements when they are abundant. "Protoplasmic life" itself, however, has variable ratios of N:P as well as C:N and C:P, owing in part to an association between rapid growth rates and high P concentrations (Elser et al. 1996).

I agree with Reiners (1986) and Sterner and Elser (2002) that together with ecological energetics, biological stoichiometry offers a solid (if incomplete) framework for understanding ecological systems. Further, I believe that linking the biological stoichiometries they discuss with the geochemical stoichiometries exhibited by element inputs to and outputs from terrestrial ecosystems can provide a common language and framework for integrating biogeochemistry as a whole. Finally, I believe that the flexibility of element cycling needs to be considered along with stoichiometry. I define "flexibility" as the ability of a particular element to cycle more rapidly within or between ecosystems than do other elements undergoing similar transformations. When they are in demand, elements with flexible cycles are transformed or transferred more rapidly than would be expected based on the underlying stoichiometry of the organisms and/or processes involved. Flexibility is important because in the long run, it is the element with the least flexible cycle rather than the one that is least abundant (relative to organisms' immediate requirements) that limits biological processes.

In this chapter I analyze stoichiometry and flexibility in forest ecosystems arrayed along a substrate age gradient in the Hawaiian Islands. Hawaii is a useful model system for this analysis (Vitousek, in press), because the islands' geological constancy, relatively low biological diversity, wide range of climates and substrate ages, and clear geochemical tracers of the sources of different elements (Crews et al. 1995; Chadwick et al. 1999; Kurtz, Derry, and Chadwick 2001; Vitousek 2002) make it possible to focus on fundamental biogeochemical processes with a clarity that is difficult to duplicate elsewhere. First, I analyze the stoichiometry and flexibility of within-system nutrient cycling, focusing on processes that could allow P to cycle through soil organic matter to organisms more rapidly than does N. Next, I discuss the stoichiometry and flexibility of element inputs to Hawaiian ecosystems. Finally, I evaluate constraints to flexibility in the case of biological N fixation, addressing why N limitation to NPP and other ecosystem processes occurs despite the apparent flexibility of N inputs that fixation provides.

Stoichiometry and Flexibility of Within-System Element Cycling

From the perspective of stoichiometry, organisms never demand one single element; their demand is rather for energy and a suite of elements in characteristic ratios that vary across groups of organisms (Sterner and Elser 2002). Nutrient cycling within ecosystems can be viewed as a consequence of groups of organisms with very different energy sources (autotroph versus heterotroph, consumer versus decomposer) and characteristic stoichiometries making their livings in the same place—with the tissues and/or waste products of each group being the fundamental resources for some or all of the others.

To date, the stoichiometric approach has been developed more fully in aquatic than terrestrial ecosystems—in large part because the slow turnover and overwhelming abundance of structural and other recalcitrant C compounds in terrestrial ecosystems makes them relatively difficult to evaluate. Nevertheless, as Sterner and Elser (2002) point out, there is a great deal to be gained from applying a stoichiometric perspective to terrestrial ecosystems—especially in that the widest differences in C:element ratios among groups of organisms anywhere occur between terrestrial plants and bacteria, the dominant producers and decomposers of terrestrial ecosystems.

I calculated the ratios of C:N and N:P in the leaves, leaf litter, roots, and wood of the dominant tree *Metrosideros polymorpha* from across the Hawaiian age gradient. Trees vary substantially across tissues and sites in their C:N ratios; they are particularly C-rich (N-poor) in the infertile young and old sites on the sequence (Figure 6.1), reflecting a greater relative investment in structural material in those

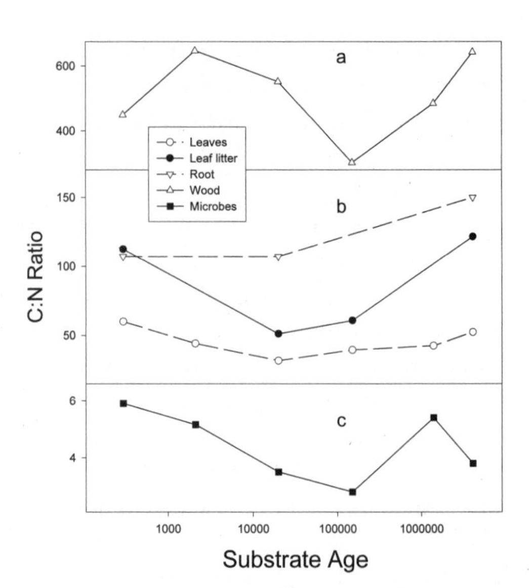

Figure 6.1. C:N ratios in the wood, roots (Ostertag and Hobbie 1999), leaves (Vitousek, Turner, and Kitayama 1995), and leaf litter (Herbert and Fownes 1999) of *Metrosideros polymorpha* and in soil microbial biomass (Torn et al. submitted) across the Hawaiian age gradient. Plant tissues have much wider C:N ratios than do microbes, particularly in the infertile young and old sites on the gradient.

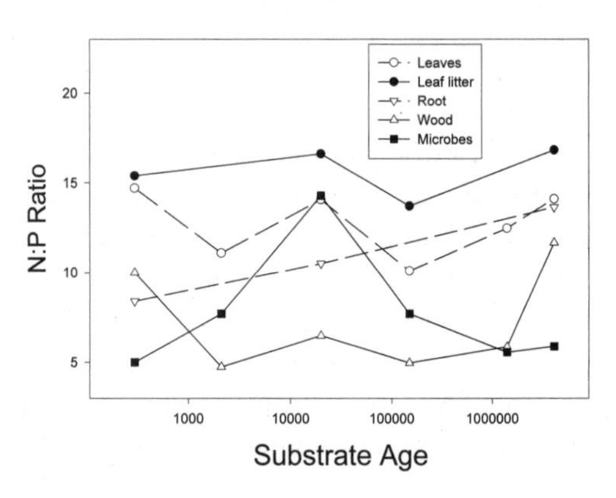

Figure 6.2. N:P ratios in *Metrosideros polymorpha* and microbial biomass across the Hawaiian age gradient; sources as in Figure 6.1. While microbes and wood have relatively low ratios, the variation in N:P ratios among plant tissues and between plants and microbes is much less than that for C:N or C:P ratios.

sites. In contrast, N:P ratios vary less across sites and tissues, displaying a relatively consistent stoichiometry (Figure 6.2) despite the fact that N limits plant productivity in the youngest site whereas P does so in the oldest (Vitousek and Farrington 1997).

The leaves of several additional plant species were sampled across the Hawaiian age gradient, again yielding substantial variation in C:N but similar N:P ratios among species and sites. Other elements follow quite different patterns. For example, Ca:P ratios vary by a factor of more than six for *Metrosideros* leaves collected across the Hawaiian age gradient, and by a factor of more than fifteen among species (Figure 6.3). The stoichiometry of ferns differs substantially from that of angiosperms; ferns are low in Mg and especially Ca, and slightly enriched

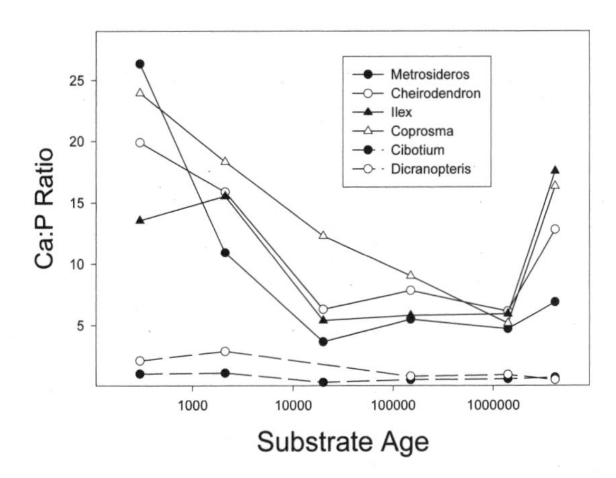

Figure 6.3. Ca:P ratios in leaves of six species that occur across the Hawaiian age gradient, from Vitousek, Turner, and Kitayama (1995). Ferns (dashed lines) have much narrower Ca:P ratios than angiosperms in all sites.

in K. I wonder if ferns' stoichiometry affects their decomposability; fungi in particular require relatively large quantities of Ca (Silver and Miya 2001), and Scowcroft (1997) observed that tree fern litter decomposes very slowly despite high concentrations of N and P and low concentrations of polyphenols and particularly lignin in their tissues (Hättenschwiler et al. in press).

Element ratios in microbial biomass were measured using chloroform fumigation/extraction (Torn et al. submitted). I am not convinced that this technique is truly quantitative, especially in organic soils. The extraordinary contrast, however, between C:N ratios in plants versus microbes—a factor of 10 to 200 (Figure 6.1)—is consistent with our understanding of plant versus microbial physiology and with observations in a wide range of ecosystems (Paul and Clark 1996; Sterner and Elser 2002). Microbial N:P ratios also are low relative to those in most plant tissues (except wood), though to a much lesser extent (Figure 6.2). Relative to their own requirements, microbes utilize C-rich and N- and P-poor resources across all of the Hawaiian sites, particularly the youngest and oldest. Under these conditions, microbial growth efficiency should be low, and the N and P that microbes do acquire should be retained effectively. Moreover, since P in plant tissues appears even less abundant than N (relative to microbial requirements) across most of the age gradient, a simple stoichiometric analysis would suggest that P should be retained within microbes more effectively than N in most of the sites.

In fact, the high C:N and C:P ratios in low-nutrient young and old sites are associated with slow rates of decomposition and even slower rates of mineralization (Figure 6.4), contributing to a plant-soil-microbial positive feedback that slows nutrient cycling substantially. Moreover, there is some evidence that the supply of P controls rates of decomposition more than does the supply of N, even in sites where N limits primary production (Vitousek 1998; Hobbie and Vitousek

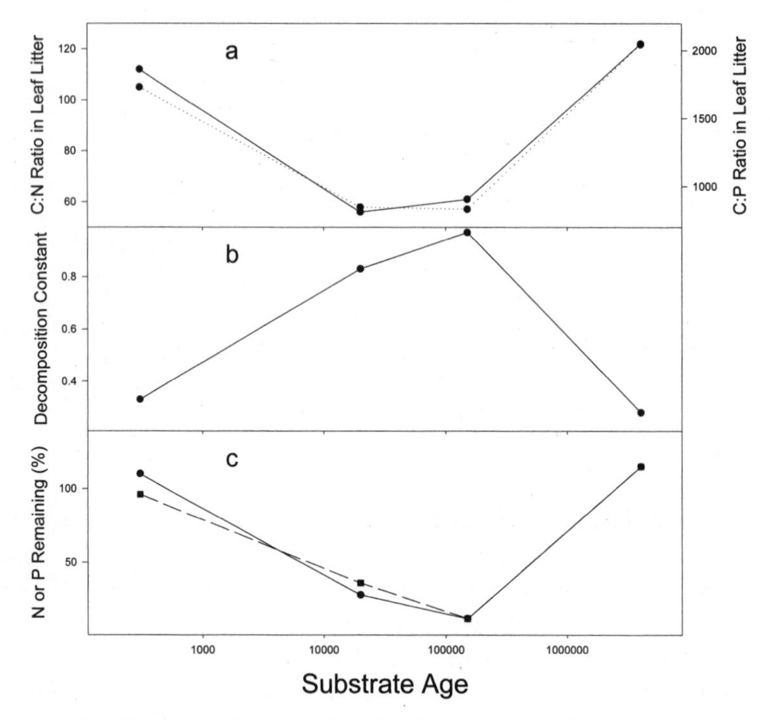

Figure 6.4. Leaf litter stoichiometry, decomposition, and N and P mineralization in sites across the Hawaiian age gradient, reanalyzed from Crews et al. (1995). (a) The C:N (solid line) and C:P (dotted line) ratios in leaf litter. (b) Decomposition rate constants, for litter decomposed *in situ.* (c) Percentage of initial N (solid line) and P (dashed line) mineralized after two years of decomposition.

2000). Several features of litter and soil organic matter and their decomposition complicate this simple analysis, however, making it difficult to argue that elemental stoichiometry proximately causes slow N and P cycling in infertile Hawaiian forests and challenging the implication that P should be in short supply relative to N in most of the sites.

First, there is abundant evidence that the C and energy richness of plant litter and soil organic matter (SOM) is more apparent than real, especially within infertile forest ecosystems. Much of the organic C is in forms that are recalcitrant to decomposition (initially lignin and polyphenols, ultimately humus and related compounds), and decomposition can be limited by the "quality" of C, rather than by the supply of N or P (Paul and Clark 1996). In Hawaii added N (and P) enhance the decomposition of high-quality litter (with low concentrations of lignin and polyphenols), whereas added nutrients have little or no effect on the

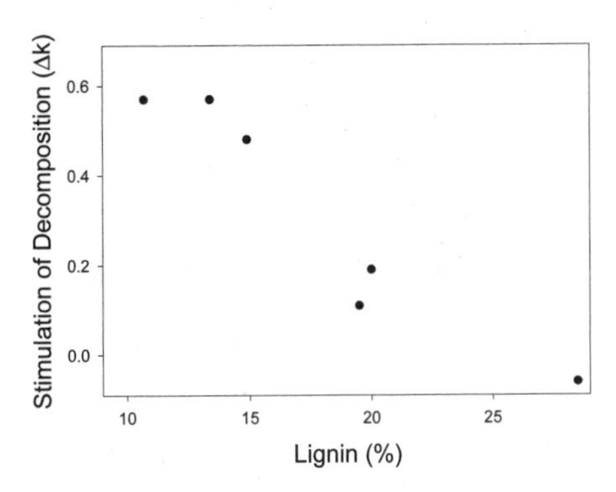

Figure 6.5. The relationship between litter lignin concentration and the stimulation of decomposition by added N, from Hobbie (2000) and Hobbie and Vitousek (2000). The decomposition rate constant for high-quality (low-lignin) litter is increased by added N, whereas the decomposition of low-quality (high-lignin) litter is not N-limited.

decomposition of lower-quality litter (Figure 6.5) (Hobbie 2000). Indeed, many recalcitrant compounds and complexes influence decomposition and nutrient cycling persistently enough to be considered factors in stoichiometry. They are not nearly as stable as elements, of course, but they are persistent relative to the lifetime of organisms involved in decomposition.

Second, although grazers may gain access to most or all of the elements in material they consume, most recalcitrant compounds are too large to be transported across decomposers' cell membranes; rather, they must be processed by extracellular enzymes until their constituents are small enough to be utilized (Sinsabaugh 1994; Carreiro et al. 2000).

Third, elements are bound differently within litter and SOM. Most N in litter and SOM bonds directly to C in protein-polyphenol complexes and other large, recalcitrant organic compounds, whereas most P is present as ester phosphates (held by a C–O–P bond) (McGill and Cole 1981). Extracellular phosphatase enzymes can release inorganic P without breaking down its organic compound or complex, in SOM if not in litter (Gressel et al. 1996), whereas the release of N generally requires the coordinated action of several enzymes. In effect, organic compounds themselves must be broken down before inorganic N or soluble amino acids can be released in biologically available forms (McGill and Cole 1981; Hunt, Stewart, and Cole 1983).

Fourth, once it is released, P may be adsorbed by a number of pathways, some of which are strong enough to remove it from circulation more or less permanently (Uehara and Gillman 1981; Sollins, Robertson, and Uehara 1988; Olander 2002).

These features of litter and SOM decomposition allow for substantial flexibility in within-system nutrient cycling, especially for P. Enzymes are built of C and N, and organisms whose growth is constrained by P can invest C and N in the prospect of obtaining P. In contrast, N-limited organisms would need to cast

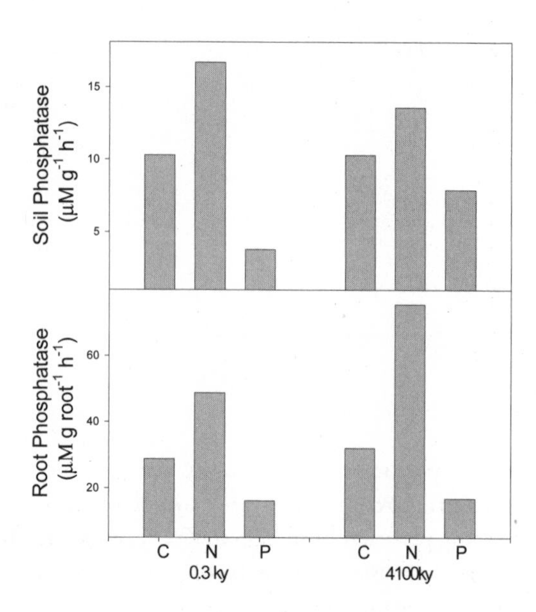

Figure 6.6. Phosphatase enzyme activity in control and fertilized plots of the young and old sites on the Hawaiian age gradient. (a) Phosphatase activity in the soil, from Olander and Vitousek (2000). (b) Root-associated phosphatase activity, from Treseder and Vitousek (2001). Both root and soil phosphatases increase in response to N additions and decrease in response to added P.

large quantities of N into the environment, in several distinct enzymes, in the prospect of obtaining more N. Accordingly, P mineralization can run ahead of the stoichiometry of SOM decomposition when P is in short supply; that flexibility is much reduced or absent in the mineralization of N, and slow rates of decomposition are more likely to drive or sustain limitation by N than limitation by P.

We evaluated the flexibility of enzyme-mediated N and P cycling along the Hawaiian age gradient by determining changes in enzyme activity following long-term fertilization (Olander and Vitousek 2000; Treseder and Vitousek 2001). Adding N to N-limited sites increases productivity and therefore the demand for P and other elements; similarly, adding P to P-limited systems increases demand for N. We found that both decomposers and mycorrhizal roots produced more extracellular phosphatase following N additions (Figure 6.6), but that decomposers did not produce more chitinase following P additions, regardless of whether P limited plant production (roots were not tested). This greater flexibility of P cycling also influenced nutrient losses, as demonstrated by the retention of P relative to N and C as dissolved organic matter leaches through soils (Figure 6.7) (Hedin, Vitousek, and Matson in press).

Stoichiometry and Flexibility of Inputs and Outputs

Biological stoichiometries and flexibilities control both element supply and demand, and so represent the dominant influence on the cycling of biophilic

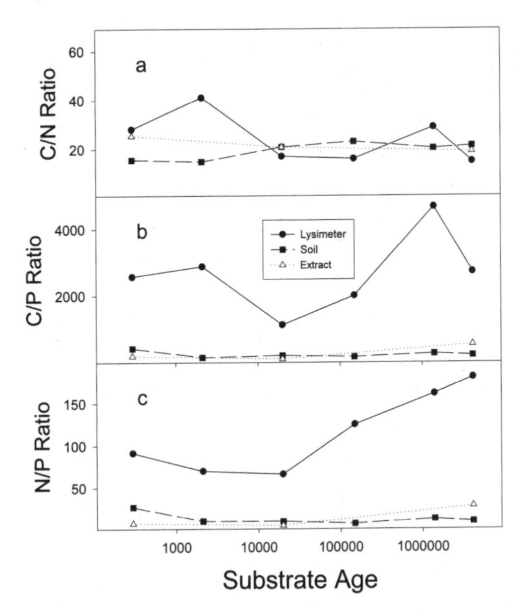

Figure 6.7. The C-N-P stoichiometry of soil organic matter (Crews et al. 1995), dissolved organic matter (DOM) in extracts of surface soil (Neff, Hobbie, and Vitousek 2000), and DOM in lysimeters deeper in the soil (Hedin et al. in press) from sites across the substrate age gradient. (a) C-N ratios do not differ systematically within or among sites. (b) C:P. (c) N:P ratios in lysimeters are much wider than those in SOM or surface-soil extracts, and wider in the low-P older sites than in nutrient-rich intermediate-aged sites, demonstrating that P is retained within these sites relative to C and N.

nutrients within ecosystems. A very different set of stoichiometries and flexibilities, however, control inputs (and to a lesser extent outputs) of elements, and so control the nutrient capital of ecosystems on longer time scales. Of the five major pathways of element input to ecosystems across the Hawaiian Islands—atmospheric deposition, basalt weathering, continental dust, volcanic sources, and biological N fixation—the first four supply elements at ratios that are primarily set by the geochemical stoichiometries of rocks and seawater rather than by the requirements of organisms.

The most consistent of these geochemical stoichiometries is that of seawater, as expressed in the marine aerosol that dominates atmospheric deposition in Hawaii (Carrillo et al. 2002). Ratios of elements in seawater are well defined; a site cannot receive 1 kilogram per hectare (kg/ha) of K from marine aerosol without also receiving 27 kg/ha of Na (Table 6.1). By measuring the flux of an element in marine aerosol and knowing its ratio to other elements in seawater, we can calculate inputs of all the major elements supplied by this pathway.

Elements in the basalt rock that makes up the Hawaiian Islands also are present at characteristic ratios, but the stoichiometry of basalt weathering is more complex than marine aerosol. First, Hawaiian basalts differ slightly within and between eruptions (Wright and Helz 1987) and more substantially between the different stages of a volcano's life cycle (Clague and Dalrymple 1987). To calculate element inputs via weathering (Chadwick et al. 1999; Vitousek, in press), we defined two basalt stoichiometries based on an average chemistry for shield-

Table 6.1. Stoichiometry of the major sources of element inputs to Hawaiian ecosystems, presented as element:P ratios

Pathway	*Ca:P*	*Mg:P*	*K:P*	*N:P*	*S:P*	*Cl:P*	*Si:P*	*Fe:P*
Basalt weathering tholeitic	106	87	3.6	<0.1	2.3	<0.1	346	109
Alkalic	20	11	3.8	<0.1	0.6	<0.1	94	27
Marine aerosol	169,000	518,000	161,000	16	1,090,000	7,800,000	23	.004
Continental dust	44	19	41	<0.1	1.4	0.9	447	38
Volcanic inputs	0	0	24	24	622	288	0	0
Plants	8	4	10	15	5	1	0.4	0.12

Source: Information from Vitousek (in press), except that ratios in plant tissue are from Bowen (1979).

building (tholeitic) eruptions versus postshield (alkalic) eruptions (Table 6.1). Second, the minerals within basalt contain elements at different ratios and weather at different rates. Third, some of the elements (e.g., P, Al) are relatively immobile, being retained within soils in secondary minerals, and weathering of these elements is shifted to later in soil development.

These differences can affect the ratios of elements added to ecosystems via basalt weathering substantially—one kg of P entering the 300-year-old site is accompanied by 85 kg of Ca, whereas in the 4.1 million-year-old site the ratio is 1 kg of P to 15 kg of Ca (Vitousek, in press). At the same time, basalt weathering declines dramatically over time, such that the young site in fact receives 0.51 kg P per hectare per year via weathering, whereas the oldest site receives only 0.008 kg P per hectare per year.

Continental dust carries elements to the Hawaiian Islands in a characteristic stoichiometry near that of average element ratios in the upper crust and quite different from that of Hawaiian basalt (Table 6.1) (Kurtz, Derry, and Chadwick 2001). Although dust inputs are small, the decline in basalt weathering over time is so dramatic that by the oldest site on the age sequence, dust supplies the majority of P inputs (Chadwick et al. 1999). Volcanic inputs represent something of a catchall, including S derived from volcanic SO_2 emissions, Cl volatilized as lava flows enter the ocean, P volatilized from hot rock, and N that is thermally fixed on flowing lava (Carrillo et al. 2002). Empirically, however, volcanic activity too provides a significant and reasonably well-characterized suite of element inputs to the youngest sites on the Hawaiian age gradient, and these inputs too are independent of organisms' requirements.

In contrast, biological N fixation can add N—and only N—when and where it is required by organisms. In this sense, N fixation represents a source of flexibility in element inputs.

The summary of input stoichiometries in Table 6.1 can be used in several ways. First, these geochemical stoichiometries are quite different from any biological stoichiometries, as illustrated here by Bowen's (1979) summary of the chemistry of angiosperms (Table 6.1). N is abundant in plants but minor in most sources of input, and without biological N fixation and/or anthropogenic N sources, N supply clearly should constrain plant growth. Second, compared with biological stoichiometries, P is depleted relative to most other rock-derived elements in all of the major pathways of element input (Table 6.1). For example, the ratio of Ca to P in plants averages 8 (Bowen 1979), but marine aerosol, basalt weathering, and continental dust have Ca:P ratios of more than 100,000, 20–105, and 44, respectively. Consequently, it makes sense that among the elements with geochemical sources, P limits the growth of plants and the productivity of ecosystems on old substrates in the Hawaiian Islands (Herbert and Fownes 1995) and over more of

Earth than any other rock-derived element. Not even the relative immobility of P or the relative flexibility of within-system P cycling can keep it from limiting plant growth on old Hawaiian soils.

Third, geochemical stoichiometries can be used to evaluate conditions other than those summarized here. How would overall element inputs differ on basalt bedrock in the center of continents, where marine aerosol might make a contribution 1–10 percent of that observed in the Hawaiian Islands, and continental dust could contribute 100–1,000 times more (Simonson 1995)?

Finally, as long as the stoichiometry of element inputs to ecosystems is controlled geochemically, then inputs of elements that limit ecosystem processes will be accompanied by inputs of nonessential elements, and of other essential elements far in excess of biological requirements. To the extent that microbes specifically weather particular minerals to obtain particular elements, weathering may be more flexible than suggested here (Blum et al. 2002; Newman and Banfield 2002). Still, it is fair to say that elements may cycle within ecosystems at ratios established by biological stoichiometry, but they flow into terrestrial ecosystems at ratios that are controlled primarily by geochemical processes.

Biological N Fixation

Whereas inputs of most elements are constrained by the stoichiometries of their sources, biological N fixation provides flexibility to N inputs. Organisms that can draw upon the inexhaustible supply of N_2 in the atmosphere should have an advantage over others where N is in short supply, and as a by-product of their activity, they should increase the supply of fixed N in the system as a whole. When N is abundant, N fixation is repressed, in effect shutting off the process when N is abundant and allowing it to be active when N is in short supply (Hartwig 1998).

This ability of biological N fixation to respond flexibly to N deficiency formed much of the basis for Redfield's (1958) analysis of marine stoichiometry. More recent analyses have come to similar conclusions, in most freshwater ecosystems (Schindler 1977), many marine ecosystems (Tyrrell 1999), and some terrestrial ecosystems (Eisele et al. 1989; Smith 1992; Crews 1993). Primary productivity in many temperate and boreal ecosystems, most grasslands, and most estuaries, however, is limited by N, despite the presence of organisms with the capacity to fix N in all these systems (Vitousek and Howarth 1991). What constrains biological N fixation and keeps it from reversing N limitation over much of the surface of the Earth?

A number of experimental studies, models, and reviews have evaluated constraints to symbiotic and cyanobacterial N fixation in terrestrial and marine ecosystems (Vitousek and Howarth 1991; Rastetter et al. 2001). Energetic constraints could be important because the cost of acquiring N via fixation is greater than that of taking up fixed N, especially as NH_4 (Gutschick 1981). Although this cost by itself is unlikely to sustain more than marginal limitation by N, it could cause plants that are dependent on N fixation to be shade-intolerant and thereby keep them from colonizing N-limited, closed-canopy ecosystems (Vitousek and Field 1999).

Second, N fixation could be constrained because the organisms that carry it out have a different stoichiometry. N fixers may require more P than do otherwise similar nonfixers (though see Sprent 1999). If so, the growth and activity of N fixers could be constrained by P while that of nonfixers would be limited by N. The nitrogenase enzyme requires Fe and Mo, and there is evidence that a low availability of one or both of these can constrain rates of N fixation (Howarth and Cole 1985; Silvester 1989; Howarth, Chan, and Marino 1999; Karl et al. 2002). Also, legumes have an N-rich stoichiometry whether or not they fix N (McKey 1994)—a distinction that probably has a deep evolutionary history and one they may share with the relatively closely related actinorrhizal N fixers (Soltis et al. 1995). In temperate and boreal regions where decomposition is slow, this greater N requirement may commit legumes to obtaining N via fixation, with all of its attendant costs, whereas the relatively high availability of N in lowland tropical environments may permit the persistence of N-demanding legumes as components of late-successional forests (Vitousek et al. 2002).

Finally, the N-rich stoichiometry of legumes in particular makes them a higher-quality food source for grazers, which have a C:N ratio in their tissues far below that of most terrestrial plants (White 1993; Ritchie, Tilman, and Knops 1998; Sterner and Elser 2002). Preferential grazing could eliminate symbiotic N fixers from N-limited communities, or it could force substantial investments in antiherbivore defense that effectively multiply the costs of N fixation.

Each of these mechanisms has been shown to be important in particular situations—and models suggest that all of them have the capacity to suppress the flexibility associated with N fixation and so contribute to sustaining N limitation. These mechanisms are not mutually exclusive (Howarth, Chan, and Marino 1999; Vitousek and Field 1999), and it should prove rewarding to understand the relative strength of these mechanisms and perhaps others as they vary in space and time and in response to anthropogenic changes such as elevated CO_2.

Challenges

I propose three areas in which we are poised to make significant progress in understanding element interactions and their implications. First, I believe that research focused on the interactions of biological and geochemical stoichiometries promises to contribute to our understanding of ecosystem dynamics in the long term and also to our understanding of the transfers of elements among ecosystems on scales from landscapes to the globe. Second, I think a consideration of flexibility as well as stoichiometry can contribute to understanding element cycling within and among ecosystems. Finally, I believe that linking the multiple resource limitation paradigm from plant and ecosystem physiology (Field et al. 1992; Rastetter, Ågren, and Shaver 1997) with the stoichiometric perspective from biogeochemistry could contribute to developing a unified perspective on energy, water, and elements in ecosystems.

Acknowledgments

I thank all of the participants in the Hawaii Ecosystem Project for the information on which this chapter is based. This work was supported by grants from the National Science Foundation and the Andrew W. Mellon Foundation. Larry Bond and Doug Turner assisted with manuscript preparation.

Literature Cited

Blum, J. D., A. Klaue, C. A. Nezat, C. T. Driscoll, C. E. Johnson, T. G. Siccama, C. Eagar, T. J. Fahey, and G. E. Likens. 2002. Mycorrhizal weathering of apatite as an important calcium source in base-poor forest ecosystems. *Nature* 417:729–731.

Bolin, B., and R. B. Cook, eds.1983. *The major biogeochemical cycles and their interactions.* SCOPE 21. Chichester, U.K.: John Wiley and Sons.

Bowen, H. J. M. 1979. *Environmental chemistry of the elements.* London: Academic Press.

Carreiro, M. M., R. L. Sinsabaugh, D. A. Repert, and D. F. Parkhurst. 2000. Microbial enzyme shifts explain litter decay responses to simulated nitrogen deposition. *Ecology* 81:2359–2365.

Carrillo, J. H., M. Galanter-Hastings, D. M. Sigman, and B. J. Huebert. 2002. Atmospheric deposition of inorganic and organic nitrogen and base cations in Hawaii. *Global Biogeochemical Cycles* 16:1–16.

Chadwick, O. A., L. A. Derry, P. M. Vitousek, B. J. Huebert, and L. O. Hedin. 1999. Changing sources of nutrients during four million years of ecosystem development. *Nature* 397:491–497.

Clague, D. A., and G. B. Dalrymple. 1987. The Hawaiian-Emperor volcanic chain. Pp. 5–73 in *Volcanism in Hawaii,* edited by R. W. Decker, T. L. Wright, and P. M. Stauffer. USGS Professional Paper 1350. Washington, D.C.: U.S. Geological Survey.

Crews, T. E. 1993. Phosphorus regulation of nitrogen fixation in a traditional Mexican agroecosystem. *Biogeochemistry* 21:141–166.

Crews, T. E., K. Kitayama, J. Fownes, D. Herbert, D. Mueller-Dombois, R. H. Riley, and P. M. Vitousek. 1995. Changes in soil phosphorus and ecosystem dynamics across a long soil chronosequence in Hawai'i. *Ecology* 76:1407–1424.

Eisele, K. A., D. S. Schimel, L. A. Kapustka, and W. J. Parton. 1989. Effects of available

P and N:P ratios on non-symbiotic dinitrogen fixation in tall grass prairie soils. *Oecologia* 79:471–474.

Elser, J. J., D. Dobberfuhl, N. A. MacKay, and J. Schampel. 1996. Organism size, life history, and N:P stoichiometry: Towards a unified view of cellular and ecosystem processes. *BioScience* 46:674–684.

Field, C. B., F. S. Chapin III, P. A. Matson, and H. A. Mooney. 1992. Responses of terrestrial ecosystems to the changing atmosphere: A resource-based approach. *Annual Review of Ecology and Systematics* 23:201–235.

Gressel, N., J. G. McColl, C. M Preston, R. H. Newman, and R. F. Powers. 1996. Linkages between phosphorus transformations and carbon decomposition in a forest soil. *Biogeochemistry* 33:97–123.

Gutschick, V. P. 1981. Evolved strategies in nitrogen acquisition by plants. *American Naturalist* 118:607–637.

Hartwig, U. A. 1998. The regulation of symbiotic N_2 fixation: A conceptual model of N feedback from the ecosystem to the gene expression level. *Perspectives in Plant Ecology, Evolution and Systematics* 1:92–120.

Hättenschwiler, S., A. E. Hagerman, and P. M. Vitousek. Polyphenols in litter from tropical montane forests across a wide range in soil fertility. *Biogeochemistry*, in press.

Hedin, L. O., P. M. Vitousek, and P. A. Matson. Pathways and implications of nutrient losses during four million years of tropical forest ecosystem development. *Ecology*, in press.

Herbert, D. A., and J. H. Fownes. 1995. Phosphorus limitation of forest leaf area and net primary productivity on a weathered tropical soil. *Biogeochemistry* 29:223–235.

Herbert, D. A., and J. H. Fownes. 1999. Forest productivity and efficiency of resource use across a chronosequence of tropical montane soils. *Ecosystems* 2: 242–254.

Hobbie, S. E. 2000. Interactions between litter lignin and soil nitrogen availability during leaf litter decomposition in a Hawaiian montane rainforest. *Ecosystems* 3:484–494.

Hobbie, S. E., and P. M. Vitousek. 2000. Nutrient limitation of decomposition in Hawaiian forests. *Ecology* 81:1867–1877.

Howarth, R. W., and J. J. Cole. 1985. Molybdenum availability, nitrogen limitation, and phytoplankton growth in natural waters. *Science* 229:653–655.

Howarth, R. W., F. Chan, and R. Marino. 1999. Do top-down and bottom-up controls interact to exclude nitrogen-fixing cyanobacteria from the plankton of estuaries? An exploration with a simulation model. *Biogeochemistry* 46:203–231.

Hunt, H. W., J. W. B. Stewart, and C. V. Cole. 1983. A conceptual model for the interactions of carbon, nitrogen, phosphorus, and sulfur in grasslands. Pp. 303–326 in *The major biogeochemical cycles and their interactions*, edited by B. Bolin and R. B. Cook. SCOPE 21. Chichester, U.K.: John Wiley and Sons.

Karl, D. M., A. Michaels, B. Bergman, D. Capone, E. Carpenter, R. Letelier, F. Lipschultz, H. Paerl, D. Sigman, and L. Stal. 2002. Dinitrogen fixation in the world's oceans. *Biogeochemistry* 57:47–98.

Kurtz, A. C., L. A. Derry, and O. A. Chadwick. 2001. Accretion of Asian dust to Hawaiian soils: Isotopic, elemental, and mineral mass balances. *Geochimica et Cosmochimica Acta* 65:1971–1983.

McGill, W. B., and C. V. Cole. 1981. Comparative aspects of cycling of organic C, N, S, and P through soil organic matter. *Geoderma* 26:267–286.

McKey, D. 1994. Legumes and nitrogen: The evolutionary ecology of a nitrogen-demanding lifestyle. Pp. 211–228 in *Advances in legume systematics,* Part 5, *The nitrogen factor,* edited by J. I. Sprent and D. McKey. Kew, England: Royal Botanic Gardens.

Neff, J. C., S. E. Hobbie, and P. M. Vitousek. 2000. Nutrient and minerological control on dissolved organic C, N and P fluxes and stoichiometry in Hawaiian soils. *Biogeochemistry* 51:283–302.

Newman, D. K., and J. F. Banfield. 2002. Geomicrobiology: How molecular-scale interactions underpin biogeochemical systems. *Science* 296:1071–1077.

Olander, L. P. 2002. Geochemical and biological control over short-term phosphorus dynamics in tropical soils. Ph.D. diss., Stanford University, Stanford, Calif.

Olander, L. P., and P. M. Vitousek. 2000. Regulation of soil phosphatase and chitinase activity by N and P availability. *Biogeochemistry* 49:175–190.

Ostertag, R., and S. E. Hobbie. 1999. Early stages of root and leaf decomposition in Hawaiian forests: Effects of nutrient availability. *Oecologia* 121:564–573.

Paul, E. A., and F. E. Clark. 1996. *Soil microbiology and biochemistry.* San Diego: Academic Press.

Rastetter, E. B., G. I. Ågren, and G. R. Shaver. 1997. Responses of N-limited ecosystems to increased CO_2: A balanced-nutrition, coupled-element-cycles model. *Ecological Applications* 7:444–460.

Rastetter, E. B., P. M. Vitousek, C. Field, G. R. Shaver, D. Herbert, and G. I. Ågren. 2001. Resource optimization and symbiotic N fixation. *Ecosystems* 4:369–388.

Redfield, A. C. 1958. The biological control of chemical factors in the environment. *American Scientist* 46:205–221.

Reiners, W. A. 1986. Complementary models for ecosystems. *American Naturalist* 127:59–73.

Ritchie, M. E., D. Tilman, and J. M. H. Knops. 1998. Herbivore effects on plant and nitrogen dynamics in oak savanna. *Ecology* 79:165–177.

Schindler, D. W. 1977. Evolution of phosphorus limitation in lakes. *Science* 195:260–262.

Scowcroft, P. G. 1997. Mass and nutrient dynamics of decaying litter from *Passiflora mollissima* and selected native species in a Hawaiian montane rainforest. *Journal of Tropical Ecology* 13:539–558.

Silver, W., and R. K. Miya. 2001. Global patterns in root decomposition: Comparisons of climate and litter quality effects. *Oecologia* 129:407–419.

Silvester, W. B. 1989. Molybdenum limitation of asymbiotic nitrogen fixation in forests of Pacific Northwest America. *Soil Biology and Biochemistry* 21:283–289.

Simonson, R. W. 1995. Airborne dust and its significance to soils. *Geoderma* 65:1-43.

Sinsabaugh, R. L. 1994. Enzymic analysis of microbial pattern and process. *Biology and Fertility of Soils* 17:69–74.

Smith, V. H. 1992. Effects of nitrogen:phosphorus supply ratios in nitrogen fixation in agricultural and pastoral systems. *Biogeochemistry* 18:19–35.

Sollins, P., G. P. Robertson, and G. Uehara. 1988. Nutrient mobility in variable- and permanent-charge soils. *Biogeochemistry* 6:181–199.

Soltis, D. E., P. S. Soltis, D. R. Morgan, S. M. Swenson, B. C. Mullin, J. M. Dowd, and P. G. Martin. 1995. Chloroplast gene sequence data suggest a single origin of the predisposition for symbiotic nitrogen fixation in angiosperms. *Proceedings of the National Academy of Science USA* 92:2647–2651.

Sprent, J. I. 1999. Nitrogen fixation and growth of non-crop legume species in diverse environments. *Perspectives in Plant Ecology, Evolution, and Systematics* 2:149–162.

Sterner, R., and J. J. Elser. 2002. *Ecological stoichiometry: The biology of elements from molecules to the biosphere.* Princeton: Princeton University Press.

Torn, M. S., P. M. Vitousek, and S. E. Trumbore. The influence of nutrient availability on soil organic matter turnover estimated by incubating and radiocarbon modeling. *Ecosystems,* submitted.

Treseder, K. K., and P. M. Vitousek. 2001. Effects of soil nutrient availability on investment in acquisition of N and P in Hawaiian rain forests. *Ecology* 82:946–954.

Tyrrell, T. 1999. The relative influences of nitrogen and phosphorus on oceanic primary production. *Nature* 400:525–531.

Uehara, G., and G. Gillman 1981. *The mineralogy, chemistry, and physics of tropical soils with variable charge clays.* Boulder, Colo.: Westview Press.

Vitousek, P. M. 1998. Foliar and litter nutrients, nutrient resorption, and decomposition in Hawaiian *Metrosideros polymorpha. Ecosystems* 1:401–407.

Vitousek, P. M. 2002. Oceanic islands as model systems for ecological studies. *Journal of Biogeography* 29:1–10.

Vitousek, P. M. *Nutrient cycling and limitation: Hawai'i as a model system.* Princeton University Press, in press.

Vitousek, P. M., and H. Farrington. 1997. Nutrient limitation and soil development: experimental test of a biogeochemical theory. *Biogeochemistry* 37:63–75.

Vitousek, P. M., and C. B. Field. 1999. Ecosystem constraints to symbiotic nitrogen fixers: A simple model and its implications. *Biogeochemistry* 46:179–202.

Vitousek, P. M., and R. W. Howarth. 1991. Nitrogen limitation on land and in the sea: How can it occur? *Biogeochemistry* 13:87–115.

Vitousek, P. M., D. R. Turner, and K. Kitayama. 1995. Foliar nutrients during long-term soil development in Hawaiian montane rain forest. *Ecology* 76:712–720.

Vitousek, P. M., K. Cassman, C. Cleveland, T. Crews, C. B. Field, N. B. Grimm, R. W. Howarth, R. Marino, L. Martinelli, E. B. Rastetter, and J. I. Sprent. 2002. Towards an ecological understanding of biological nitrogen fixation. *Biogeochemistry* 57:1–45.

White, T. C. R. 1993. *The inadequate environment: Nitrogen and the abundance of animals.* Berlin: Springer-Verlag.

Wright, T., and R. T. Helz. 1987. Recent advances in Hawaiian petrology and geochemistry. Pp. 625–640 in *Volcanism in Hawaii,* edited by R. W. Decker, T. L. Wright, and P. M. Stauffer. USGS Professional Paper 1350. Washington, D.C.: U.S. Geological Survey.

7

Element Interactions: Theoretical Considerations

Göran I. Ågren, Dag O. Hessen, Thomas R. Anderson, James J. Elser, and Peter De Reuter

Carbon, nitrogen, and phosphorus are central among the elements that are required in the largest quantities by living organisms for growth and reproduction. It is therefore unsurprising that both autotroph and heterotroph growth rates are often identified as being limited by one of these elements. The identity of the limiting element is generally considered to depend on the system studied, with nitrogen mostly identified as the limiting element in terrestrial and marine systems (Vitousek and Howarth 1991) and phosphorus in freshwater systems (Schindler 1977). Carbon, on the other hand, seems less often to be directly limiting to terrestrial plant growth, although growth responses are commonly obtained when plants are grown in elevated CO_2 (Poorter 1993; Norby 1996; Mooney et al. 1999). The reason why a particular element is limiting is still enigmatic because of the enormous quantities of CO_2 and N_2 in the atmosphere and the existence of nitrogen-fixing plants, which should make these elements essentially freely available. Living organisms, however, generally have element compositions that differ widely from their nearest environment (e.g., Ågren and Bosatta 1998). Human interventions in the global cycles of carbon, nitrogen, and phosphorus (Vitousek et al. 1997; Falkowski et al. 2000) are now changing the relative availability of the elements, with global effects on climate and ecosystems.

Short-term experiments using controlled environments have shown that plants grown at doubled CO_2 exhibit photosynthetic rates that are typically 20–50 percent higher than under ambient conditions (e.g., Ceulemans and Mousseau 1994; Idso and Idso 1994; Mooney et al. 1999), often leading to significant biomass increase. On the other hand, by regulating the growth rate of plants through the rate of addition of mineral elements, it is possible to achieve changes in relative growth rates from a few percent per day to over 50 percent per day (Ingestad and Ågren 1992, 1995). This result points to the importance of the mode of supply (concentration versus rate) for interaction between elements (Rastetter and Shaver 1992). Elements that are supplied as a concentration cannot be exhausted, and allocation of uptake effort by an organism will increase its uptake of this element. Elements that are supplied as a rate, on the other hand, can be exhausted and an organism may lack control of the uptake of such an element. For this reason, the whole ecosystem response is complex, and the effective mode of supply may also depend on the time scale. Long-term experiments suggest that at time scales of decades the uptake of CO_2 is regulated by the availability of other elements like N (Oren et al. 2001). On sufficiently long time scales, it has been argued that P will ultimately be the limiting element and that biological and geological processes such as the weathering rate and flux of P to oceans ultimately will regulate the global CO_2-O_2 balance (Redfield, Ketchum, and Richards 1963; Schindler 1977; Lenton and Watson 2000; Lenton 2001).

Ecological stoichiometry is a tool for analyzing how the balance of elements in organisms affects production, nutrient cycling, and food web dynamics (Sterner and Elser 2002). The origins of stoichiometry are rooted in agricultural studies involving Liebig's law of the minimum. In its first formulation this law was restricted to supply of mineral elements as fertilizers, but later formulations have also attempted to include other factors such as temperature. In a pure stoichiometric context, however, it is our view that the law of the minimum should be applied only to the elements that are used as components in growth; with this use the law of the minimum has a direct connection to more recent ideas related to the chemical stoichiometry of living biomass. Other factors that influence growth, such as temperature, do this by changing the rate at which all the elemental resources are utilized but without necessarily altering the underlying stoichiometry. Ingestad and Ågren (1995) suggested that the interaction between these different kinds of factors be called orthogonal to indicate that they can act on growth independently of each other. Nevertheless the concept of a single limiting element is open to question—the complexity of organism physiology means that there may be circumstances where co-limitation by more than one resource is possible. For example, cultures of yeast can be co-limited by both oxygen and glucose supply rates (Duboc and von Stockar 1998), shifting between different degrees of aerobicity in response to the oxygen:glucose supply rate.

In this chapter we will examine basic stoichiometric constraints on organisms at different trophic levels and indicate the consequences of organisms with different stoichiometries having to serve as food sources for each other. We will also discuss feedbacks from the availability of one element to fluxes of another element in ecosystems.

Stoichiometry of Cells and Organisms

The value of stoichiometry in ecological studies comes in identifying the consequences of and in understanding the controls of systematic variations in C:element stoichiometry in organisms and ecosystems. One key contrast is found when considering photoautotrophs (autotrophs hereafter) versus heterotrophs, especially metazoans. Although animals generally tend to maintain body elemental composition within relatively narrow bounds (see below), autotrophs have great flexibility in their C:nutrient ratios, and these ratios vary systematically with the growth rate of the plant.

Plants

In general, the overall range of relative requirements for different elements seems to be similar for all photoautotrophic organisms (i.e., unicellular and colonial algae and cyanobacteria in aquatic systems, vascular plants in terrestrial systems) except that terrestrial autotrophs normally contain relatively more carbon than aquatic autotrophs (Elser et al. 2000), owing to extensive investment in C-rich, low-nutrient, structural components (Table 7.1). Both N:C and P:C ratios respond strongly to limitation by N and P, respectively, and may also respond more weakly to limitation by the alternate nutrient (N:C responds weakly to P limitation and P:C responds weakly to N limitation), indicating, in relative terms, a larger allocation to active N- and P-based compounds than C-based structural compounds at higher growth rates. This variability in stoichiometry is characteristic of the overall plasticity of autotrophic organisms and has its physiological base in the ability of autotrophs to maintain a balance between the types of compounds required for growth (Droop 1974).

Another cause of variability in stoichiometry comes from excess uptake of elements—that is, autotrophs have the capacity to accumulate elements in excess of what is immediately needed for growth. Nitrogen can be stored in various amino acids, in inactive Rubisco, or as nitrate, whereas P can be stored as inorganic phosphate (orthophosphate or polyphosphate) or in organic phosphate-containing compounds. The reasons behind, and the consequences of, these two sources of variability in element ratios are, of course, entirely different and should not be confused. When comparing C:element ratios in various autotrophs, in different habitats, or in different seasons, it is therefore necessary to know whether the element

Table 7.1. Element concentrations in plants

Macronutrients						*Micronutrients*					
N	*P*	*K*	*Ca*	*S*	*Mg*	*Mn*	*Fe*	*Zn*	*Cu*	*B*	*Mo*
100	10	35	2.5	5	4	0.05	0.2	0.05	0.03	0.05	0.007

Note: Element concentrations as percent (by weight) of nitrogen in Norway spruce (*Picea abies*) foliage, considered to be optimal. Optimal nitrogen concentration is considered to be 20 mg g^{-1} (Linder 1995).

in question is limiting or not. Thingstad (1987) provided an elegant example of this when he demonstrated how heterotrophic bacteria could deplete their environment of C, N, or P or combinations of two or all three elements depending on growth rate and element ratios in the supply. Although the environment could be depleted of all three elements, it was still only one of them that determined growth rate.

The ability to take up an excess of available elements can clearly be seen when relative growth rates of a terrestrial autotroph and two aquatic autotrophs are plotted as a function of nutrient:C ratios (Figure 7.1). The terrestrial plant, *Betula pendula*, took up little N or P above what was required to achieve its maximum relative growth rate (cf. open and solid circles in Figure. 7.1). On the other hand, an aquatic autotroph (*Dunaliella tertiolecta*) took up N to an excess of 50 percent, and another autotroph (*Monochrysis lutheri*) took up P to a concentration five times that required for maximum relative growth rate. The ratio between a limiting and a nonlimiting element may therefore be highly variable. The ratio of element concentrations at the same relative growth rate when the element is limiting, however, gives the stoichiometry at co-limitation. It is possible that this ratio will vary with growth rate (Ågren 1988). At very high element availabilities, autotrophs do not take up any further quantities of an element despite its presence in the environment (this maximum can be exceeded if external concentrations become very high and toxic internal concentrations are reached, e.g., Ingestad and Lund 1979). When the element availability goes down, the plant can still maintain its maximum relative growth rate while decreasing its element concentration (the horizontal portion of the relative growth rate versus element concentration relations in Figure 7.1). In this region there can be considerable changes in the form in which the elements are stored within the plant, but there is no limitation on growth rate. It is only when element availability has decreased to the sloping part of the relation that there is a growth limitation. The existence of an excess uptake of nonlimiting elements also means that nutrient use efficiencies can only be used for the limiting element.

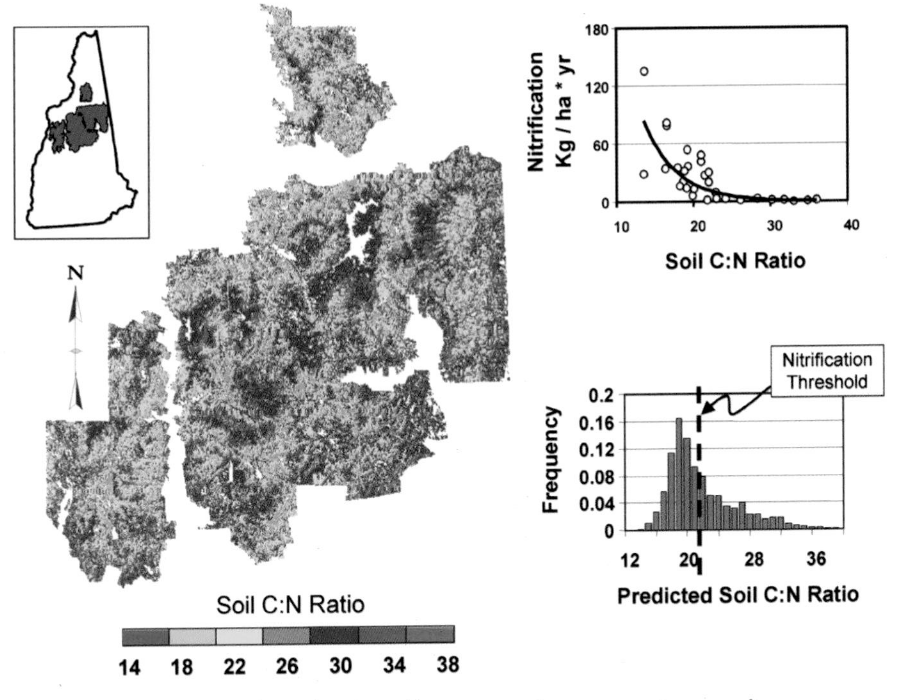

Colorplate 1. An example application of hyperspectral remote sensing to scaling ecosystem biogeochemistry from plots to complex landscapes (after Ollinger et al. 2002). Data from NASA's Airborne Visible and InfraRed Imaging Spectrometer (AVIRIS) were obtained for the White Mountain National Forest in New Hampshire, USA (shown in upper left inset). Image data were calibrated to field measurements of canopy nitrogen from eighty-six plots. Estimates of soil C:N ratios were derived by combining the resulting canopy N coverage with an observed relationship between canopy N and soil C:N ($R^2 = 0.72$). The potential utility of spatial data for soil C:N stems from its role in a variety of processes, including net nitrification. The panel at upper right shows a trend between soil C:N ratio and net nitrification, indicating a threshold at approximately C:N = 24. Analysis of a histogram of estimated soil C:N for the region (lower right) indicates that 62 percent of the region's land area is expected to fall below this threshold.

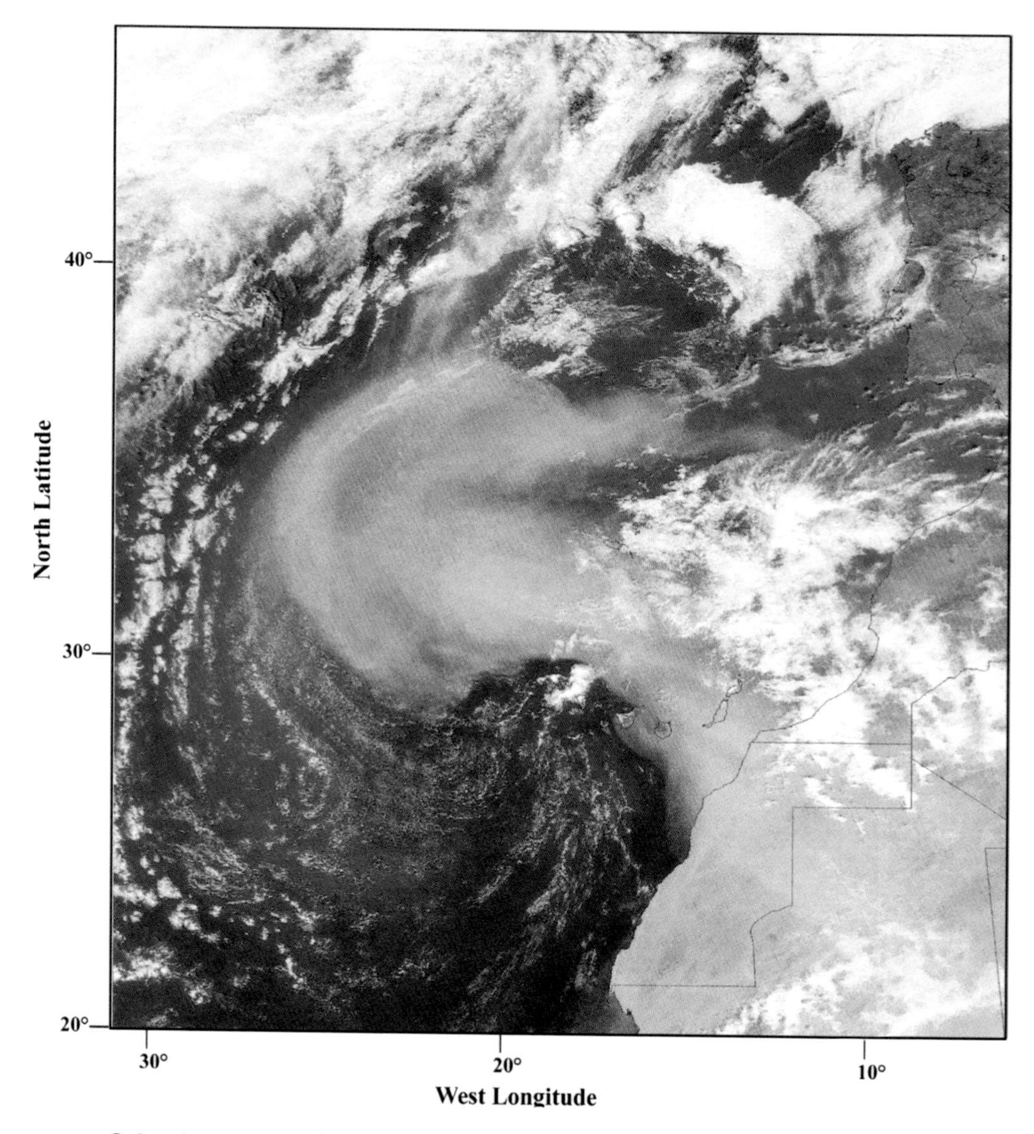

Colorplate 5. Dust plume from a sandstorm over the Sahara Desert, 26 February 2000. Image courtesy of the SeaWiFS Project, NASA/Goddard Space Flight Center and ORBIMAGE Corp.

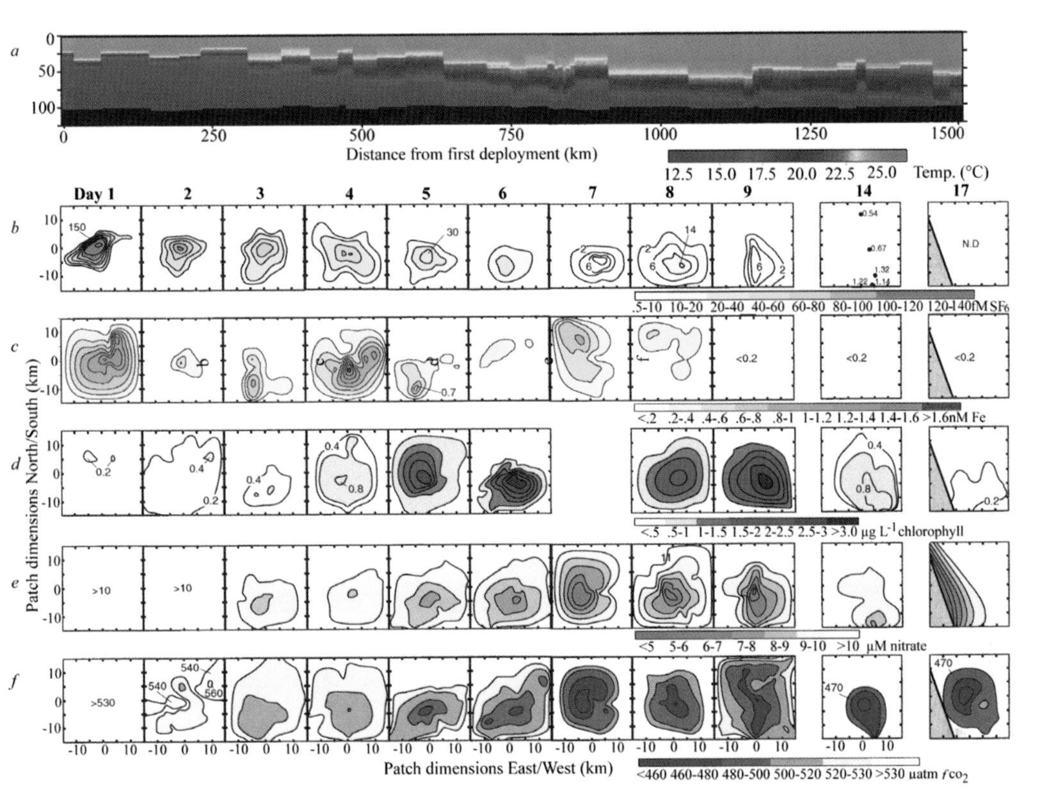

Colorplate 6. Distribution of responses to iron enrichment in time and space during IronEx-II. The squares in each panel follow the seeded, fertilized water mass through time as it drifted to the southwest. (a) vertical section of temperature; (b) SF_6 tracer concentration (fM); (c) dissolved iron (nM); (d) chlorophyll µg/l); (e) nitrate (µM); (f) fCO_2 (µatm). Figure reproduced courtesy of *Nature* (Macmillan Journals, Ltd.).

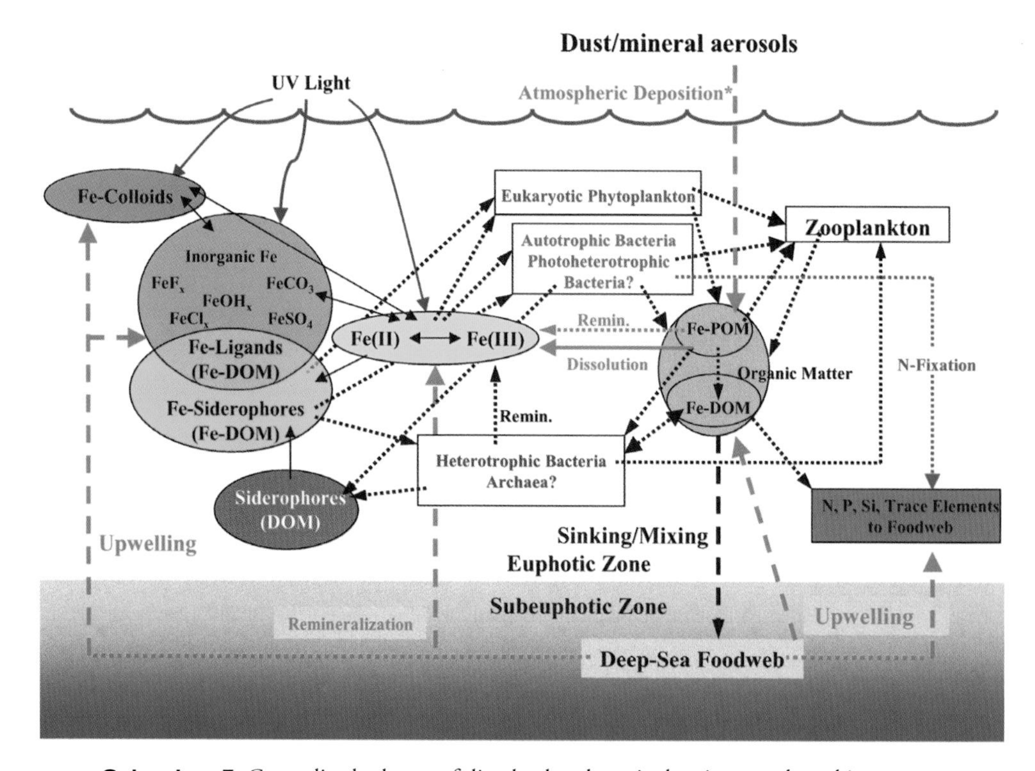

Colorplate 7. Generalized scheme of dissolved and particulate iron pools and iron cycles within the oceanic surface layer. Red lines denote new sources of iron. Black lines denote iron transformation processes. Solid lines represent chemical or photochemical transformations. Dashed lines represent physical transports. Dotted lines represent biochemical processes or transformations. POM = particulate organic matter; DOM = dissolved organic matter; Remin. = remineralization. *Atmospheric deposition includes both wet and dry deposition. See Turner and Hunter (2001) for details of iron transformations.

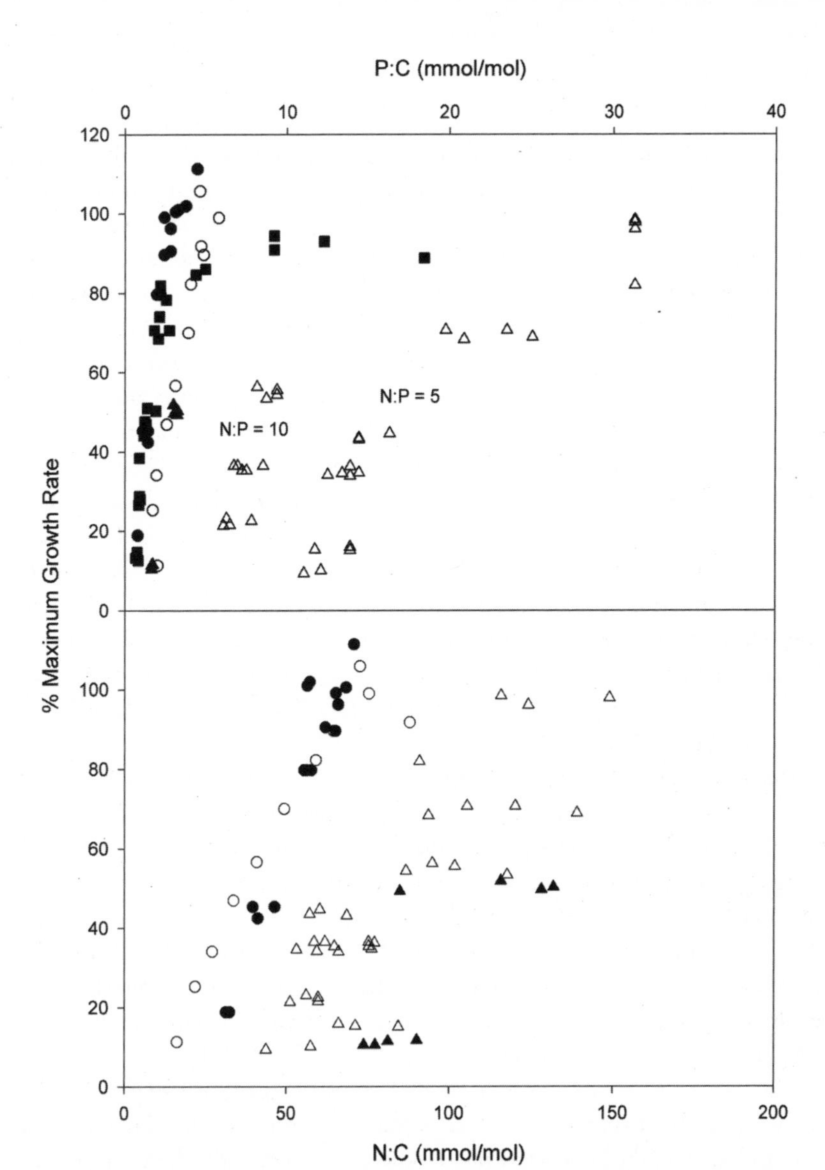

Figure 7.1. Relative growth rates as functions of element ratios for a terrestrial plant (*Betula pendula*) and two aquatic plants (*Dunaliella tertiolecta* and *Monochrysis lutheri*) when grown under P limitation or N limitation. There are no values for *D. tertiolecta* grown under P limitation and at maximum growth rate. P:C ratios for *D. tertiolecta* under N limitation form two clusters indicated by N:P = 10 and N:P = 5. These two N:P ratios correspond to the N:P ratios in the nutrient supply. Data for *B. pendula* from Ingestad (1979) and Ericsson and Ingestad (1988) and for *D. tertiolecta* and *M. lutheri* from Goldman, McCarthy, and Peavey (1979). The scatter in data points for *D. tertiolecta* is partly a result of the data points being extracted from a graph.

Key: ● = *Betula pendula* grown under P limitation; ○ = *Betula pendula* grown under N limitation; ▲ = *Dunaliella tertiolecta* grown under P limitation; Δ = *Dunaliella tertiolecta* grown under N limitation; ■ = *Monochrysis lutheri* grown under P limitation

It is well known that the N content of leaves and other organs of plants grown under elevated CO_2 is lower than that in plants cultivated at ambient CO_2 (Ceulemans and Mousseau 1994; Curtis and Wang 1998). Most studies have only looked at effects on N. In a compilation of results of various studies, Cotrufo, Ineson, and Scott (1998) report a mean plant tissue N concentration in elevated CO_2 of 86 percent of the concentration in ambient CO_2 concentration. Variability in the extent of N content response is, however, large among studies, but there are some systematic responses: N-fixing plants and C_4 plants show little response to altered CO_2 regime. Data regarding elements other than nitrogen are too limited to allow any general conclusions. A study by Overdieck (1993) finds that tissue N:P ratio might decrease under elevated CO_2 but the ratio of N to other elements remains unchanged. The extent to which the C:element ratio of woody tissues might respond to CO_2 is, however, less clear. The C:N and C:P ratios in stem wood were much less variable than in foliage when several Nordic fertilization experiments with *Picea abies* (Andersson, Brække, and Hallbäcken 1998) were analyzed (G. I. Ågren, pers. comm.), but C:N varied more in stem wood than in foliage in stands of *Pinus radiata* in New Zealand (Beets and Madgwick 1988).

Decreases in the N:C ratio of live tissue in CO_2 fertilization experiments with plants do not reliably translate to similar changes in plant litter (Norby and Cotrufo 1998). Thus, consequences for turnover rates of litters as a result of a changing N:C ratio are not clear. Although it has frequently been shown that the N:C and P:C ratios correlate with the decomposability of a litter (e.g. Melillo, Aber, and Muratore 1982; Enriquez, Duarte, and Sand-Jensen 1993), this finding could also reflect the importance of differences in carbon chemistry to the decomposer community rather than the importance of differences in stoichiometry per se (Joffre and Ågren 2001).

The flexibility in the C:element ratio in terrestrial plants is also closely associated with shifting tissue allocation, such that with an increasing C:nutrient ratio more and more of the plant growth is allocated to root growth when N or P is limiting (Andrews, Raven, and Sprent 2001). The allocation of growth in response to an increasing C:nutrient ratio can go in the other direction, however, when other elements, such as Mg, are limiting (Ericsson 1995). The response to N and P limitation is understandable as a strategy to allocate resources to acquisition of the scarcest resource (Bloom, Chapin, and Mooney 1985), whereas Mg deficiency might lead to a shortage of carbon because of limitation of photosynthesis and hence the plant perceives a need to allocate more to shoots. The extent to which strategies for coping with element deficiencies have been elaborated may differ in terrestrial and aquatic plants.

Animals

In contrast to the great physiological plasticity in growth form, tissue allocation, and thus biochemical and elemental composition exhibited by autotrophic organisms, metazoans exhibit far less physiological variation, a relatively narrow range of even interspecific variation in C:N:P ratios, and a generally nutrient-rich biomass compared with autotrophs (Elser et al. 2000; Sterner and Elser 2002). Thus, when consumers meet autotroph biomass with low nutrient content, constraints on animal growth by low nutrient content seem inevitable. Over the past decade stoichiometric theory has also been used extensively as a tool to investigate the growth of heterotrophic organisms and recycling of elements in response to the elemental composition of food (McNaughton 1988; Sterner and Hessen 1994; Elser et al. 2000). Ratios of C, N, and P are compared in consumers and prey, and, using simple mass balance principles, the element in shortest supply relative to demand is identified. Potentially limiting elements are expected to be utilized for growth and transferred in food chains with high efficiency, while nonlimiting elements, by definition present in excess, must be disposed of and recycled.

If there are shortages of elements other than C, growth efficiency of C will decrease. That is, more of ingested C will go "to waste" as feces or respiration or will be stored. The fate of this C is not trivial for either the individual organism or the ecosystem as a whole. The stoichiometric approach to production dynamics of individual organisms is centered using Liebig's principle to identify the primary limiting element, the one in least supply relative to demand. In practice, it has been assumed that limiting nutrient elements (N, P) can potentially be used with a growth efficiency approaching 100 percent—that is, there is total reclaim of ingested elements and zero excretion (e.g., Andersen and Hessen 1991; Urabe and Watanabe 1992). Carbon, however, is always lost as CO_2 in meeting the demands of respiration, so for this reason alone, growth efficiency of C is less than 100 percent.

Decomposers

Stoichiometric imbalance also plays a key role in affecting decomposer/detritivore guilds that encounter nutrient-poor plant detritus. Indeed, such imbalances are likely to be even more severe than in herbivory, at least in the initial stages of detrital processing, as it is common in terrestrial plants for considerable nutrients to be reclaimed from leaves and other tissues before loss (Aerts 1996; Killingbeck 1996). Unfortunately, less is known about the flexibility in stoichiometric relations in decomposers than in plants, although changes between fungi and bacteria as the principal decomposer should be important. Nevertheless, stoichiometric relations are important for maintaining the element cycles in the ecosystem, and a consequence is that soils are relatively richer in N than C relative to plants (Ågren and Bosatta 1998).

Decomposition of soil organic matter, considered a key process in soil ecosystem functioning (Griffiths et al. 2000), may be limited by C, N, or P depending on soil type, management, and vegetation. Invertebrate decomposers have typically one order of magnitude higher N and P concentrations than the fresh plant litter (Swift, Heal, and Anderson 1979), and bacterial mineralization of organic C is enhanced by additions of inorganic P or N (Melillo, Aber, and Muratore 1982; Enriquez, Duarte, and Sand-Jensen 1993; Hessen et al. 1994). Nonetheless, indications that decomposition is C limited are provided by the frequently observed phenomenon that microbial activity and soil respiration increase after physical disturbance of the soil, such as through tillage (Van Veen and Kuikman 1990). The mechanism behind this phenomenon is that decomposition has become substrate-C limited because of the nonaccessibility of the soil organic matter to the microbes through the micro-pore structure of the soil and that the "protected" soil organic matter becomes available by the physical disturbance (Van Veen and Kuikman 1990). C limitation is also indicated for relatively nutrient-rich agricultural soils, in which microbes and soil fauna are found to mineralize N when processing food and producing new microbial biomass (Verhoef and Brussaard 1990; Van Veen et al. 1991). N limitation may occur in situations in which the soil organic matter that the microbes use as their substrate has a much higher C:N ratio than that of the microbes, which is a common feature for most natural soil ecosystems.

When decomposing soil organic matter with a high C:N ratio, the decomposers require extra N to produce new microbial biomass. In situations in which the pool of available inorganic N is limited, the energy conversion efficiency (yield), and hence decomposition, will decrease (Bloem et al. 1994; Ågren, Bosatta, and Magill 2001). This is, however, not a universal feature, and sometimes addition of N will increase the decomposition rate (e.g., Hobbie 2000; Resh, Binkley, and Parrotta 2002). Interaction between C and N use by decomposers might explain these contradictory results. Ågren, Bosatta, and Magill (2001) describe the fraction of carbon remaining of a litter cohort with the following function of time

$$g(t) = \frac{1}{(1 + a u_0 t)^{\frac{1-e_0}{b e_0}}}$$

where e_0 is the energy conversion efficiency, u_0 the decomposers' growth rate, and a and b parameters of no interest for this discussion. Ågren, Bosatta, and Magill

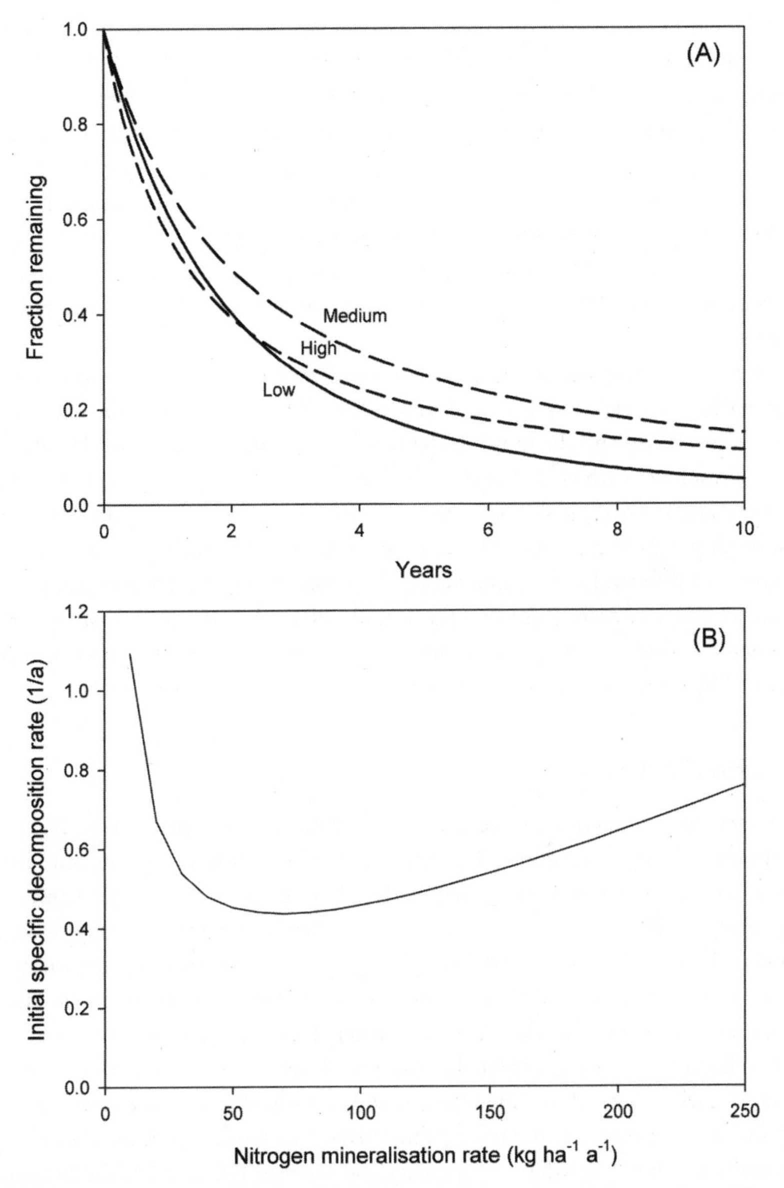

Figure 7.2. Effect of N availability on the decomposition rate of a litter cohort. The curves in this figure have been calculated by assuming that the decomposer efficiency as a function of the rate of N mineralization is $e_0(m) = e_{max}(1 - \exp\{-km\})$ and the decomposer growth rate is $u_0(m) = u_{00} + u_{01}m$. (A) Fraction of the litter remaining as a function of time. (B) Initial specific decomposition rate as a function of N mineralization rate.

(2001) and Franklin et al. (in press) observed that e_0 is an increasing function of inorganic N availability. On the other hand, a closer examination of the results by Ågren, Bosatta, and Magill (2001) indicates that u_0 might be an increasing function of inorganic N availability. Combining these two observations can lead to the results presented in Figure 7.2, where the decomposition rate is lowest at intermediate levels of N availability. The decomposition rate is high at low N availability despite a slow growth rate of decomposers because they are inefficient in their C use because of lack of N. At high N availability the abundant N allows the decomposers to grow rapidly to the extent that this outweighs the more efficient C use.

A distinguishing feature of the decomposition system is that the decomposers start with a material rich in C and hence N or P limiting, but as the substrate is used, C is released and the limiting element incorporated until a critical C:element ratio is reached at which C becomes limiting. The decomposers can then release N and P and make these nutrients available for uptake by plants again. These critical break points in N:C and P:C ratios are well-defined stoichiometric quantities (Ågren and Bosatta 1998). Their values, however, are coupled to inorganic nutrient availability (Ågren, Bosatta, and Magill 2001). A consequence of the use of inorganic N and P is that the decomposer trophic links may be much less constrained by C:nutrient ratios than herbivore linkages.

Terrestrial Ecosystems

The element interactions in terrestrial ecosystems are shaped not only by the properties of the plants and soil communities of which they consist, but also by the interactions between the plant and soil systems (with animals sometimes appearing as an important linking agent; Loreau 1995; Daufresne and Loreau 2001). Plants feed the soil with litter of varying element ratios. It is clear that different plant species have differing capacities to retranslocate elements down to a minimal level before shedding tissues as litter (Killingbeck 1996). It is, however, not well studied to what extent this minimal level is achieved or to what extent retranslocation depends on the nutrient status of the tissue before senescence. In the soil, decomposers modify the stoichiometry. The resulting C:element ratios are generally much smaller than in the litter and seem to reflect more the properties of the decomposers than of the litter (Ågren and Bosatta 1998). The reason that different ecosystems, such as forest soils versus agricultural soils, have soils with different C:element ratios is therefore more a property of the type of decomposers that the system allows than the C:element ratio of the litter per se. One can also speculate that there is a nonlinear feedback from the productivity level to the C:element ratios such that higher productivities lead to lower C:element ratios (see section on decomposers). The existence of such a feedback also indicates that there will be interactions between C:element ratios and external inputs of elements, and the openness of the system to element fluxes becomes a key issue (Rastetter, Ågren, and Shaver 1997). Since different elements will have different openness—for example, most ecosystems are more open with respect to N than to P—the time scale at which different elements operate will differ, with N being more important at shorter time scales and P at longer ones (Kirschbaum et al. 1998).

Marine Ecosystems

Marine seston stoichiometry has been widely assumed to closely conform to the Redfield ratio of 106C:16N:1P (Redfield, Ketchum, and Richards 1963). Conventional wisdom is that changes in pCO_2 in marine systems will not affect the elemental composition of plankton, although the matter is unresolved (e.g., Burkhardt and Riebesell 1997; Burkhardt, Zondervan, and Riebesell 1999). The C:N ratio of sinking zooplankton fecal pellets is variable (Anderson 1994). This ratio should in principle be inversely related to the supply of limiting nitrogen to zooplankton—that is, excess C may remain unassimilated in the gut.

In general it is to be expected that export fluxes are proportional to biological production in the sunlit euphotic zone. Various nutrients are thought to play limiting roles in marine systems. Nitrogen is generally considered to be the "proximate limiting nutrient" representing local limitation according to Liebig's law, while phosphate supplied by continental weathering and fluvial discharge is the "ultimate limiting nutrient" influencing system productivity on long time scales (Tyrell 1999). Shortfalls of N can be compensated for in the long term through N_2 fixation, although low availability of trace elements, notably iron, and light may limit this process (Martin 1990). Thus, this distinction between P and N as ultimate and proximate limiting factors is far from clear cut (Falkowski 1997).

Although Redfield ratios have become an accepted empirical observation, a conceptual understanding of their biological basis is lacking (Falkowski 2000). Still, feedback mechanisms involving the regulation of total ocean N content involving nitrogen-fixing and denitrifying organisms are undoubtedly important.

Discussion

Ecosystem responses reflect the integration of processes at the cellular or organismal level and their interaction with abiotic factors. In both organisms and ecosystems the production efficiency in terms of C uptake depends heavily on the balance between intake and demand of key nutrient elements. Nutrients like N or P will be recycled and reused by autotrophs, while excess C may either be oxidized to inorganic forms by the organisms themselves or heterotrophic decomposers or be more permanently buried in sediment or soils. These similarities in feedbacks arise not because ecosystems mimic organismal allocation rules for C, N, or P, but because the interacting biota themselves must obey stoichiometric rules at the level of their individual cells and bodies. Thus, in the simplest scenarios, the fate of C will be determined by the absolute and relative demands for C, N, and P of each organism relative to its resources. In food webs the large discrepancy in elemental ratios found between autotrophs and herbivores causes a low efficiency in the use of C (i.e., much C "in excess"). An evolutionary explanation for such a mismatch in elemental ratios between autotrophs and grazers may, from the autotrophs' perspective, be seen as an adaptation to cause lower grazer fitness (at some expense to its own growth, however; cf. Moran and Hamilton 1980), whereas the grazers' dilemma appears to be a trade-off between rapid growth and the risk of direct mineral limitation. Thus, high and low body levels of and demands for P and N in consumers may be seen as evolutionary adaptations along the classic r–K axis. In any case, these inexorable constraints mean that herbivores rarely find optimal food and that this autotroph production is allowed to go elsewhere.

To extrapolate from organisms to ecosystems may be a dubious process, but the processes described here may also reflect large-scale ecosystem properties. The unavoidable imbalances between autotrophs and consumers appear to have the net effect that overall transfer of primary production to higher trophic levels is impaired under conditions of nutrient-limited autotroph production. As a result, much C can be sequestered in standing biomass of nutritionally unsuitable autotroph biomass with a high C:nutrient ratio. This result further affects the fate of production by increasing the percentage of production entering detrital pools to be entrained into slow detrital food webs, where the efficiency of C processing is low because of the detritus's low nutrient content. This situation is illustrated by a cross-ecosystem comparison (Cebrian 1999). Systems dominated by autotrophs with low C:element ratios and fast turnover rates such as phytoplankton communities are effectively grazed, and little C is sequestered to recalcitrant detrital pools, whereas terrestrial systems dominated by high C:element ratios and low turnover rates sequester most of the fixed C to detritus.

Acknowledgment

This chapter is an abbreviated and modified version of a paper by D. O. Hessen, G. I. Ågren, T. R. Anderson, J. J. Elser, and P. De Reuter, Carbon sequestration in ecosystems: The role of stoichiometry, submitted for publication in *Ecology*.

Literature Cited

Aerts, R. 1996. Nutrient resorption from senescing leaves of perennials: Are there general patterns? *Journal of Ecology* 84:597–608.

Ågren, G. I. 1988. The ideal nutrient productivities and nutrient proportions. *Plant, Cell and Environment* 11:613–620.

Ågren, G. I., and E. Bosatta. 1998. *Theoretical ecosystem ecology: Understanding element cycles.* Cambridge: Cambridge University Press.

Ågren, G. I., E. Bosatta, and A. M. Magill. 2001. Combining theory and experiment to understand effects of inorganic nitrogen on litter decomposition. *Oecologia* 128:94–98.

Andersen, T., and D. O. Hessen. 1991. Carbon, nitrogen, and phosphorus content of fresh-water zooplankton. *Limnology and Oceanography* 36:807–814.

Anderson, T. R. 1994. Relating C:N ratios in zooplankton food and faecal pellets using a biochemical model. *Journal of Experimental Marine Biology and Ecology* 184:183–199.

Andersson F. O., F. H. Brække, and F. Hallbäcken, eds. 1998. Nutrition and growth of Norway spruce forests in a Nordic climatic and deposition gradient. *TemaNord*: 566. Copenhagen: Nordic Council of Ministers.

Andrews, M., J. A. Raven, and J. J. Sprent. 2001. Environmental effects on dry matter partitioning between shoot and root of crop plants: Relations with growth and shoot protein concentration. *Annals of Applied Biology* 138:57–68.

Beets, P. N., and H. A. I. Madgwick. 1988. Above-ground dry matter and nutrient content of Pinus radiata as affected by lupin, fertiliser, thinning, and stand age. *New Zealand Journal of Forestry Science* 18:43–64.

Bloem, J., G. Lebbink, K. B. Zwart, L. A. Bouwman, S. L. G. E. Burgers, J. A. Devos, and P. C. de Ruiter. 1994. Dynamics of microorganisms, microbivores, and nitrogen mineralization in winter-wheat fields under conventional and integrated management. *Agriculture Ecosystems and Environment* 51:129–143.

Bloom, A. J., F. S. Chapin III, and H. A. Mooney. 1985. Resource limitation in plants: An economic analogy. *Annual Review of Ecology and Systematics* 16:363–392.

Burkhardt, S., and U. Riebesell. 1997. CO_2 availability affects elemental composition (C:N:P) of the marine diatom *Skeletonema costatum. Marine Ecology Progress Series* 155:67–76.

Burkhardt, S., I. Zondervan, and U. Riebesell. 1999. Effect of CO_2 concentration on C:N:P ratio in marine phytoplankton: A species comparison. *Limnology and Oceanography* 44:683–690.

Cebrian, J. 1999. Patterns in the fate of production in plant communities. *American Naturalist* 154:449–468.

Ceulemans, R., and M. Mousseau. 1994. Effects of elevated CO_2 on woody plants. *New Phytologist* 127:425–446.

Cotrufo, M. F., P. Ineson, and A. Scott. 1998. Elevated CO_2 reduces the nitrogen concentration of plant tissues. *Global Change Biology* 4:43–54.

Curtis, P. S., and X. Wang. 1998. A meta-analysis of elevated CO_2 effects on woody plant mass, form, and physiology. *Oecologia* 113:299–313.

Daufresne, T., and M. Loreau. 2001. Plant-herbivore interactions and ecological stoichiometry: When do herbivores determine plant nutrient limitation? *Ecology Letters* 4:196–206.

Droop, M. R. 1974. The nutrient status of algal cells in continuous culture. *Journal of the Marine Biological Association of the United Kingdom* 54:825–855.
Duboc, P., and U. von Stockar. 1998. Dual limitations: Kinetic and stoichiometric analysis. *Thermochimica Acta* 309:121–128.
Elser, J. J., W. F. Fagan, R. F. Denno, D. R. Dobberfuhl, A. Folarin, A. Huberty, S. Interlandi, S. S. Kilham, E. McCauley, K. L. Schulz, E. H. Siemann, and R. W. Sterner. 2000. Nutritional constraints in terrestrial and freshwater foodwebs. *Nature* 408:578–580.
Enriquez, S., C. M. Duarte, and K. Sand-Jensen. 1993. Patterns in decomposition rates among photosynthetic organisms: The importance of detritus C:N:P content. *Oecologia* 94:457– 471.
Ericsson, T. 1995. Growth and shoot:root ratio of seedlings in relation to nutrient availability. *Plant and Soil* 168/169:205–214.
Ericsson, T., and T. Ingestad. 1988. Nutrition and growth of birch seedlings at varied relative phosphorus addition rates. *Physiologia Plantarum* 7:227–235.
Falkowski, P .G. 1997. Evolution of the nitrogen cycle and its influence on the biological sequestration of CO_2 in the ocean. *Nature* 387:272–275.
———. 2000. Rationalizing elemental ratios in unicellular algae. *Journal of Phycology* 36:3–6.
Falkowski, P. G., R. J. Scholes, E. Boyle, J. Canadell, D. Canfield, J. Elser, N. Gruber, K. Hibbard, P. Högberg, S. Linder, F. T. MacKenzie, B. Moore, T. Pedersen, Y. Rosenthal, S. Seitzinger, V. Smetacek, and W. Steffen. 2000. The global carbon cycle: A test of our knowledge of Earth as a system. *Science* 290:291–296.
Franklin, O., P. Högberg, A. Ekblad, and G. I. Ågren. Pine forest floor carbon accumulation in response to N and PK additions: Bomb 14C modelling and respiration studies. *Ecosystems*, in press.
Goldman, J. C., J. J. McCarthy, and D. G. Peavey. 1979. Growth rate influence on the chemical composition of phytoplankton in oceanic waters. *Nature* 279:210–214.
Griffiths, B. S., K. Ritz, R. D. Bardgett, R. Cook, S. Christensen, F. Ekelund, S. J. Sørensen, E. Bååth, J. Bloem, P. C. de Ruiter, J. Dolfing, and B. Nicolardot. 2000. Ecosystem response of pasture soil communities to fumigation-induced microbial diversity reductions: An examination of the biodiversity-ecosystem function relationship. *Oikos* 90:279–294.
Hessen, D. O., K. Nygaard, K. Salonen, and A. Vähätalo. 1994. The effect of substrate stoichiometry on microbial activity and carbon degradation in humic lakes. *Environment International* 20:67–76.
Hobbie, S. E. 2000. Interactions between litter lignin and soil nitrogen availability during leaf litter decomposition in a Hawaiian montane forest. *Ecosystems* 3:484–494.
Idso, K. E., and S. B. Idso. 1994. Plant responses to atmospheric CO_2 enrichment in the face of environmental constraints: A review of the past 10 years' research. *Agricultural and Forest Meteorology* 69:153–203.
Ingestad, T. 1979. Nitrogen stress in birch seedlings. II. N, K, P, Ca, and Mg nutrition. *Physiologia Plantarum* 45:149–157.
Ingestad, T., and G. I. Ågren. 1992. Theories and methods on plant nutrition and growth. *Physiologia Plantarum* 84:177–184.
———. 1995. Plant nutrition and growth: Basic principles. *Plant and Soil* 169:15–20.

Ingestad, T., and A. B. Lund. 1979. Nitrogen stress in birch seedlings. I. Growth technique and growth. *Physiologia Plantarum* 45:137–148.

Joffre, R., and G. I. Ågren. 2001. From plant to soil: Litter production and decomposition. Pp. 83–99 in *Terrestrial Global Productivity*, edited by J. Roy, B. Saugier, and H. A. Mooney. San Diego: Academic Press.

Killingbeck, K. T. 1996. Nutrients in senescing leaves: Key to the search for potential resorption and resorption proficiency. *Ecology* 77:1716–1727.

Kirschbaum, M. U. F., B. E. Medlyn, D. A. King, S. Pongracic, D. Murty, H. Keith, P. K. Khanna, P. Snowdon, and J. R. Raison. 1998. Modelling forest-growth response to increasing CO_2 concentration in relation to various factors affecting nutrient supply. *Global Change Biology* 4:23–41.

Lenton, T. 2001. The role of land plants, phosphorus weathering, and fire in the rise and regulation of atmospheric oxygen. *Global Change Biology* 7:613–629.

Lenton, T., and A. J. Watson. 2000. Redfield revisited: 1. Regulation of nitrate, phosphate, and oxygen in the ocean. *Global Biochemical Cycles* 14:225–248.

Linder, S. 1995. Foliar analysis for detecting and correcting nutrient imbalances in Norway spruce. *Ecological Bulletins* 44:178–190.

Loreau, M. 1995. Consumers as maximizers of matter and energy flow in ecosystems. *American Naturalist* 145:22–42.

Martin, J. 1990. Glacial-interglacial CO_2 change: The iron hypothesis. *Paleoceanography* 5:1–13.

McNaughton, S. J. 1988. Mineral nutrition and spatial concentration of African ungulates. *Nature* 334:343–345.

Melillo, J. M., J. D. Aber, and J. F. Muratore. 1982. Nitrogen and lignin control of hardwood leaf litter decomposition dynamics. *Ecology* 63: 621–626.

Mooney, H. A., J. Canadell, F. S. Chapin III, J. R. Ehleringer, C. Körner, R. E. McMurtrie, W. J. Parton, L. F. Pitelka, and E.-D. Schulze.1999. Ecosystem physiology responses to global change. Pp. 141–189 in *The Terrestrial Biosphere and Global Change: Implications for Natural and Managed Ecosystems,* edited by B. H. Walker, W. L. Steffe, J. Canadell, and J. S. I. Ingram. Cambridge: Cambridge University Press.

Moran, N., and W. D. Hamilton. 1980. Low nutritive quality as defense against herbivores. *Journal of Theoretical Biology* 86:247–254.

Norby, R. J. 1996. Forest canopy productivity index. *Nature* 381:564.

Norby, R. J., and M. F. Cotrufo. 1998. Global change: A question of litter quality. *Nature* 396:17–18.

Oren, R., D. S. Ellsworth, K. H. Johnsen, N. Phillips, B. E. Ewers, C. Maier, K. V. R. Schafer, H. McCarthy, G. Hendrey, S. G. McNulty, and G. G. Katul. 2001. Soil fertility limits carbon sequestration by forest ecosystems in a CO_2-enriched atmosphere. *Nature* 411:469–472.

Overdieck, D. 1993. Elevated CO_2 and the mineral content of herbaceous and woody plants. *Vegetatio* 104/105:403–411.

Poorter, H. 1993. Interspecific variation in the growth response of plants to an elevated ambient CO_2 concentration. *Vegetatio* 104/105:77–97.

Rastetter, E. B., and G. R. Shaver. 1992. A model of multiple-element limitations for acclimating vegetation. *Ecology* 73:1157–1174.

Rastetter, E.B., G. I. Ågren, and G. R. Shaver. 1997. Responses of N-limited ecosystems to increased CO_2: A balanced-nutrition, coupled-element-cycles model. *Ecological Applications* 7:444–460.

Redfield, A. C., B. C. Ketchum, and F. A. Richards. 1963. The influence of organisms on the composition of sea-water. Pp. 26–77 in *The Sea,* Vol. 2, edited by M. N. Hill. New York: Interscience.

Resh, S. C., D. Binkley, and J. A. Parrotta. 2002. Greater soil carbon sequestration under nitrogen-fixing trees compared with Eucalyptus species. *Ecosystems* 5:217–231.

Schindler, D. W. 1977. Evolution of phosphorus limitation in lakes. *Science* 195:260–262.

Sterner, R. W., and D. O. Hessen. 1994. Algal nutrient limitation and the nutrition of aquatic herbivores. *Annual Review of Ecology and Systematics* 25:1–29.

Sterner, R. W., and J. J. Elser. 2002. *Ecological stoichiometry: The biology of elements from molecules to the biosphere.* Princeton: Princeton University Press.

Swift, M. J., O. W. Heal, and J. M. Anderson. 1979. *Decomposition in terrestrial ecosystems.* Oxford: Blackwell.

Thingstad, T. F. 1987. Utilization of N, P, and organic C by heterotrophic bacteria. I. Outline of a chemostat theory with a consistent concept of 'maintenance' metabolism. *Marine Ecology Progress Series* 35:99–109.

Tyrrell, T. 1999. The relative influences of nitrogen and phosphorus on oceanic primary production. *Nature* 400:525–531.

Urabe, J., and Y. Watanabe. 1992. Possibility of N or P limitation for planktonic cladocerans: An experimental test. *Limnology and Oceanography* 37:244–251.

Van Veen, J. A., and P. J. Kuikman.1990. Soil structural aspects of decomposition of organic matter by microbes. *Biogeochemistry* 11:213–233.

Van Veen, J. A., E. Liljeroth, L. J. A. Lekkerkerk, and S. C. van de Geijn. 1991. Carbon fluxes in plant-soil systems at elevated atmospheric CO_2 levels. *Ecological Applications* 1:175–181.

Verhoef, H. A., and L. Brussaard. 1990. Decomposition and nitrogen mineralization in natural and agro-ecosystems: The contribution of animals. *Biogeochemistry* 11:175–211.

Vitousek, P. M., and R.W. Howarth. 1991. Nitrogen limitation on land and in the sea: How can it occur? *Biogeochemistry* 13:87–115.

Vitousek, P. M., J. D. Aber, R. W. Howarth, G. E. Likens, P. A. Matson, D. W. Schindler, W. H. Schlesinger, and D. G. Tilman. 1997. Human alterations of the global nitrogen cycle: Sources and consequences. *Ecological Applications* 7:737–750.

8

Biogeochemical Interactions and Biodiversity

Valerie T. Eviner and F. Stuart Chapin III

There has been considerable recent interest in the ecosystem consequences of global and regional shifts in biogeochemical cycling and the loss of species diversity. Since biogeochemical cycling is the result of an interaction of numerous organisms, the link between diversity and biogeochemistry is a natural one. Changes in either diversity or biogeochemical cycling can dramatically influence one another, and the large human impact on both has the capacity to compromise the ecosystem functions on which society relies.

A number of significant changes are occurring in the Earth's biogeochemistry (Austin et al., Chapter 2, this volume; Galloway, Chapter 14, this volume; Chapin et al. 2000). Because of these shifts in global and regional biogeochemistry, as well as land use changes and the introduction of exotic species, we are in the midst of one of the largest extinction events in the history of life on Earth, with extinction rates 100–1,000 times greater than prehuman rates. The ecosystem response to these multiple biogeochemical shifts depends largely on the response of the biotic community. For example, shifts in vegetation in response to these human-induced changes can have ecosystem impacts that are larger and different in direction than the direct biogeochemical response to these changes (Hobbie 1996; Shaw and Harte 2001). Beyond a shift in composition, loss of diversity can compromise the capacity of a community to perform ecosystem functions, particularly in response to multiple stresses (Griffiths et al. 2000; Degens et al. 2001; Muller et al. 2002).

In this chapter we explore the feedbacks between changes in biogeochemical interactions and biodiversity. Plant species richness is the component of biodiversity that is most frequently studied in relationship to ecosystem function, but we include other examples where they are available.

Effects of Changes in Biogeochemical Cycling on Diversity

Shifts in the biotic community in response to a variety of biogeochemical changes have been well documented (Table 8.1). Deposition of toxic levels of heavy metals and SO_2 can decrease the diversity of plants, soil microbes, and several other types of organisms. These decreases in species richness are directional changes through selection for species that can tolerate levels of pollutants that are potentially toxic to many organisms (Table 8.1). The differential responses of components of the biotic community to these pollutants can have significant impacts on biogeochemical cycling. For example, heavy metals suppress N_2O production more than NO production (Holtan-Hartwig et al. 2002).

Plant diversity substantially decreases in response to chronic N inputs across a large number of ecosystems through selection for fast-growing plant species and canopy dominants and selection against legumes and small-statured species. N deposition also causes shifts in microbial community composition and diversity (Table 8.1). Additions of other limiting nutrients in terrestrial systems do not have as strong an effect on community diversity as does N. For example, even though P was the most limiting nutrient to plant growth, P additions did not alter plant species richness in a chalk grassland, whereas N additions substantially decreased plant species diversity (Willems, Peet, and Bik 1993). Similarly, in an annual grassland, additions of P and K had very little effect on plant community composition, whereas additions of N decreased species richness (Goldberg and Miller 1990). These results show that N additions tend to have strong impacts on plant species diversity, whereas community composition has variable responses to additions of other nutrients, even if they are limiting.

Eutrophication by N and P in aquatic systems also tends to decrease the diversity of aquatic organisms and selects for plants that are good competitors at high nutrient levels and for animals that tolerate low oxygen levels. In both aquatic and terrestrial habitats, these directional changes in community composition are likely to determine the ecosystem effects of resource additions (Fridley 2002).

The effects of elevated CO_2 on plant and microbial community composition are variable. Elevated CO_2 often results in a change in plant and microbial species composition, but there are no generalizable responses in the nature of species that are selected for. Several studies, however, have found that elevated CO_2 increases plant species evenness (proportionate distribution of species) (Table 8.1).

Although climate is not specifically a biogeochemical change, increased fluxes of CO_2, CH_4, and N_2O into the atmosphere are resulting in warming and possible changes in precipitation patterns that will likely impact the biological community. Changing weather patterns will cause directional shifts in species composition and diversity, based on the life history of the species (Sternberg et al.

Table 8.1. Shifts in the biotic community in response to a variety of biogeochemical changes

Biogeochemical change	*Diversity change*	*Selection for species*
SO_2 deposition	Decreased species richness of:	Selection against sensitive species
	plants[1]	
	soil microarthropods[2]	
	macroinvertebrates[3]	
	ants[4]	
Heavy metals	Decreased species richness of:	
	plants[5]	Selection for tolerant plants[6]
	microbes	Selection for resistant microbes with fast growth[7]
		Change in microbial community structure and function[8]
	aquatic macroinvertebrates[9]	Selection for tolerant dominants[10]
N additions	Terrestrial plants—decreased richness[11]	Selection for fast-growing, tall species[12]
		Selection against short-statured species, legumes[13]
	Aquatic organisms—decreased plant diversity[14]	Select for fast-growing plants and against native species
		Select for animals that tolerate low oxygen[15]
	Mycorrhizal diversity decreases [16]	Select for species that are small-spored, less mutualistic, do well at high N[16]
	Soil microbes	Increase fungi[17]
		Selection for microbial communities dependent on plant compositional shifts [18]

(continued on next page)

Table 8.1. *(continued)*

Biogeochemical change	*Diversity change*	*Selection for species*
Elevated CO_2	Plants—increased evenness, but no change in richness[19]	Shifts in plant species composition, but no consistent groups of species selected for, some evidence of selection against dominants[20]
	Microbes	Larger effects on composition than diversity Shifts in composition but not consistent[21]
Warming	Plants	Shifts in plant composition, selection for species depends on other environmental conditions[22]
	Microbes—increased diversity[23]	Preferential stimulation of methanogens over methane-oxidizers[24]
	Soil mesofauna—increased diversity if moisture is not limiting[25]	

[1] Singh, Agrawal, and Narayan 1994; [2] Bressan and Paoletti 1997; [3] Carcamo and Parkinson 2001; [4] Hoffman, Griffiths, and Andersen 2000; [5] Salemaa, Vanha-Majamaa, and Derome 2001; [6] Del Rio et al. 2002, Lasat 2002; [7] Muller et al. 2001; [8] Kozdroj and van Elsas 2001, Muller et al. 2001, Muller et al. 2002; [9] Clements and Kiffney 1995; [10] Clements 1994, Clements et al. 2000; [11] Reed 1977, Inouye et al. 1987, Tilman 1988, Goldberg and Miller 1990, Huenneke et al. 1990, Wilson and Tilman 1991, Berendse, Aerts, and Bobbink 1993, Willems, Peet, and Bik 1993, Inouye and Tilman 1995; [12] Dukes and Mooney 1999; [13] Jones and Winans 1967, Willems, Peet, and Bik 1993; [14] Tilman 1999, Mosier et al. 2001; [15] Byers 2002; [16] Arnolds 1991, Egerton-Warburton and Allen 2000, Lilleskov, Fahey, and Lovett 2001, Egerton-Warburton et al. 2001, Lilleskov et al. 2002; [17] Arnolds 1991; [18] Bardgett et al. 1999; [19] Leadley et al. 1999, Niklaus et al. 2001a, He, Bazzaz, and Schmid 2002; [20] Leadley and Korner 1996, Chiariello and Field 1996, Potvin and Vassuer 1997, Owensby et al. 1999, Dukes and Mooney 1999; [21] O'Neill 1994, Kandeler et al. 1998, Hungate et al. 2000, Phillips et al. 2002, Dhillion, Roy, and Abrams 1996, Rillig et al. 1997, Grayston et al. 1998; [22] Harte and Shaw 1995, Chapin and Korner 1996, Hobbie 1996, Saleska, Harte, and Torn 1999, Price and Waser 2000; [23] Andrews et al. 2000; [24] Schlesinger 1997; [25] Harte et al. 1995.

1999). Several studies have shown that warming influences plant species composition through selection for certain plant species (Harte and Shaw 1995; Chapin and Korner 1996; Hobbie 1996), although this is not always the case (Price and Waser 2000). These shifts in vegetation composition are largely responsible for ecosystem responses to warming (Hobbie 1996; Saleska, Harte, and Torn 1999; Shaw and Harte 2001). Similarly, an increase in microbial diversity in response to warming increases soil C respiration (Andrews et al. 2000), while preferential stimulation of methanogenic bacteria over methane-oxidizers could stimulate increased CH_4 flux from wetland soils (Schlesinger 1997).

Clearly, changes in biotic diversity in response to biogeochemical factors are not random additions or deletions of species; they are selective changes in species composition. Resulting changes in biogeochemical processes may be better explained by these changes in species composition than by changes in species richness.

The Ecosystem Effects of Species Diversity

Mechanisms

Many components of biological diversity could be important in determining ecosystem processes. Most studies have focused on the ecosystem consequences of species richness within a group of organisms (e.g., plant species diversity or microbial diversity). The most commonly proposed mechanism for a diversity effect on ecosystem function is complementarity. Organisms differ in their niches, including how, when, and where they play their role, and the combination of species with different niches can affect ecosystem function in ways that differ from the effects of any single species. For example, species differences in the type, timing, or spatial zone of resource uptake can enhance overall community resource utilization, leading to an increase in plant production. The more unique the niches of the species are, the greater the diversity effect.

The observed relationships between diversity and ecosystem functions may also be explained by the "sampling effect" (Aarrsen 1997; Huston 1997). The sampling effect is based on one or a subset of species with strong ecosystem effects. In this case, the relationship between species richness and ecosystem function is not a mechanistic effect of diversity per se, but rather, it is due to an increased statistical probability of including a particular species in a community of higher diversity.

A third mechanism, "interspecific facilitation" (Cardinale, Palmer, and Collins 2002), has received less attention in the literature. Like complementarity, this ecosystem effect is a result of the activity of multiple organisms. Whereas complementarity is a function of the accumulated independent effects of the component species, in interspecific facilitation, the function and sometimes the existence of these species complexes rely on close interactions among the component species. Some examples include symbiotic N fixation and substrate degradation that can only be achieved by a closely associated community of microbes (consortia). Extreme examples are eukaryotic cells and their intracellular units that are endosymbiotic prokaryotes or the descendents of these (e.g., mitochondria, chloroplasts, S-oxidizing bacteria, methanogens, methane-oxidizing bacteria) (Caldwell et al. 1997).

A fourth alternative is that there is no mechanistic effect of diversity on particular ecosystem processes; rather diversity and an ecosystem process co-vary because they are controlled by the same environmental factor. This is not a mechanism that can explain direct experimental manipulations of diversity, but it can be an important consideration when comparing natural systems or experimental treatments in which species richness is manipulated through a change in environmental factors (e.g., Huston 1997; Wardle 2001). For example, when comparing community diversity and productivity in unmanipulated natural plots in very low-productivity environments, the correlation between productivity and diversity may not be causative but due to the fact that both are limited by the same resources (Wardle 2001).

At a mechanistic level, sampling effects, complementarity, and interspecific facilitation all depend on a mixture of species that differ in a key trait or combinations of traits. These mechanisms are not mutually exclusive, and the importance of these multiple mechanisms highlights a critical feature of biological diversity: a given function can be attained through multiple pathways.

The Effects of Biological Diversity on Different Ecosystem Functions

Species diversity can influence many ecosystem processes, but ecosystem processes differ in their relationships to diversity and/or composition of the biotic community. The strongest relationships between species richness and ecosystem processes are for those processes that are directly related to resource utilization (reviewed in Eviner and Chapin, in review). A number of studies have observed a positive relationship between plant richness (Naeem et al. 1994; Tilman et al. 1997; Hooper and Vitousek 1998; Mulder, Uliassi, and Doak 2001) or mycorrhizal richness (van der Heijden et al. 1998) and plant productivity. Increased productivity in response to increased plant species richness can be mediated through complementarity in the use of various resources, but this relationship is often weak and is usually better explained by the presence of highly productive species (Naeem et al. 1994; Tilman et al. 1997; Hooper and Vitousek 1998; Spehn et al. 2000a; Mulder, Uliassi, and Doak 2001). Plant species richness also contributes to nutrient and water retention, but again, species composition usually plays a larger role than species number (Naeem et al. 1994, 1995; Tilman, Wedin, and Knops 1996; Tilman et al. 1997; Hooper and Vitousek 1998).

Much as plant species richness can enhance productivity through complementarity of resource use, a diverse microbial community may enhance decomposition of litter or soil organic matter through the same mechanism (Solanius 1981; McGrady-Steed, Harris, and Morin 1997; Dobranic and Zak 1999). Decomposition rates are not, however, always enhanced by increased diversity of microbial communities because the identity of species plays a larger role than their number (Finlay, Maberly, and Cooper 1997). For example, selection for certain microbes can enhance decomposition of plant residues despite decreases in microbial community diversity (Griffiths et al. 2000). Microbial community composition can also affect the degradation of specific compounds (Janzen, Dormaar, and McGill 1995; Colombo, Cabello, and Arambarri 1996; De Boer, Klein Gunnewiek, and Parkinson 1996). There are many substrates, particularly environmental contaminants, that cannot be degraded by isolated populations of microbes, but only by consortia (Rozgaj and Glancersoljan 1992; Caldwell et al. 1997), particularly in anaerobic environments (Palmer et al. 1997). Composition of the microbes within consortia can dramatically alter degradation rates (Wolfaardt et al. 1994), indicating that the effects of a community on degradation are due not only to the number of species, but also to their composition.

Similar to decomposition, rates of nitrification, denitrification (Martin, Trevors, and Kaushik 1999; Griffiths et al. 2000), and methane oxidation (Willison et al. 1997; Griffiths et al. 2000) relate to microbial diversity and are also strongly influenced by the composition of the microbial community (Landi et al. 1993). Overall, it is clear that microbial community composition can have strong impacts on ecosystem function (Colombo, Cabello, and Arambarri 1996; De Boer, Klein Gunnewiek, and Parkinson 1996; Balser, Kinzig, and Firestone 2001).

While microbial diversity can enhance decomposition and nutrient cycling, plant composition generally has a stronger effect on these processes, because the species composition of the plant community determines the substrates available for microbial utilization. Although the identity of plant species has strong effects on these processes, there is little reason to hypothesize an effect of plant species richness on these processes. All experimental evidence indicates that decomposition rates are not related to plant species richness, but to plant composition (Naeem et al. 1995; Wardle, Bonner, and Nicholson 1997; Bardgett and Shine 1999; Hector et al. 2000; Spehn et al. 2000b; Knops, Wedin, and Tilman 2001). Several of these studies found that decomposition rates of litter mixtures could not be predicted based on rates of the component species alone (nonadditive effect), suggesting the importance of substrate mixtures, but these were both positive and negative effects and were largely due to specific combinations of substrates (species composition), not an increase in the number of litter types (Wardle and Nicholson 1996; Wardle, Bonner, and Nicholson 1997; Hector et al. 2000). There have been a few mechanistic hypotheses linking plant species richness and decomposition rates. Higher diversity treatments tend to have higher percent cover and soil water use, resulting in lower temperature and moisture for decomposition. Although these directional microclimate changes would be predicted to have a negative effect on decomposition, the magnitude of these microclimate effects on decomposition rate was extremely small (Hector et al. 2000). Similarly, plant substrate quality has a larger impact on nutrient mineralization than does plant richness. Although plant species richness can stimulate N cycling through an increase in substrate quantity (Spehn et al. 2000b) or decrease nitrification rates through higher plant NH_4 uptake (Niklaus et al. 2001b), these effects tend to be weaker than the influence of substrate quality on N cycling. This accounts for the frequent observation that the role of richness in ecosystem function is overwhelmed by species composition and environmental factors (Fridley 2002; Loreau 1998; Loreau 2000; Wardle, Bonner, and Barker 2000).

Although species composition is key to a mechanistic understanding of the role of the biotic community in determining ecosystem processes, diversity per se is important to ensuring stability of these species effects. Different functions and environmental tolerances are usually distributed independently among species. There may be redundancy in a function, but it is unlikely that there will be overlap in the suite of functions associated with any species, including functional roles, environmental tolerances, physiological requirements, and microhabitat preferences (Beare et al. 1995). Species diversity is critical for the maintenance of these roles under fluctuating conditions. Diversity provides a community with multiple ways of performing a function, providing stability in biogeochemical processes despite shifts in environmental conditions (McNaughton 1977; Rao and Willey 1980; Chapin and Shaver 1985; McGrady-Steed, Harris, and Morin 1997; Naeem and Li 1997; Griffiths et al. 2000; Degens et al. 2001).

Effects of Species Composition on Interactions among Biogeochemical Cycles

Biogeochemical cycling is largely mediated through the biotic community, which both supplies and metabolizes substrates. Species composition largely controls the ratios at which elements cycle in systems and the potential of these element cycles to interact as a result of the suite of traits, functions, and environmental preferences of the species present in a community.

C, N, S, and P are not cycled in simple stoichiometric ratios because these elements are mobilized and stabilized by different mechanisms (McGill and Cole

1981). N is bonded directly to the C skeleton of organic compounds, and thus mineralization of N is a by-product of the intracellular breakdown of organic compounds by microbes for energy (C). This accounts for the strong relationship between litter C:N and rates of decomposition and net N mineralization. Because P is not directly bonded to the C skeleton of organic matter but is associated with C through ester bonds, mineralization of P occurs through enzymatic activity, which cleaves P from organic matter. Thus mineralization of P from organic matter is not directly related to litter C:P ratio but is controlled more directly by plant and microbial need for P. P can also be accessed by organisms through production of charged organic compounds that can free PO_4 from binding sites in the soil or through accumulation of calcium oxalates that increase P availability by decreasing soil calcium. The control of sulfur release from organic matter is intermediate between N and P because it occurs in organic matter in both C-bonded and ester-bonded forms. Thus, organic S can be released through both microbial C metabolism and directly by biotic demand for S.

Because these elements are stabilized by different mechanisms, the effect of an organism on any one of these elements can be independent of its effect on other elements. For example, grassland plant species with very similar effects on N cycling can differ substantially in their effects on P cycling (Hooper and Vitousek 1998; Eviner 2001), water fluxes, and soil C (Eviner and Chapin 2001). Similarly, the influence of the plant community on nutrient retention differs for different elements (Naeem et al. 1994, 1995). The functions of species can also vary independently from their environmental tolerances (Eviner and Chapin 2001), so shifts in environmental conditions are not likely to select for species with specific suites of functional effects. A notable exception to this is that soil fertility levels often select for plant species that reinforce these levels of fertility (Chapin 1980). Species not only differ in their effects on multiple elements, but also in how they couple multiple elements. The ratios of elements required for optimal growth can differ among species, and many organisms can store nutrients that are in excess at one time and later use them for growth.

The stoichiometry of decomposers can greatly influence rates of N mineralization. Fungi, with a high C:N ratio (15:1–45:1), will mineralize relatively more N than bacteria, which have a low C:N ratio (3:1–5:1) and higher requirement for N (Paul and Clark 1995). For both herbivores and decomposers, the ratio of nutrient release from a given substrate will be determined by the requirements of those organisms for the different nutrients. A species with a higher N:P requirement will recycle N at a relatively lower concentration than a species with a lower N:P requirement (Sterner, Elser, and Hessen 1992; Elser et al. 2000). The balance of nutrients released will change because of shifts in both the species providing the substrate (producers) and the consumers.

The ratios of element uptake and resorption by plants can have substantial effects on biogeochemical interactions, since plant litter is the major substrate for element recycling and fuels many biogeochemical processes. Plant species differ in the ratios of elements taken up to produce new biomass and resorbed prior to litter fall (Aerts and Chapin 2000), so species differences in these properties alter rates at which different elements cycle. For example, many studies have linked plant species effects on decomposition and N cycling to litter C:N ratios (Taylor, Parkinson, and Parsons 1989; Wedin and Tilman 1990; Scott and Binkley 1997; Maithani et al. 1998; Mueller et al. 1998; Mafongaya, Barak, and Reed 2000).

Species differences in C quality of litter also have strong effects on rates of decomposition and mineralization (Melillo, Aber, and Muratore 1982; Aber, Melillo, and McClaugherty 1990; Palm and Sanchez 1991; Taylor et al. 1991; Stump and Binkley 1993; Schimel et al. 1996), because secondary compounds can inhibit microbial activity (Lodhi and Killingbeck 1980; Thibault, Fortin, and Smirnoff 1982). These secondary compounds can be more important than C:N ratios in explaining plant species effects on biogeochemical cycling (Eviner 2001).

Although plant regulation of N and C dynamics has largely focused on litter quality, up to a tenfold difference in rates of net N mineralization can occur because of a threefold difference in labile C inputs by different species (Wedin and Pastor 1993). These labile C inputs are generally simple C compounds that are readily available as an energy source to microbes and can accelerate decomposition of recalcitrant litter and soil organic matter (Sallih and Bottner 1988; Mueller et al. 1998; Bottner, Pansu, and Sallih 1999). These recalcitrant C compounds are often not used as an energy source, and microbes often require labile C to provide energy for the metabolism of these recalcitrant compounds in their search for N. Labile C will stimulate decomposition of a recalcitrant compound under N-limited conditions but not under high N availability (van Ginkel, Gorissen, and Van Veen 1996). These examples clearly demonstrate that the biogeochemistry of C is affected not only by an interaction of nutrients, but also by the interaction of different types of C compounds that are produced by different plant species.

Plant composition can also influence biogeochemical cycling by altering microclimate conditions such as temperature, pH, soil moisture, and O_2 concentrations (van Vuuren et al. 1992; Mack 1998; Eviner 2001; Caraco and Cole 2002). For example, in waterlogged soils or sediments, plant species that enhance soil O_2 concentrations have been responsible for enhanced decomposition of organic matter (Allen et al. 2002), increased nitrification rates (Engelaar et al. 1995), decreased CH_4 formation (Grosse, Jovy, and Tiebel 1996), and lower consumption of SO_4, leading to decreased mobilization of Fe and mineralization of P (Christensen 1999). Interactions of plant inputs also determine the fate of nitrate in anaerobic environments by regulating the concentrations of NO_3 and organic C available to denitrifiers. The amount of C in relation to NO_3 is positively correlated to the amount of N_2 versus N_2O produced by denitrification. At high levels of labile C, however, the fate of NO_3 can shift from denitrification to dissimilatory nitrate reduction to ammonium (DNRA) (Fazzolari, Nicolardot, and Germon 1998; Silver, Herman, and Firestone 2001; Yin et al. 2002).

Ultimately, a plant species effect on biogeochemical cycling reflects its suite of traits. For example, an interaction of litter, microclimate effects, and labile C inputs determines plant species effects on net N mineralization and nitrification, soil C dynamics, and decomposition, and these traits vary independently among species (Eviner 2001). This independent distribution of multiple ecosystem functions among species is a key driver of the effect of the biotic community on biogeochemical interactions.

Biogeochemical interactions are also related to species-specific responses of microbes to combinations of substrates and environmental conditions. For example, the effect of N in limiting lignin decomposition depends on the species of white rot fungus present, because N inhibits some species less than others (Schimel 1995). Similarly, variation in methane oxidation with N concentrations depends on the composition of the microbial community (Groffman and Bohlen 1999). Ultimately, the suite of functions and environmental tolerances of a given species will determine the interaction of elements, their stoichiometry, and the interactions of biogeochemical cycling with environmental conditions. By altering environmental conditions and the ratios of available elements, any given species can also influence the distribution and activity of other organisms, further altering biogeochemical interactions. Thus, it is not surprising that the stoichiometry of elements is highly uncoupled in both terrestrial and aquatic ecosystems.

Biogeochemical Interactions and Species Mixtures

The interactions between biogeochemical cycles are mediated by the suite of organismal traits being expressed, and the nature of traits has a much larger role in determining biogeochemical cycling than does the diversity of traits. There is no clear relationship between species richness and biogeochemical interactions because the organismal traits that determine the dynamics of different elements do not consistently occur together when species are compared. The ecosystem effects of species mixtures often cannot be predicted based on the patterns of species effects in monocultures (nonadditive effects), as seen in processes ranging from decomposition (Naeem et al. 1995; Wardle and Nicholson 1996; Wardle, Bonner, and Nicholson 1997) to N cycling (Blair, Parmelee, and Beare 1990; Wardle, Bonner, and Nicholson 1997; Finzi and Canham 1998; Nilsson, Wardle, and

Dahlberg 1999; Eviner 2001) to plant growth (Newman et al. 1979; Lawley, Newman, and Campbell 1982; Lawley, Campbell, and Newman 1983; Nilsson, Wardle, and Dahlberg 1999). Just as the ecosystem effects of a species depends on an interaction of traits that a given species exhibits, these trait combinations have the same effect on biogeochemical cycling when they are due to the mixture of multiple species, each of which exhibits some of these traits (Eviner 2001). Some trait combinations, however, are unique to species mixtures, seldom occurring within a single plant species (e.g., litter with recalcitrant C and high N content). Nonadditive effects of species mixtures are mediated by an interaction of particular traits of particular species, not simply by a diversity of traits. Nonadditive effects of species mixtures can also be due to shifts in species traits resulting from the identity of the neighboring species. For example, nonadditive effects of plant mixtures on N cycling in California annual grasslands relate to shifts in litter quantity and quality, labile C inputs, and microclimate effects when species are grown in mixture versus monoculture (Eviner 2001). Nonadditive effects of plant mixtures can also be mediated through shifts in the activity and distribution of other organisms that play key roles in biogeochemical cycling (Blair, Parmelee, and Beare 1990; Williams and Alexander 1991). For example, N and C dynamics in plant mixtures were nonadditive functions of the component monocultures because earthworms responded nonlinearly to the plant mixtures (Saetre 1998).

All of these examples demonstrate that the rates and interactions of biogeochemical cycles can best be understood by focusing on the composition of organisms and the traits they exhibit, not simply on species number, because combinations of specific organismal traits mediate elemental fluxes.

Interactions between Biodiversity and Biogeochemical Interactions

Ultimately, an interaction of species and biogeochemical cycles determines ecosystem processes, and species diversity can influence ecosystem responses to biogeochemical changes by influencing the pool of species that can respond to varying conditions. For example, high-diversity plant communities have a higher enhancement of net primary production (NPP) in response to elevated CO_2 (Niklaus et al. 2001a; Reich et al. 2001; He, Bazzaz, and Schmid 2002) and N (Fridley 2002; He, Bazzaz, and Schmid 2002) than do low-diversity communities. Plant richness, however, does not always enhance ecosystem CO_2 assimilation (Stocker et al. 1999), and species composition plays an important role in determining these responses. The diversity-induced enhancement of NPP in response to elevated CO_2 in a calcareous grassland only occurred in the early years of the experiment, when CO_2 increased growth of stress-tolerant, noncompetitive species. These species decreased over time in the elevated CO_2 plots (Niklaus et al. 2001a), and with this shift in plant composition, the diversity-CO_2 interaction disappeared.

These studies clearly show that communities with high species richness are more likely to contain species that can positively respond to enhanced resource availability. A natural extension of this logic would suggest that a loss of species richness could compromise ecosystem function in response to environmental stresses owing to the presence of fewer species able to tolerate the stressful conditions. Many studies have shown that decreases in diversity in response to a single environmental stress do not compromise ecosystem function, because of the ability of other species to compensate for the function of the lost species. Several studies have shown, however, that these less diverse microbial communities cannot maintain ecosystem function in response to additional environmental stresses (Griffiths et al. 2000; Degens et al. 2001; Muller et al. 2002) and are also less able to respond to subsequent positive changes, such as additions of labile substrates (Muller et al. 2002). These results are particularly striking because all three of these examples are from soil microbial communities, which are assumed to contain ample diversity to buffer any change and still maintain function.

These studies emphasize that, although several species may be able to play the same ecosystem role under stable conditions, a diversity of organisms is critical for maintaining that ecosystem role under changing conditions, and loss of diversity can dramatically compromise ecosystem function. This decreased function occurs because species that have similar functions can have very different environmental tolerances. Thus one environmental stress can decrease diversity through selection for species with specific tolerances, compromising the ability of the remaining community to respond to another stress.

It is essential to consider the relative importance of changes in biogeochemical processes versus shifts in diversity on ecosystem processes. For example, there has been much debate over the importance of plant species richness to productivity. Although experimental evidence has indicated that species richness can positively affect production, cross-site comparisons of natural communities usually indicate that the productivity-diversity relationship is a bell-shaped curve (Huston 1997; Wardle 2001). In extreme resource-limited environments, both productivity and species richness tend to be low and are limited by the same resources. Both species richness and productivity increase in ecosystems with moderate resource availability. In systems with extremely high resource availability, productivity is high, but species richness tends to be low. In low-resource or extreme environments, there are not many organisms that can cope with the conditions that limit production. At the other extreme, high N availability in terrestrial environments strongly selects for species that are fast growing and highly competitive. A similar trend would be expected in many aquatic systems in response to P availability. Over fluctuating conditions, a diversity of plants allows production to be somewhat stabilized by providing species that can tolerate these varying conditions. Sustained differences in environmental conditions, however, can strongly select for species that are best adapted to those conditions, decreasing diversity. These communities may be optimized to their current conditions but may also have less flexibility in the face of other environmental changes.

Conclusion

In this chapter we have given numerous examples where species composition and richness alter biogeochemical processes. In processes influenced by resource use, complementarity of species functions can account for the effects of species richness, but the strength of this effect is determined by the extent to which species differ in their functional traits. Overall, the composition of the biotic community plays a much stronger role than species richness in determining biogeochemical cycling. The importance of composition is further highlighted because shifts in the biotic community due to biogeochemical changes alter species richness through selection for certain species. Suites of organismal traits determine elemental stoichiometry and environmental tolerances of species, which then determine how element cycles interact. For example, plant species can differ significantly in their effects on environmental conditions and the stoichiometry of elements. These suites of traits greatly affect the activity and composition of the consumer communities. The elemental composition of the consumer community dictates the stoichiometry and rates of nutrient release from plant litter, and the processes mediated by the microbial community are sensitive to the relative availability of elements. Thus, organism-induced variation in nutrient forms and concentrations can differentially affect almost every biogeochemical flux.

The vast majority of research linking species diversity and ecosystem processes has focused on this relationship under a relatively constant set of conditions. In these experimental communities the effect of species richness on ecosystem processes saturates at a low species number, and species composition tends to have a larger effect than species richness. Ecological theory, and a handful of experiments that have tested it, show that over changing conditions, the maintenance of ecosystem processes depends on a diversity of species. Species have unique combinations of functions, environmental preferences, and species interactions, and this suite of traits determines how a species responds and maintains itself, its functions, and its interactions under different conditions. There is very little true redundancy in the overall role played by species when accounting for these suites of traits.

Acknowledgments

Our thanks go to Jill Baron and Osvaldo Sala for helpful comments on an earlier draft. This work is a contribution to the program of the Institute of Ecosystem Studies.

Literature Cited

Aarssen, L. 1997. High productivity in grassland ecosystems: Effected by species diversity or productive species? *Oikos* 80:183–184.

Aber, J., J. Melillo, and C. McClaugherty. 1990. Predicting long-term patterns of mass loss, nitrogen dynamics and soil organic matter formation from initial fine litter chemistry in temperate forest ecosystems. *Canadian Journal of Botany* 68:2201–2208.

Aerts, R., and F. S. Chapin. 2000. The mineral nutrition of wild plants revisited: A re-evaluation of processes and patterns. *Advances in Ecological Research* 30:1–67.

Allen, W. C., P. B. Hook, J. A. Biederman, and O. R. Stein. 2002. Temperature and wetland plant species effects on wastewater treatment and root zone oxidation. *Journal of Environmental Quality* 31:1010–1016.

Andrews, J. A., R. Matamala, K. M. Westover, and W. H. Schlesinger. 2000. Temperature effects on the diversity of soil heterotrophs and the delta C-13 of soil-respired CO_2. *Soil Biology and Biochemistry* 32:699–706.

Arnolds, S. 1991. Decline of ectomycorrhizal fungi in Europe. *Agriculture Ecosystems and Environment* 35:209–244.

Balser, T., A. P. Kinzig, and M. K. Firestone. 2001. Linking microbial communities and ecosystem function. Pp. 265–357 in *The functional consequences of biodiversity*, edited by A. Kinzig, S. Pacala, and D. Tilman. Princeton: Princeton University Press.

Bardgett, R. D., J. L. Mawdsley, S. Edwards, P. J. Hobbs, J. S. Rodwell, and W. J. Davies. 1999. Plant species and nitrogen effects on soil biological properties of temperate upland grasslands. *Functional Ecology* 13:650–660.

Bardgett, R. D., and A. Shine. 1999. Linkages between plant litter diversity, soil microbial biomass and ecosystem function in temperate grasslands. *Soil Biology and Biochemistry* 31:317–321.

Beare, M., D. Coleman, D. Crossley, P. Hendrix, and E. Odum. 1995. A hierarchical approach to evaluating the significance of soil biodiversity to biogeochemical cycling. *Plant and Soil* 170:5–22.

Berendse, F., R. Aerts, and R. Bobbink. 1993. Atmospheric nitrogen deposition and its impact on terrestrial ecosystems. Pp. 104–121 in *Landscape ecology of a stressed environment,* edited by C. Vos and P. Opdam. London: Chapman and Hall.

Blair, J., R. Parmelee, and M. Beare. 1990. Decay rates, nitrogen fluxes, and decomposer communities of single- and mixed-species foliar litter. *Ecology* 71:1976–1985.

Bottner, P., M. Pansu, and Z. Sallih. 1999. Modelling the effect of active roots on soil organic matter turnover. *Plant and Soil* 2166:15–25.

Bressan, M., and M. G. Paoletti. 1997. Leaf litter decomposition and soil microarthropods affected by sulphur dioxide fallout. *Land Degradation and Development* 8:189–199.

Byers, J. E. 2002. Physical habitat attribute mediates biotic resistance to non-indigenous species invasion. *Oecologia* 130:146–156.
Caldwell, D. E., G. M. Wolfaardt, D. R. Korber, and J. R. Lawrence. 1997. Do bacterial communities transcend Darwinism? *Advances in Microbial Ecology* 15:105–191.
Caraco, N. F., and J. J. Cole. 2002. Contrasting impacts of a native and alien macrophyte on dissolved oxygen in a large river. *Ecological Applications* 12:1496–1509.
Carcamo, H. A., and D. Parkinson. 2001. Localized acidification near sour gas processing plants: Are forest floor macro-invertebrates affected? *Applied Soil Ecology* 17:199–213.
Cardinale, B. J., M. A. Palmer, and S. L. Collins. 2002. Species diversity enhances ecosystem functioning through interspecific facilitation. *Nature* 415:426–429.
Chapin, F. I., and C. Korner. 1996. Arctic and alpine biodiversity: Its patterns, causes, and ecosystem consequences. Pp. 7–32 in *Functional role of biodiversity: a global perspective*, edited by H. A. Mooney, J. Cushman, E. Medina, O. Sala, and E. Schulze. New York: John Wiley and Sons.
Chapin, F. I., E. Zavaleta, V. Eviner, R. Naylor, P. Vitousek, H. Reynolds, D. U. Hooper, S. Lavorel, O. Sala, S. Hobbie, M. Mack, and S. Diaz. 2000. Consequences of changing biodiversity. *Nature* 405:234–242.
Chapin III, F. S. 1980. The mineral nutrition of wild plants. *Annual Review of Ecology and Systematics* 11:233–260.
Chapin, F. S., and G. Shaver. 1985. Individualistic growth response of tundra plant species to environmental manipulations in the field. *Ecology* 66:564–576.
Chiariello, N., and C. Field. 1996. Annual grassland responses to elevated CO_2 in multiyear community microcosms. Pp. 139–155 in *Carbon dioxide, populations, and communities*, edited by C. Korner and F. A. Bazzaz. San Diego: Academic Press.
Christensen, K. K. 1999. Comparison of iron and phosphorus mobilization from sediments inhabited by Littorella uniflora and Sphagnum sp. at different sulfate concentrations. *Archiv für Hydrobiologie* 145:257–275.
Clements, W. 1994. Benthic invertebrate community responses to heavy metals in the upper Arkansas River basin, Colorado. *Journal of the North American Benthological Society* 13:30–44.
Clements, W., and P. Kiffney. 1995. The influence of elevation on benthic responses to heavy metals in Rocky Mountain streams. *Canadian Journal of Fisheries and Aquatic Science* 52:1966–1977.
Clements, W., D. Carlisle, J. Lazorchak, and P. Johnson. 2000. Heavy metals structure benthic communities in Colorado mountain streams. *Ecological Applications* 10:626–638.
Colombo, J., M. Cabello, and A. Arambarri. 1996. Biodegradation of aliphatic and aromatic hydrocarbons by natural soil microflora and pure cultures of imperfect and lignolitic fungi. *Environmental Pollution* 94:355–362.
De Boer, W., P. Klein Gunnewiek, and D. Parkinson. 1996. Variability of N mineralization and nitrification in a simple, simulated microbial forest soil community. *Soil Biology and Biochemistry* 28:203–211.
Degens, B. P., L. A. Schipper, G. P. Sparling, and L. C. Duncan. 2001. Is the microbial community in a soil with reduced catabolic diversity less resistant to stress or disturbance? *Soil Biology and Biochemistry* 33:1143–1153.

Del Rio, M., R. Font, C. Almela, D. Velez, R. Montoro, and A. D. Bailon. 2002. Heavy metals and arsenic uptake by wild vegetation in the Guadiamar River area after the toxic spill of the Aznalcollar mine. *Journal of Biotechnology* 98:125–137.

Dhillion, S. S., J. Roy, and M. Abrams. 1996. Assessing the impact of elevated CO_2 on soil microbial activity in a Mediterranean model ecosystem. *Plant and Soil* 187:333–342.

Dobranic, J. K., and J. C. Zak. 1999. A microtiter plate procedure for evaluating fungal functional diversity. *Mycologia* 91:756–765.

Dukes, J. S., and H. A. Mooney. 1999. Does global change increase the success of biological invaders? *Trends in Ecology and Evolution* 14:135–139.

Egerton-Warburton, L., and E. Allen. 2000. Shifts in arbuscular mycorrhizal communities along an anthropogenic nitrogen deposition gradient. *Ecological Applications* 10:484–496.

Egerton-Warburton, L., R. Graham, E. Allen, and M. Allen. 2001. Reconstruction of historical changes in mycorrhizal fungal communities under anthropogenic nitrogen deposition. *The Royal Society of London Proceedings B. Biological Sciences* 1484:2479–2848.

Elser, J. J., R. W. Sterner, E. Gorokhova, W. F. Fagan, T. A. Markow, J. B. Cotner, J. F. Harrison, S. E. Hobbie, G. M. Odell, and L. J. Weider. 2000. Biological stoichiometry from genes to ecosystems. *Ecology Letters* 3:540–550.

Engelaar, W. M. H. G., J. C. Symens, H. J. Laanbroek, and C. W. M. Blom. 1995. Preservation of nitrifying capacity and nitrate availability in waterlogged soils by radial oxygen loss from roots of wetland plants. *Biology and Fertility of Soils* 20:243–248.

Eviner, V. T. 2001. Linking plant community composition and ecosystem dynamics: Interactions of plant traits determine the ecosystem effects of plant species and plant species mixtures. Ph.D. diss., Department of Integrative Biology, University of California, Berkeley.

Eviner, V. T., and F. I. Chapin. 2001. The effects of California grassland species on their ecosystems: Implications for sustainable agriculture and rangeland management. *California Agriculture* 55:254–259.

———. 2003. The role of species richness and composition in determining ecosystem processes. *American Naturalist,* in review.

Fazzolari, E., B. Nicolardot, and J. C. Germon. 1998. Simultaneous effects of increasing levels of glucose and oxygen partial pressures on denitrification and dissimilatory nitrate reduction to ammonium in repacked soil cores. *European Journal of Soil Biology* 34:47–52.

Finlay, B., S. Maberly, and Cooper J. I. 1997. Microbial diversity and ecosystem function. *Oikos* 80:209–213.

Finzi, A., and C. Canham. 1998. Non-additive effects of litter mixtures on net N mineralization in a southern New England forest. *Forest Ecology and Management* 105:129–136.

Fridley, J. D. 2002. Resource availability dominates and alters the relationship between species diversity and ecosystem productivity in experimental plant communities. *Oecologia* 132:271–277.

Goldberg, D., and T. Miller. 1990. Effects of different resource additions on species-diversity in an annual plant community. *Ecology* 71:213–225.

Grayston, S. J., C. D. Campbell, J. L. Lutze, and R. M. Gifford. 1998. Impact of

elevated CO_2 on the metabolic diversity of microbial communities in N-limited grass swards. *Plant and Soil* 203:289–300.
Griffiths, B. S., K. Ritz, R. D. Bardgett, R. Cook, S. Christensen, F. Ekelund, S. J. Sorensen, E. Baath, J. Bloem, P. C. de Ruiter, J. Dolfing, and B. Nicolardot. 2000. Ecosystem response of pasture soil communities to fumigation-induced microbial diversity reductions: An examination of the biodiversity-ecosystem function relationship. *Oikos* 90:279–294.
Groffman, P. M., and P. J. Bohlen. 1999. Soil and sediment biodiversity: Cross-system comparisons and large-scale effects. *BioScience* 49:139–148.
Grosse, W., K. Jovy, and H. Tiebel. 1996. Influence of plants on redox potential and methane production in water-saturated soil. *Hydrobiologia* 340:93–99.
Harte, J., M. S. Torn, F. Chang, B. Feifarek, A. Kinzig, R. Shaw, and K. Shen. 1995. Global warming and soil microclimate—results from a meadow-warming experiment. *Ecological Applications* 5:132–150.
Harte, J., and R. Shaw. 1995. Shifting dominance within a montane vegetation community—results of a climate-warming experiment. *Science* 267:876–880.
He, J. S., F. A. Bazzaz, and B. Schmid. 2002. Interactive effects of diversity, nutrients and elevated CO_2 on experimental plant communities. *Oikos* 97:337–348.
Hector, A., A. J. Beale, A. Minns, S. J. Otway, and J. H. Lawton. 2000. Consequences of the reduction of plant diversity for litter decomposition: Effects through litter quality and microenvironment. *Oikos* 90:357–371.
Hobbie, S. 1996. Temperature and plant species control over litter decomposition in Alaskan tundra. *Ecological Monographs* 66:503–522.
Hoffman, B. D., A. D. Griffiths, and A. N. Andersen. 2000. Responses of ant communities to dry sulfur deposition from mining emissions in semi-arid tropical Australia, with implications for the use of functional groups. *Austral Ecology* 25:653–663.
Holtan-Hartwig, L., M. Bechmann, T. R. Hoyas, R. Linjordet, and L. R. Bakken. 2002. Heavy metals tolerance of soil denitrifying communities: N_2O dynamics. *Soil Biology and Biochemistry* 34:1181–1190.
Hooper, D., and P. Vitousek. 1998. Effects of plant composition and diversity on nutrient cycling. *Ecological Monographs* 68:121–149.
Huenneke, L., S. Hamburg, R. Koide, H. A. Mooney, and P. Vitousek. 1990. Effects of soil resources on plant invasion and community structure in Californian serpentine grassland. *Ecology* 71:478–491.
Hungate, B. A., C. H. Jaeger, G. Gamara, F. S. Chapin, and C. B. Field. 2000. Soil microbiota in two annual grasslands: Responses to elevated atmospheric CO_2. *Oecologia* 124:589–598.
Huston, M. A. 1997. Hidden treatments in ecological experiments: Re-evaluating the ecosystem function of biodiversity. *Oecologia* 110:449–460.
Inouye, R., N. Huntly, D. Tilman, J. Tester, M. Stillwell, and K. Zinnel. 1987. Old-field succession on a Minnesota sand plain. *Ecology* 68:12–26.
Inouye, R., and D. Tilman. 1995. Convergence and divergence of old-field vegetation after 11 years of nitrogen addition. *Ecology* 76:1872–1887.
Janzen, R., J. Dormaar, and W. McGill. 1995. A community-level concept of controls on decomposition processes: Decomposition of barley straw by *Phanerochaete chrysosporium* or *Phlebia radiata* in pure or mixed culture. *Soil Biology and Biochemistry* 27:173–179.

Jones, M. and S. Winans. 1967. Subterranean clover vs. nitrogen-fertilized annual grasslands: Botanical composition and protein content. *Journal of Range Management* 20:8–12.

Kandeler, E., D. Tscherko, R. D. Bardgett, P. J. Hobbs, C. Kampichler, and T. H. Jones. 1998. The response of soil microorganisms and roots to elevated CO_2 and temperature in a terrestrial model ecosystem. *Plant and Soil* 202:251–262.

Knops, J. M. H., D. Wedin, and D. Tilman. 2001. Biodiversity and decomposition in experimental grassland ecosystems. *Oecologia* 126:429–433.

Kozdroj, J., and J. D. van Elsas. 2001. Structural diversity of microorganisms in chemically perturbed soil assessed by molecular and cytochemical approaches. *Journal of Microbiological Methods* 43:197–212.

Landi, L., L. Badalucco, F. Pomare, and P. Nannipieri. 1993. Effectiveness of antibiotics to distinguish the contributions of fungi and bacteria to net nitrogen mineralization, nitrification, and respiration. *Soil Biology and Biochemistry* 25:1771–1778.

Lasat, M. M. 2002. Phytoextraction of toxic metals: A review of biological mechanisms. *Journal of Environmental Quality* 31:109–120.

Lawley, R., E. Newman, and R. Campbell. 1982. Abundance of endomycorrhizas and root-surface microorganisms on three grasses grown separately and in mixtures. *Soil Biology and Biochemistry* 14:237–240.

Lawley, R., R. Campbell, and E. Newman. 1983. Composition of the bacterial flora of the rhizosphere of three grassland plants grown separately and in mixtures. *Soil Biology and Biochemistry* 15:605–607.

Leadley, P. W., and C. Korner. 1996. Effects of elevated CO_2 on plant species dominance in a highly diverse calcareous grassland. Pp. 159–175 in *Carbon dioxide, populations, and communities,* edited by C. Korner and F. A. Bazzaz. San Diego: Academic Press.

Leadley, P. W., P. A. Niklaus, R. Stocker, and C. Korner. 1999. A field study of the effects of elevated CO_2 on plant biomass and community structure in a calcareous grassland. *Oecologia* 118:39–49.

Lilleskov, E., T. Fahey, and G. Lovett. 2001. Ectomycorrhizal fungi aboveground community changes over an atmospheric deposition gradient. *Ecological Applications* 11:397–401.

Lilleskov, E., T. Fahey, T. Horton, and G. Lovett. 2002. Belowground ectomycorrhizal fungal community change over a nitrogen deposition gradient in Alaska. *Ecology* 83:104–115.

Lodhi, M., and K. Killingbeck. 1980. Allelopathic inhibition of nitrification and nitrifying bacteria in ponderosa pine (*Pinus ponderosa* Dougl.) community. *American Journal of Botany* 67:1423–1429.

Loreau, M. 1998. Biodiversity and ecosystem functioning: A mechanistic model. *Proceedings of the National Academy of Sciences of the United States of America* 95:5632–5636.

Loreau, M. 2000. Biodiversity and ecosystem functioning: Recent theoretical advances. *Oikos* 91:3–17.

Mack, M. C. 1998. Effects of exotic grass invasion on ecosystem nitrogen dynamics in a Hawaiian woodland. Ph.D. diss., University of California, Berkeley.

Mafongoya, P., P. Barak, and J. Reed. 2000. Carbon, nitrogen, and phosphorus mineralization of tree leaves and manure. *Biology and Fertility of Soils* 30:298–305.

Maithani, K., A. Arunachalam, R. Tripathi, and H. Pandey. 1998. Influence of leaf litter

quality on N mineralization in soils of subtropical humid forest regrowths. *Biology and Fertility of Soils* 27:44–50.
Martin, T. L., J. T. Trevors, and N. K. Kaushik. 1999. Soil microbial diversity, community structure, and denitrification in a temperate riparian zone. *Biodiversity and Conservation* 8:1057–1078.
McGill, W., and C. Cole. 1981. Comparative aspects of cycling of organic C, N, S and P through soil organic matter. *Geoderma* 26:267–286.
McGrady-Steed, J., P. Harris, and P. J. Morin. 1997. Biodiversity regulates ecosystem predictability. *Nature* 390:162–165.
McNaughton, S. 1977. Diversity and stability of ecological communities: A comment on the role of empiricism in ecology. *American Naturalist* 111:515–525.
Melillo, J., J. Aber, and J. Muratore. 1982. Nitrogen and lignin control of hardwood leaf litter decomposition dynamics. *Ecology* 63:621–626.
Mosier, A. R., M. A. Bleken, P. Chaiwanakupt, E. C. Ellis, J. R. Freney, R. B. Howarth, P. A. Matson, K. Minami, R. Naylor, K. N. Weeks, and Z. L. Zhu. 2001. Policy implications of human-accelerated nitrogen cycling. *Biogeochemistry* 52:281–320.
Mueller, T., L. Jensen, E. Nielsen, and J. Magid. 1998. Turnover of carbon and nitrogen in a sandy loam soil following incorporation of chopped maize plants, barley straw, and blue grass in the field. *Soil Biology and Biochemistry* 30:561–571.
Mulder, C. P. H., D. D. Uliassi, and D. F. Doak. 2001. Physical stress and diversity-productivity relationships: The role of positive interactions. *Proceedings of the National Academy of Sciences of the United States of America* 98:6704–6708.
Muller, A. K., K. Westergaard, S. Christensen, and S. J. Sorensen. 2001. The effect of long-term mercury pollution on the soil microbial community. *Fems Microbiology Ecology* 36:11–19.
Muller, A. K., K. Westergaard, S. Christensen, and S. J. Sorensen. 2002. The diversity and function of soil microbial communities exposed to different disturbances. *Microbial Ecology* 44:49–58.
Naeem, S., L. Thompson, S. Lawler, J. H. Lawton, and R. Woodfin. 1994. Declining biodiversity can alter the performance of ecosystems. *Nature* 368:734–737.
Naeem, S., K. Hakansson, J. Lawton, M. Crawley, and L. Thompson. 1995. Biodiversity and plant productivity in a model assemblage of plant species. *Oikos* 76:259–264.
Naeem, S., and S. B. Li. 1997. Biodiversity enhances ecosystem reliability. *Nature* 390:507–509.
Newman, E., R. Campbell, P. Christie, A. Heap, and R. Lawley. 1979. Root microorganisms in mixtures and monocultures of grassland plants. Pp. 161–173 in *The soil-root interface,* edited by J. Harley and R. Russell. London: Academic Press.
Niklaus, P. A., P. W. Leadley, B. Schmid, and C. Korner. 2001a. A long-term field study on biodiversity x elevated CO_2 interactions in grassland. *Ecological Monographs* 71:341–356.
Niklaus, P. A., E. Kandeler, P. W. Leadley, B. Schmid, D. Tscherko, and C. Korner. 2001b. A link between plant diversity, elevated CO_2, and soil nitrate. *Oecologia* 127:540–548.
Nilsson, M., D. Wardle, and A. Dahlberg. 1999. Effects of plant litter species composition and diversity on the boreal forest plant-soil system. *Oikos* 86:16–26.
O'Neill, E. 1994. Responses of soil biota to elevated atmospheric carbon dioxide. *Plant and Soil* 165: 55–65.

Owensby, C. E., J. M. Ham, A. K. Knapp, and L. M. Auen. 1999. Biomass production and species composition change in a tallgrass prairie ecosystem after long-term exposure to elevated atmospheric CO_2. *Global Change Biology* 5:497–506.

Palm, C., and P. Sanchez. 1991. Nitrogen release from the leaves of some tropical legumes as affected by their lignin and polyphenolic contents. *Soil Biology and Biochemistry* 23:83–88.

Palmer, M., A. Covich, B. Finlay, J. Gilbert, K. Hyde, R. Johnson, T. Kairesalo, S. Lake, C. Lovell, R. Naiman, C. Ricci, F. Sabater, and D. Strayer. 1997. Biodiversity and ecosystem processes in freshwater sediments. *Ambio* 26:571–577.

Paul, E., and F. Clark. 1995. *Soil microbiology and biochemistry.* San Diego: Academic Press.

Phillips, R. L., D. R. Zak, W. E. Holmes, and D. C. White. 2002. Microbial community composition and function beneath temperate trees exposed to elevated atmospheric carbon dioxide and ozone. *Oecologia* 131:236–244.

Potvin, C., and L. Vassuer. 1997. Long-term CO_2 enrichment of a pasture community: Species richness, dominance, and succession. *Ecology* 78:666–677.

Price, M. V., and N. M. Waser. 2000. Responses of subalpine meadow vegetation to four years of experimental warming. *Ecological Applications* 10:811–823.

Rao, M., and R. Willey. 1980. Evaluation of yield stability in intercropping: Studies on sorghum/pigeonpea. *Experimental Agriculture* 16:105–116.

Reed, F. 1977. Plant species number, biomass accumulation, and productivity of a differentially fertilized Michigan old-field. *Oecologia* 30:45–53.

Reich, P. B., J. Knops, D. Tilman, J. Craine, D. Ellsworth, M. Tjoelker, T. Lee, D. Wedin, S. Naeem, D. Bahauddin, G. Hendrey, S. Jose, K. Wrage, J. Goth, and W. Bengston. 2001. Plant diversity enhances ecosystem responses to elevated CO_2 and nitrogen deposition. *Nature* 410:809–812.

Rillig, M. C., K. M. Scow, J. N. Klironomos, and M. F. Allen. 1997. Microbial carbon-substrate utilization in the rhizosphere of Gutierrezia sarothrae grown in elevated atmospheric carbon dioxide. *Soil Biology and Biochemistry* 29:1387–1394.

Rozgaj, R., and M. Glancersoljam. 1992. Total degradation of 6-aminonapthalene-2-sulfonic acid by a mixed culture consisting of different bacterial genera. *Fems Microbiology Ecology* 86:229–235.

Saetre, P. 1998. Decomposition, microbial community structure, and earthworm effects along a birch-spruce soil gradient. *Ecology* 79: 843–846.

Salemaa, M., I. Vanha-Majamaa, and J. Derome. 2001. Understory vegetation along a heavy-metal pollution gradient in SW Finland. *Environmental Pollution* 112:339–350.

Saleska, S. R., J. Harte, and M. S. Torn. 1999. The effect of experimental ecosystem warming on CO_2 fluxes in a montane meadow. *Global Change Biology* 5:125–141.

Sallih, Z., and P. Bottner. 1988. Effect of wheat (*Triticum aestivum*) roots on mineralization rates of soil organic matter. *Biology and Fertility of Soils* 7:67–70.

Schimel, J. 1995. Ecosystem consequences of microbial diversity and community structure. Pp. 239–254 in *Arctic and alpine biodiversity: Patterns, causes and ecosystem consequences,* edited by F. Chapin and C. Koerner. Berlin: Springer-Verlag.

Schimel, J., K. Van Cleve, R. Cates, T. Clausen, and P. Reichardt. 1996. Effects of balsam poplar (*Populus balsamifera*) tannins and low molecular weight phenolics on microbial activity in taiga floodplain soil: Implications for changes in N cycling during succession. *Canadian Journal of Botany* 74:84–90.

Schlesinger, W. H. 1997. *Biogeochemistry.* San Diego: Academic Press.

Scott, N., and D. Binkley. 1997. Foliage litter quality and annual net N mineralization: Comparison across North American forest sites. *Oecologia* 111:151–159.

Shaw, M. R., and J. Harte. 2001. Control of litter decomposition in a subalpine meadow-sagebrush steppe ecotone under climate change. *Ecological Applications* 11:1206–1223.

Silver, W. L., D. J. Herman, and M. K. Firestone. 2001. Dissimilatory nitrate reduction to ammonium in upland tropical forest soils. *Ecology* 82:2410–2416.

Singh, J., M. Agrawal, and D. Narayan. 1994. Effect of power-plant emissions on plant community structure. *Ecotoxicology* 3:110–122.

Solanius, P. 1981. Metabolic capabilities of forest soil microbial populations with reduced species diversity. *Soil Biology and Biochemistry* 13:1–10.

Spehn, E. M., J. Joshi, B. Schmid, M. Diemer, and C. Korner. 2000a. Aboveground resource use increases with plant species richness in experimental grassland ecosystems. *Functional Ecology* 14:326–337.

Spehn, E. M., J. Joshi, B. Schmid, J. Alphei, and C. Korner. 2000b. Plant diversity effects on soil heterotrophic activity in experimental grassland ecosystems. *Plant and Soil* 224:217–230.

Sternberg, M., V. K. Brown, G. J. Masters, and I. P. Clarke. 1999. Plant community dynamics in a calcareous grassland under climate change manipulations. *Plant Ecology* 143:29–37.

Sterner, R. W., J. J. Elser, and D. O. Hessen. 1992. Stoichiometric relationships among producers, consumers, and nutrient cycling in pelagic ecosystems. *Biogeochemistry* 17:49–67.

Stocker, R., C. Korner, B. Schmid, P. A. Niklaus, and P. W. Leadley. 1999. A field study of the effects of elevated CO_2 and plant species diversity on ecosystem-level gas exchange in a planted calcareous grassland. *Global Change Biology* 5:95–105.

Stump, L., and D. Binkley. 1993. Relationships between litter quality and nitrogen availability in Rocky Mountain forests. *Canadian Journal of Forest Research* 23:492–502.

Taylor, B., D. Parkinson, and W. Parsons. 1989. Nitrogen and lignin content as predictors of litter decay rates: A microcosm test. *Ecology* 70:97–104.

Taylor, B., C. Prescott, W. Parsons, and D. Parkinson. 1991. Substrate control of litter decomposition in four Rocky Mountain coniferous forests. *Canadian Journal of Botany* 69:2242–2250.

Thibault, J., J. Fortin, and W. Smirnoff. 1982. In vitro allelopathic inhibition of nitrification by balsam poplar and balsam fir. *American Journal of Botany* 69:676–679.

Tilman, D. 1988. *Plant strategies and the dynamics and structure of plant communities.* Monographs in Population Biology. Princeton: Princeton University Press.

Tilman, D., D. Wedin, and J. Knops. 1996. Effects of biodiversity on nutrient retention and productivity in grasslands. *Nature* 379:718–720.

Tilman, D., J. Knops, D. Wedin, P. Reich, M. Ritchie, and E. Siemann. 1997. The influence of functional diversity and composition on ecosystem processes. *Science* 277:1300–1302.

Tilman, D. 1999. Global environmental impacts of agricultural expansion: The need for

sustainable and efficient practices. *Proceedings of the National Academy of Sciences of the United States of America* 96:5995–6000.
van der Heijden, M., J. N. Klironomos, M. Ursic, P. Moutoglis, R. Streitwolf-Engel, T. Boller, A. Wiemken, and I. Sanders. 1998. Mycorrhizal fungal diversity determines plant biodiversity, ecosystem variability, and productivity. *Nature* 396:69–72.
van Ginkel, J. H., A. Gorissen, and J. A. Van Veen. 1996. Long-term decomposition of grass roots as affected by elevated atmospheric carbon dioxide. *Journal of Environmental Quality* 25:1122–1128.
van Vuuren, M., R. Aerts, F. Berendse, and W. de Visser. 1992. Nitrogen mineralization in heathland ecosystems dominated by different plant species. *Biogeochemistry* 16:151–166.
Wardle, D. A., K. I. Bonner, and G. M. Barker. 2000. Stability of ecosystem properties in response to aboveground functional group richness and composition. *Oikos* 89:11–23.
Wardle, D. A. 2001. Experimental demonstration that plant diversity reduces invisibility: Evidence of a biological mechanism or a consequence of sampling effect? *Oikos* 95:161–170.
Wardle, D., and K. Nicholson. 1996. Synergistic effects of grassland plant species on soil microbial biomass and activity: Implications for ecosystem-level effects of enriched plant diversity. *Functional Ecology* 10:410–416.
Wardle, D., K. Bonner, and K. Nicholson. 1997. Biodiversity and plant litter: Experimental evidence which does not support the view that enhanced species richness improves ecosystem function. *Oikos* 79:247–258.
Wedin, D., and D. Tilman. 1990. Species effects on nitrogen cycling: A test with perennial grasses. *Oecologia* 84:433–441.
Wedin, D., and J. Pastor. 1993. Nitrogen mineralization dynamics in grass monocultures. *Oecologia* 96:186–192.
Willems, J. H., R. K. Peet, and L. Bik. 1993. Changes in chalk grassland structure and species richness resulting from selective nutrient additions. *Journal of Vegetation Science* 4:203–212.
Williams, B., and C. Alexander. 1991. Interactions on mixing litters from beneath sitka spruce and scots pine and the effects on microbial activity and N-mineralization. *Soil Biology and Biochemistry* 23:71–75.
Willison, T. W., M. S. O'Flaherty, P. Tlustos, K. W. T. Goulding, and D. S. Powlson. 1997. Variations in microbial populations in soils with different methane uptake rates. *Nutrient Cycling in Agroecosystems* 49:85–90.
Wilson, S. D., and D. Tilman. 1991. Interactive effects of fertilization and disturbance on community structure and resource availability in an old-field plant community. *Oecologia* 88:61–71.
Wolfaardt, G. M., J. R. Lawrence, R. Robarts, and D. E. Caldwell. 1994. The role of interactions, sessile growth, and nutrient amendments on the degradative efficiency of a microbial consortium. *Canadian Journal of Microbiology* 40:331–340.
Yin, S. X., D. Chen, L. M. Chen, and R. Edis. 2002. Dissimilatory nitrate reduction to ammonium and responsible microorganisms in two Chinese and Australian paddy soils. *Soil Biology and Biochemistry* 34:1131–1137.

PART III
Lithosphere

9

Keystone Molecules and Organic Chemical Flux from Plants

Manuel T. Lerdau

The ghost of Alfred Redfield looms large over the study of elemental interactions and plant ecology. The paradigm of elemental ratios has driven most studies of the roles of organisms in affecting biogeochemical cycling, and explicit emphasis on stoichiometry has proven a powerful method for understanding features of elemental cycling that would otherwise make little sense. By considering organisms as fixed entities, similar to minerals in terms of having consistent chemical compositions within a taxonomic unit, ecologists and biogeochemists have been able to link particular taxa to biogeochemical cycles and element interactions. To a large extent the power of this conceptual framework stems from the fact that fixed entities (the organisms and rocks) exist at certain stoichiometric ratios (SRs), and those SRs change when the entities are acted upon by other forces, such as decomposition or weathering. The significance of a particular taxon (or rock type) can then be directly inferred from its SR, with only minor accounting for the particular forms in which elements are held, such as carbon in the form of lignin versus carbon in the form of cellulose or calcium as gypsum or plagioclase.

This approach ignores, however, many of the most exciting, important, and poorly understood ways in which organisms affect elemental interactions and biogeochemistry. Through their physiological activities, particularly through the flux of especially reactive compounds from leaves and roots to the atmosphere and soil, plants affect elemental interactions far out of proportion to the SR of the particular compounds. These compounds can be considered "keystone molecules" in biogeochemical cycles and element interactions because their importance cannot be predicted from their abundance within an ecosystem or the atmosphere. Perhaps the best known of these keystone molecules are signaling compounds given off by certain plant taxa that induce bacteria of particular genera to infect

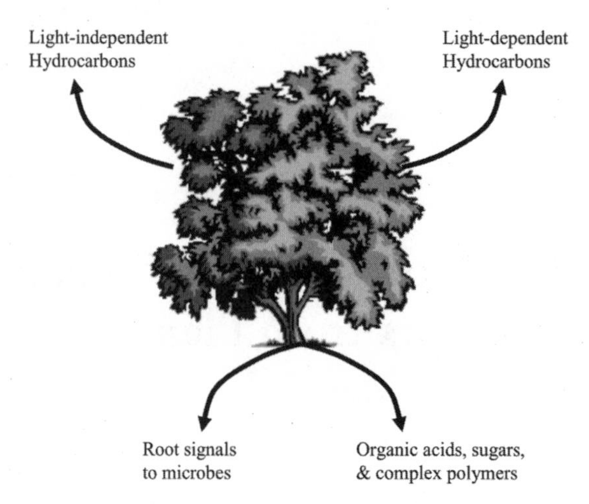

Figure 9.1. Schematic of the main classes of Keystone Molecules discussed in this chapter

roots, form nodules, and begin symbiotic biological nitrogen fixation (BNF) (Long 2001). Symbiotic BNF is the major mechanism by which atmospheric dinitrogen is converted to a biologically available form.

This chapter considers two broad classes of keystone molecules: organic molecules produced in and discharged from roots into soils and organic molecules produced in and emitted from leaves to the atmosphere (Figure 9.1). Together, the fluxes of these molecules, though tiny in quantity when compared with the amount of carbon that moves between plants and the environment in the form of carbon dioxide, are key determinants of redox potential and pollutant dynamics in the troposphere as well as element cycling in soils and ecosystem carbon balance. Understanding the biological regulation of these fluxes is essential for determining how they will respond to global environmental change and how these responses will, in turn, influence such change.

Two main processes lead to the loss of organic carbon (not including organ loss, e.g., leaf fall and organism death, both of which lead to decomposition and carbon mineralization) (Ågren et al., Chapter 7, this volume). Organic carbon is lost from leaves in the form of volatile compounds that are tied to leaf expansion, stress physiology, and herbivore defense. Global estimates of these losses are in the range of one petagram (Pg) of carbon per year, an amount equivalent to the sum of biogenic and anthropogenic methane and roughly equal to 1 percent of terrestrial net primary productivity (Guenther et al. 1995). The second major pathway involves the deposition of organic molecules from roots into the soil solution (Marschner 1995). Estimates of the amount of organic carbon loss from roots are much less well constrained than losses from leaves, primarily because of the large number of physiological processes involved with these losses, the wide diversity of compounds that are lost, and the difficulty of measuring these fluxes under realistic conditions. Despite these uncertainties, it is likely that root organic carbon fluxes are as large or larger than fluxes from leaves.

Regulation of Fluxes

Leaves

The fluxes from leaves fall into two major categories—those whose emissions are under active physiological regulation by the plant and those that appear to be emitted in a manner solely related to vapor pressure and resistance to diffusion. There is also a large phylogenetic component to the fluxes of most phytogenic volatile organic carbon (VOC) compounds. That is, certain taxa emit large amounts of certain compounds and very little of others, whereas other taxa emit few if any species of VOC compounds. The two most important, in terms of their abundance and impacts on the atmosphere, regulated VOC compounds are isoprene (C_5H_8) and methyl butenol (C_5H_9OH); among the unregulated VOC compounds, the most important are the mono- and sesqui-terpenes, and methanol ($C_{10}H_{16}$, $C_{15}H_{24}$, and CH_3OH).

Numerous studies over the past twenty-five years have concentrated on elucidating the physiological controls over the production and emission of isoprene, methyl butenol (MBO), and the monoterpenes. Isoprene and MBO are produced in a light-dependent manner in chloroplasts, and they diffuse from the chloroplast to the atmosphere, exiting the leaf through the stomata (Monson et al. 1995). Unlike water vapor, which exists in the intercellular air spaces at vapor pressures close to saturation, isoprene and MBO in leaves are far below their saturation vapor pressures, so decreases in stomatal conductance are balanced by increases in vapor pressure. That is, under a broad range of steady-state conditions, there is no effect of conductance on flux. In addition, the lack of a significant stored pool of either of these compounds and the relatively short diffusion pathway from chloroplast to stoma mean that flux rates are accurate estimations of production rates.

Both isoprene and MBO production/emission show hyperbolic responses to light intensity when sun leaves are measured. In shade leaves the response approximates a rectangular hyperbola, suggesting control by energy or substrate availability. The temperature responses approximate Michaelis-Menton kinetics, with typical maxima in the range 35°–40° C. Simple physiochemical models suggest that the temperature response is too steep to be explainable entirely by vapor pressure effects; short-term biosynthetic activity must also increase in response to temperature (Lerdau and Keller 1997).

On a longer time scale, the nitrogen content of leaves correlates well with their capacity for VOC emission (Litvak et al. 1996). This nitrogen relationship probably stems from both a direct effect of nitrogen on the content and activity of the enzymes responsible for isoprene/MBO synthesis and an indirect effect on photosynthesis and thus the availability of fixed carbon for the production of isoprene/MBO. An additional control that operates at the scale of the leaf is the developmental state of the leaf. In angiosperms it appears that isoprene emission does not begin until leaf expansion is completed; in contrast, isoprene is emitted from expanding conifer needles (Lerdau and Gray 2003).

Perhaps the most important, and least understood, scale of regulation is at the level of taxon. Certain taxa produce or emit a particular VOC, whereas others do not. The highest variance in isoprene production or emission appears to occur among, rather than within, genera (Harley, Monson, and Lerdau 1999). In contrast, MBO emission variability is primarily intrageneric, and monoterpenes vary at the scale of families (Lerdau and Gray 2003). Although there have been attempts to understand the patterns of VOC emission in an explicitly phylogenetic framework, these have yielded success only for the terpenes and MBO (Loreto et al. 1998; Trapp and Croteau 2001). Not enough taxa have been sampled for conclusive statements to be made regarding the evolutionary origins of isoprene. Understanding the phylogenetics of isoprene emission is, however, crucial for developing predictions for ecosystems where plant taxa are known but their isoprene emission status is not.

A crucial, and still unanswered, question regarding isoprene and MBO is the function(s) of these compounds. Early studies suggested that isoprene was part of the pathway involving a second carboxylase in leaves or that isoprene was linked to photorespiration, but these suggestions have not received support (reviewed in Lerdau and Gray 2003). More recently several authors have argued that isoprene is linked to protection against damage from either high temperatures or oxidants, but again there is controversy surrounding these proposals (Hewitt and Salter 1992; Logan, Monson, and Potosnak 2000; Loreto et al. 2001; Sharkey and Yeh 2001). Elucidation of function will require detailed physiological and biochemical studies that are conducted within a phylogenetically informed framework. Evolutionary analysis of isoprene production suggests that it evolved multiple times within land plants, and putative functions must fit with the observed taxonomic distribution of production (Harley, Monson, and Lerdau et al. 1999; Lerdau and Gray 2003).

In contrast to the complex physiological regulation of isoprene and MBO emission, the short-term regulation of most mono- and sesquiterpene fluxes are explainable entirely by the vapor pressure of the compounds within the foliage and the resistance to diffusion posed by the foliar structure and anatomy (Dement, Tyson, and Mooney 1975; Lerdau 1991). The vapor pressures of the terpenes in solution are predictable by their individual vapor pressures and their concentrations (according to Henry's Law). Because most mono- and sesquiterpenes are stored in specialized canals, glands, or ducts, their major resistance is a fixed value, based on the anatomy and chemistry of the duct itself. The key exception to this rule occurs when foliage is damaged by herbivores and the storage organelles are exposed directly to the atmosphere (Litvak and Monson 1998). When this occurs, resistance goes to zero and flux rates become very large. Recent detailed enzymological studies demonstrate that when this sort of damage occurs the plants are capable of upregulating terpene synthesis to maintain high intercellular concentrations in the face of rapid volatilization (Litvak and Monson 1998).

An interesting and regionally important exception to the rule that monoterpene flux occurs in a light-independent manner from stored pools involves several of the oaks *(Quercus)* native to the Mediterranean basin (Peñuelas and Llusia 1998). Unlike other oaks, these trees produce no terpenes in a light-dependent manner; in addition, they do not produce isoprene. Physiological studies of terpene production in Mediterranean oaks suggest that terpenes may serve in stress-tolerance physiology in a manner similar to isoprene in other plants (Loreto et al. 2001).

The mono- and sesquiterpenes whose fluxes occur in a light-independent manner from stored pools serve functions that are well understood by plant ecologists. They act as defensive compounds, deterring herbivores and acting as toxins against many fungal pathogens. This well-understood function has allowed the development of terpene production-emission models that scale from ecological control over concentration to short-term regulation over emission (Litvak and Monson 1998). Because of the high degree of taxonomic specificity of terpene production, such models must always parameterize species composition.

Roots

Like fluxes from leaves, fluxes from roots can be divided into those whose loss from the plant is under active physiological regulation and those whose loss appears to be an unavoidable consequence of activity such as growth (Marschner 1995). Among the regulated fluxes, sugars, complex polymers, organic acids, and amino acids appear to be the most important, while most of the organic carbon lost during root growth is in the form of complex polymers (Bret-Harte and Silk 1994). Again, as with leaf-derived compounds, the specific organic carbon produced by roots also has a large phylogenetic component, although much less is known about the full range of root fluxes.

In contrast to the leaf fluxes already discussed, the physiological regulation of root fluxes is very poorly understood. There is little doubt that root organic carbon flux is linked to growth, with faster-growing roots tending to have higher fluxes, but in general we know too little to make certain statements regarding direct influences of environmental factors. Just as important, we know little about how individual compound flux is regulated. That is, we do not understand the cues that turn on the fluxes of certain compounds and turn off others. Given the compound-specific nature of their effects in soils, such qualitative understanding is crucial.

Perhaps the most important conceptual distinction between leaf and root fluxes results from what happens to these compounds after they exit the plant. In the case of leaf volatiles, convection and turbulence move the compounds away from the leaf in a matter of seconds so that by the time they react they are far from the plant. Leaf organic carbon tends to have a very low ratio of "time spent near plant:lifetime of organic molecule." In contrast, the organic carbon that exits from roots is discharged into the mixed liquid/solid medium that is soil, and this carbon usually reacts when it is still quite close to the plant. That is, it has a very high ratio of "time spent near plant:lifetime of organic molecule."

This high ratio makes it much more likely that the impacts of root organic carbon flux affect the plant that is responsible for the flux. Thus there is a much greater opportunity for natural selection to act upon the flux as well as the production of root organic carbon. Indeed, many of the putative adaptive functions of root organic carbon flux involve modulation of the rhizosphere, and most of the impacts on element interactions and biogeochemical processes (discussed below) of root discharges can be considered adaptive mechanisms by which individual plants modify both the physio/chemical and biotic environments around their roots.

Impacts of Organic Carbon Fluxes

Fluxes from Leaves

Organic carbon losses from plants have direct effects on element interactions in both the atmosphere and soils. These effects are so great that models of each must include organic carbon-input terms or they will be unable to capture key processes that determine the turnover time and ultimate fate of both carbon and nutrients. Until recently these impacts have largely been ignored because researchers have focused on CO_2 fluxes, which are, after all, two orders of magnitude higher than organic carbon fluxes. Such an emphasis on a simple mass balance approach has ignored, however, the reactivity of the particular molecules moving from plants to the environment and the particular importance of the reactions governed by these molecules.

The atmospheric chemical impacts of VOC emissions from leaves include both homogeneous (gas-gas) and heterogeneous (gas-particle and gas-liquid) reactions. The best-studied VOC in terms of atmospheric impact and element interactions is isoprene. Isoprene reacts, during the daytime, with the hydroxyl radical to form a series of short-lived intermediates. This photochemical oxidation continues until carbon monoxide is produced. Carbon monoxide is quite stable, with tropospheric lifetimes on the order of six months. Global photochemical model estimates suggest that isoprene oxidation contributes approximately one-third of the total CO burden on an annual scale (Fuentes et al. 2000). In addition, because isoprene competes with methane for OH, isoprene increases the atmospheric lifetime of CH_4 by lowering OH concentration. This increase in lifetime can be as high as 25 percent, which translates into a significant increase in CH_4 concentration and importance as a greenhouse gas (Jacob and Wofsy 1988).

Recently, attention has focused on the heterogeneous reactions that dominate the chemistry of the larger (lower vapor pressure) VOCs. These heterogeneous reactions are complex and difficult to model mechanistically. A great deal of progress has been made, however, through theoretical and smog chamber studies that have been used to develop and parameterize empirical models (e.g., Fuentes et al. 2000). Surprisingly, and contrary to initial modeling studies, the aerosols formed during these heterogeneous reactions appear to play a very small role in the formation of cloud condensation nuclei; in contrast, most organic aerosols that lead to cloud condensation nuclei appear to be derived from primary biogenic microparticles such as pollen and spores (Andreae et al. 2002, but see Dowd et al. 2002).

The most important atmospheric roles of these larger VOC compounds appear to lie in their tendency to form organic acids that are deposited in precipitation (and thus influence precipitation pH in rural areas) and to form organic nitrates that allow for the long-distance transport of reactive nitrogen (Lerdau, Guenther, and Monson 1997; Fuentes et al. 2000). This effect on soil pH in rural areas may be extremely important in affecting both cation and phosphorus availability. The production, transport, and disassociation of organic nitrates have strong impacts on nitrogen deposition and cycling because they allow for long-distance transport of pollutant nitrogen. These nitrates then disassociate and the released nitrogen can be deposited onto ecosystems hundreds of kilometers away from the original source. The recent suggestion that these larger VOCs may play a direct role in surface radiative balance under certain conditions requires field verification before its importance can be accepted (Fuentes et al. 2001).

One of the more novel, and controversial, suggested impacts of leaf-derived VOCs is their effect on carbon balance (Crutzen et al. 1999). The logic of this argument is simple and, superficially, appealing. As noted, there is general agreement that global fluxes of leaf-derived VOCs are approximately one Pg of carbon per year. This value is about 50 percent of "missing carbon sink" that has attracted much attention from people measuring and modeling the global carbon cycle (Schimel et al. 2001; Ollinger et al., Chapter 4, this volume; Moldan et al., Chapter 5, this volume). Crutzen et al. (1999) suggested that, because models of carbon flux typically consider only CO_2, leaf-level fluxes of organic carbon could explain up to half of the missing sink because the models do not consider the efflux of carbon from terrestrial systems to the atmosphere.

Two issues complicate, and perhaps invalidate, the argument that VOCs are important components of ecosystem carbon balance. First, the ultimate fate of much of the volatile carbon lost in organic forms is CO_2, and the time period until that organic carbon is oxidized to CO_2 is often less than a year. On an annual time scale, then, the removal of CO_2 from the atmosphere by photosynthesis and conversion to VOCs will be balanced by the oxidation of organic carbon in the atmosphere. Estimates of the missing carbon sink are based on models that run on annual time steps (Schimel et al. 2001), so VOCs cannot be the missing sink as defined in these models. Another component of this first problem is that those larger volatile organics that often enter ecosystems as partially oxidized compounds such as organic acids either remain in the ecosystem or are lost as CO_2 during soil oxidation. In other words, even though VOC fluxes to the atmosphere are approximately equivalent to the missing sink, on an annual time scale almost all of their carbon ends up as CO_2 or as organic carbon in soils.

The second problem is a classic of biogeochemical budget calculations. Annual terrestrial carbon fluxes into ecosystems by carbon dioxide uptake (photosynthesis) are approximately 100 Pg of carbon. Fluxes of carbon in the form of carbon dioxide to the atmosphere (respiration) are of a similar order of magnitude. Annual organic carbon fluxes are approximately 1 percent of these. The *uncertainty* in the carbon dioxide flux estimates dwarfs the magnitude of the organic carbon flux estimates, and much more progress in resolving the carbon balance will be made through reducing these uncertainties than could possibly be made through improving organic carbon estimates. Given these two issues, it is hard to imagine how VOCs could play a major role in balancing the global carbon budget.

Fluxes from Roots

While leaf-derived organic carbon may have a small direct impact on the biogeochemical cycle of carbon, the indirect impacts of root-derived fluxes are probably enormous. These impacts occur because of the "priming effect" that these high-quality (highly reduced and thus energy-rich) compounds have on microbial activity and nutrient cycling (Clarholm 1985). The release by roots of sugars, amino acids, and other organic acids leads to a rapid stimulation of microbial activity and subsequent increase in the mineralization of organically bound nutrients, and there have been recent advances in detecting these fluxes in nonsterile soils (e.g., Jaeger et al. 1999; Bringhurst, Cardon, and Gage 2001). In addition, for reasons not well understood, these energy-rich molecules sometimes lead to immobilization of nitrogen and a suppression of nitrogen turnover.

Although ecosystem models do not yet explicitly consider either priming or suppression caused by root-associated carbon fluxes, recent empirical studies suggest that carbon fluxes from roots on scales from days, to weeks, to seasons can influence soil respiration and ecosystem carbon exchange. For example, episodic movement of carbon from red oak shoots to roots is linked to matching episodic increases in soil respiration; these changes in soil respiration are correlated with both root and microbial biomass and activity over time scales of weeks (Cardon et al. 2002). As biogeochemical modelers incorporate more biological mechanisms and attempt to predict fluxes across larger temporal and spatial scales, they will need to include both priming and suppression processes as well as seasonal dynamics of root activity and fine root turnover.

Another mechanism by which organic carbon losses from roots influence nutrient and biogeochemical cycling is through organic acid impacts on soil pH and the subsequent sorption and desorption of bound phosphorus. Much mineral-bound phosphorus is held in a pH-dependent manner, and the impacts of organic acids, particularly in tropical soils where P is often limiting, can be profound. In addition, citrate, a common soil exudate, participates in chelation reactions that increase P availability, and many tropical plants appear to exude particularly large amounts of citrate in response to P deficits (Marschner 1995).

Responses of Organic Carbon Fluxes to Global Environmental Change

One of the challenges facing attempts to place organic carbon fluxes from plants within the context of element interactions and biogeochemical cycles is to understand how these fluxes will respond to some of the broad-scale changes occurring in the global environment. The first step in developing this understanding is to examine the direct effects of environmental changes on the ecophysiological processes underpinning organic carbon flux from plants to the atmosphere and soils. Two of the most important changes occurring on a global scale that are directly affecting these processes are increases in the partial pressure of carbon dioxide and increases in temperature (Melillo et al., Chapter 1, this volume). An indirect, though possibly very large, effect is the change in species composition occurring as a result of anthropogenic activities and the subsequent changes in organic carbon fluxes because many of the physiological processes are highly species-specific (Lerdau and Slobodkin 2002). These species-specific processes may provide the most important link between population and community ecology and biogeochemistry.

Only a few studies have examined the effects of elevated CO_2 on VOC emissions from leaves, and the results have not been consistent. In the case of monoterpenes, it appears that elevated CO_2 has no impact on emissions at the leaf level. Of course, as much as elevated CO_2 leads to an increase in biomass, it will also lead to an increase in VOC flux. The few isoprene-elevated CO_2 studies undertaken have found either a depression or an increase in isoprene emission, with a hint that whole-plant carbon balance may explain the observed differences. Isoprene emission from *Populus*, which shows a dramatic growth response to elevated CO_2, declines in response to increases in CO_2; in contrast, *Quercus* shows a much weaker growth response to elevated CO_2, and isoprene emission increases (Sharkey, Loreto, and Delwiche 1991). Other studies on whole plant carbon balance in *Populus* also suggest that there may exist a trade-off between growth responses and isoprene emission rates (Funk, Jones, and Lerdau 1999). This subject, however, still requires further investigation.

Quite a few researchers have considered the question of elevated CO_2 effects on organic carbon losses from roots; however, most empirical efforts have focused on larger-scale questions of carbon dynamics in soils (e.g., Cardon et al. 2001). Higher fluxes of labile carbon cause an increase in microbial activity, soil respiration, and nitrogen turnover on short time scales (Hamilton and Frank 2001). Until recently, it was assumed that these effects included an overall increase in decomposition rate. Detailed studies using stable isotopes, however, have shown that this increase in decomposition of labile matter is accompanied by a decrease in the decomposition of more recalcitrant carbon; the net effect is that the turnover time of soil carbon may increase as the ecosystem arrives at a new equilibrium state (Cardon et al. 2001). Studies of these complex relationships across years and in different soil types are essential to determine the generality of these results.

While elevated CO_2 impacts on organic carbon losses from both leaves and roots are complex and difficult to model, the impacts of temperature appear to be much more straightforward (at least in the case of leaf-derived fluxes). Larger VOCs such as monoterpenes and sesquiterpenes volatilize from stored pools according to their vapor pressures, and temperature has a simple exponential relationship with vapor pressure. Litvak and Monson (1998) have shown that as monoterpene losses due to volatilization increase, the rate of monoterpene synthesis is up regulated, suggesting that plants will maintain higher loss rates in the face of higher temperatures for extended periods of time. These larger VOCs are such a small fraction of the plant's carbon budget that increases in synthesis are likely not to be costly (Lerdau and Gershenzon 1997). The smaller VOCs such as isoprene and methyl butenol are emitted directly upon synthesis, and numerous detailed studies have shown that, over the range of temperatures that are considered realistic in climate models, biosynthetic and emission rates increase exponentially (Harley, Monson, and Lerdau 1999; Lerdau and Gray 2003). As higher temperatures lead to an increase in root growth, it is likely that they will also lead to an increase in the loss of organic carbon from roots, but it is also possible that higher temperatures may cause a depression in root growth and organic carbon discharge. On this crucial topic the empirical database is very small.

Changes in species composition are likely to have dramatic effects on organic carbon losses from plants because the processes underlying these losses are usually highly species-specific (Lerdau and Slobodkin 2002). Whereas photosynthesis and respiration tend to be very similar across plants when one accounts for differences in specific leaf area and nitrogen content, VOC losses from leaves can differ by four orders of magnitude across species (a point first noted by Zimmerman 1979). The magnitude of variability of root losses is not as well known but may well be just as large. This profound difference—the difference between the exchange of carbon as CO_2 and the exchange of organically bound carbon—fits with broader theories regarding the regulation of plant metabolism and the selective pressures on primary metabolism, such as photosynthesis and respiration, versus those on secondary metabolism, such as allocation to defensive or protective compounds.

Briefly, these theories state that all plants face the same challenges in terms of acquiring carbon, and they must adapt to the differential availability of water, nutrients, and light. This requirement leads to the prediction that allocation should occur in such a manner that all resources are equally limiting to growth when growth is considered across the entire life span of the plant (Bloom, Chapin, and Mooney 1985; Lerdau 1992). Because all plants require basically the same resources to acquire carbon, the selective pressures on the physiological processes involved with carbon acquisition should be fairly uniform. In contrast, there has been much greater variation among species in selection for defensive compounds because of the high diversity of pest and pathogen threats that plants face. In addition, because the selective pressures stem from organisms that can, themselves, evolve, the likelihood exists for plants to respond and the complexity to grow rapidly (coevolution, as defined by Ehrlich and Raven 1964).

This immense variability makes organic carbon losses from ecosystems highly sensitive to changes in species composition. In particular, fluxes of VOCs from leaves to the atmosphere will change in response to changes in land use. For example, the two broad-leafed taxa most important in current efforts at commercial forestry, *Populus* and *Eucalyptus*, are both isoprene emitters (Lerdau, Guenther, and Monson 1997). As plantations of these two taxa increase across the Earth's surface, so will the flux of highly reduced compounds. In contrast, most crop species emit low levels of VOCs, so agricultural extensification could lead to decline in VOC fluxes at the ecosystem scale and subsequent changes in the redox potential of the troposphere (Lerdau and Slobodkin 2002).

The challenge facing the research community is to understand how these changes in keystone molecule flux will alter element interactions and biogeochemical cycles. A key point to bear in mind is that these changes in keystone molecule flux will occur within the context of the driving global environmental change itself. For example, global warming may cause an increase in monoterpene flux, and the impacts of that increase in terpene flux on tropospheric redox potential may be temperature-dependent. Increases in carbon dioxide may cause an increase in root-derived keystone molecules, and the effects of these molecules on nutrient cycling and nutrient relations of plants may depend upon the concentration of CO_2. Such complexities dictate that process-based models be used to try and predict the overall impacts of interacting direct and indirect effects.

Future Priorities

Future efforts on organic carbon keystone molecule losses from plants should focus on both the regulation of these losses and the impacts they have on biogeochemical cycles. With respect to fluxes from both roots and leaves, the biggest lacunae surround our understanding of regulation at phylogenetic, life-form, and ecosystem scales. The magnitude of unexplained variability that occurs across taxa dwarfs the unexplained variability occurring at organismal and physiological scales. We know far more about how light, temperature, or moisture influence fluxes from any one plant than we do about why two unrelated plants growing in similar conditions may have wildly different fluxes. Developing prognostic models that are useful across ecosystems and biomes will be severely limited until we have a better handle on the mechanisms underlying this unexplained variation. In the case of fluxes from roots, we also know surprisingly little about the regulation of these fluxes under realistic conditions and how these fluxes will respond to large-scale environmental changes. A major challenge facing those studying root fluxes is to understand the controls over fluxes that are typically considered "unregulated" at the physiological scale. Natural selection almost certainly acts on the amount and quality of resources shed from roots (as it acts on the tissue quality of senesced leaves), but few investigators have examined the role selection may play on these organic losses from roots. With leaf fluxes the physiochemical, physiological, and biochemical regulation are surprisingly well understood (Sharkey and Yeh 2001; Lerdau and Gray 2003) and of lower priority for future research.

One potentially important aspect of organic carbon emissions from plants explicitly not considered in this paper because too little is known about the regulation, magnitude, or impacts involves emissions from marine algae. Compared with our understanding of terrestrial fluxes, studies of marine fluxes are in their infancy. Recent work suggests, however, that these fluxes may be important in local and regional atmospheric chemistry (Chuck, Turner, and Liss 2002). A recent review of VOC fluxes from algae argues for the ubiquity of these fluxes and their importance from an organismal perspective (Steinke, Malin, and Liss 2002). We have no idea yet, however, about the importance of these fluxes in global atmospheric chemistry or climate, nor do we have a clear picture from a phylogenetic perspective of which chemicals are emitted from which ecosystems.

The impacts of these organic carbon losses from plants to both soils and the atmosphere stand as one of the major growth fields in biogeochemistry. We have only scratched surfaces regarding the role that soil fluxes play in priming microbial activity and stimulating both carbon and element cycling. Indeed, there are no reliable estimates of the magnitudes of these fluxes. It may well turn out that in most terrestrial ecosystems, the majority of element cycling occurs in the rhizosphere and is stimulated or suppressed by organic carbon fluxes from roots. With respect to VOC losses from leaves, we have a good handle on how isoprene affects homogeneous chemistry in the troposphere and the dynamics of ozone, carbon monoxide, and hydroxyl radicals, but we know very little about what roles the larger organic compounds such as aldehydes, sesquiterpenes, and ketones may play in aerosol formation, heterogeneous chemistry, organic acid deposition, and radiative balance. Investigating these issues requires collaborative efforts among atmospheric scientists, micrometeorologists, and biologists. Together, these groups can conduct research that brings together the bio-, geo-, and chem- of biogeochemistry and address problems that will otherwise remain largely ignored and, as they have been in the past, unsolved.

Acknowledgments

I am grateful for a Bullard Fellowship and a Hrdy Visiting Professorship (both from Harvard University) for support during preparation of this manuscript. I also thank Geeta Bharathan, Zoe Cardon, Deborah Clark, Chris Field, Jeff Levinton, Kenichi Satake, Mary Scholes, and an anonymous reviewer for comments on earlier drafts.

Literature Cited

Andreae, M. O., P. Artaxo, C. Brandao, F. E. Carswell, P. Ciccioli, A. L. da Costa, A. D. Culf, J. L. Esteves, J. H. C. Gash, J. Grace, P. Kabat, J. Lelieveld, Y. Malhi, A. O. Manzi, F. X. Meixner, A. D. Nobre, C. Nobre, M. D. L. P. Ruivo, M. A. Silva-Dias, P. Stefani, R. Valentini, J. von Jouanne, and M. J. Waterloo. 2002. Biogeochemical cycling of carbon, water, energy, trace gases, and aerosols in Amazonia: The LBA-EUSTACH experiments. *Journal of Geophysical Research* 107 (D20): art. no. 8066.

Bloom, A., F. Chapin, and H. Mooney. 1985. Resource limitation in plants: An economic analogy. *Annual Review of Ecology and Systematics* 16:363–392.

Bret-Harte, M. S., and W. Silk. 1994. Fluxes and deposition rates of solutes in growing roots of *Zea mays*. *Journal of Experimental Botany* 45:1733–1742.

Bringhurst, R. M., Z. G. Cardon, and D. J. Gage. 2001. Galactosides in the rhizosphere: Utilization by Sinorhizobium meliloti and development of a biosensor. *Proceedings of the National Academy of Sciences* 98:4540–4545.

Cardon, Z. G., A. D. Czaja, J. L. Funk, and P. L. Vitt. 2002. Periodic carbon flushing to roots of Quercus rubra saplings affects soil respiration and rhizosphere microbial biomass. *Oecologia* 133:215–223.

Cardon, Z. G., B. A. Hungate, C. A. Cambardella, F. S. Chapin III, C. B. Field, E. A. Holland, and H. A. Mooney. 2001. Contrasting effects of elevated CO_2 on old and new soil carbon pools. *Soil Biology and Biochemistry* 33:365–373.

Chuck, A. L., S. M. Turner, and P. S. Liss. 2002. Direct evidence for a marine source of C-1 and C-2 alkyl nitrates. *Science* 297:1151–1154.

Clarholm, M. 1985. Interactions of bacteria, protozoa, and plants leading to mineralization of nitrogen. *Soil Biology and Biochemistry* 17:181–187.

Crutzen, P., R. Fall, I. Galbally, and W. Lindinger. 1999. Parameters for global ecosystem models. *Nature* 399:535.

Dement, W. A., B. Tyson, and H. Mooney. 1975. Mechanism of monoterpene volatilization in *Salvia mellifera*. *Phytochemistry* 14:2555–2557.

Dowd, D., P. Aalto, K. Hameri, M. Kulmala, and T. Hoffmann. 2002. Atmospheric particles from organic vapours. *Nature* 416:497–498.

Ehrlich, P. R., and P. H. Raven. 1964. Butterflies and plants: A study in coevolution. *Evolution* 18:586–608.

Fuentes, J. D., B. P. Hayden, M. Garstang, M. T. Lerdau, D. Fitzjarrald, D. D. Baldocchi, R. Monson, B. Lamb, and C. Geron. 2001. New directions: VOCs and biosphere-atmosphere feedbacks. *Atmospheric Environment* 35:189–191.

Fuentes, J. D., M. T. Lerdau, R. Atkinson, D. Baldocchi, J. W. Bottenheim, P. Ciccioli, B. Lamb, C. Geron, L. Gu, A. Guenther, T. D. Sharkey, and W. Stockwell. 2000. Biogenic hydrocarbons in the atmospheric boundary layer: A review. *Bulletin of the American Meteorological Society* 81:1537–1575.

Funk, J., C. Jones, and M. T. Lerdau. 1999. Whole plant controls on isoprene emission. *Oecologia* 118:333–339.

Guenther, A., N. Hewitt, D. Erickson, R. Fall, C. Geron, T. Graedel, P. Harley, L. Klinger, M. T. Lerdau, W. McKay, T. Pierce, B. Scholes, R. Steinbrecher, R. Tallamraju, J. Taylor, and P. Zimmerman. 1995. A global model of natural volatile organic compound emissions. *Journal of Geophysical Research* 100:8873–8892.

Hamilton, E. W., and D. A. Frank. 2001. Can plants stimulate soil microbes and their own nutrient supply? Evidence from a grazing tolerant grass. *Ecology* 82:2397–2402.

Harley, P. C., R. K. Monson, and M. T. Lerdau. 1999. Ecological and evolutionary aspects of isoprene emission from plants. *Oecologia* 118:109–123.

Hewitt, C. N., and L. Salter. 1992. Ozone-hydrocarbon interactions in plants. *Phytochemistry* 31:4045–4050.

Jacob, D., and S. Wofsy. 1988. Photochemistry of biogenic emissions over the Amazon forest. *Journal Geophysical Research* 93:1477–1486.

Jaeger, C. H., S. E. Lindow, S. Miller, E. Clark, and M. K. Firestone. 1999. Mapping of sugar and amino acid availability in soil around roots with bacterial sensors of sucrose and Tryptophan. *Applied and Environmental Microbiology* 65:2685–2690.

Lerdau, M. T., and L. Slobodkin. 2002. Trace gas emissions and species-dependent ecosystem services. *Trends in Ecology and Evolution* 17:309–312.

Lerdau, M. T. 1992. Future discounts and resource allocation in plants. *Functional Ecology* 6:371–375.

Lerdau, M. T., and D. Gray. 2003. The ecology and evolution of light dependent and light independent phytogenic volatile organic carbon. *New Phytologist* 157:199–211.

Lerdau, M. T., and J. Gershenzon. 1997. Allocation theory and the costs of chemical defenses in plants. Pp. 265–277 in *Resource allocation in plants and animals,* edited by F. Bazzaz and J. Grace. San Diego: Academic Press.

Lerdau, M. T. 1991. Plant function and biogenic terpene emission. Pp. 121–134 in *Trace gas emissions by plants,* edited by T. D. Sharkey, E. A. Holland, and H. A. Mooney. San Diego: Academic Press.

Lerdau, M. T., A. Guenther, and R. Monson. 1997. Production and emission of volatile organic compounds by plants. *BioScience* 47:373–383.

Lerdau, M. T., and M. Keller. 1997. Controls over isoprene emission from trees in a sub-tropical dry forest. *Plant, Cell and Environment* 20:569–578.

Litvak, M. E., and R. K. Monson. 1998. Patterns of induced and constitutive monoterpene production in conifer needles in relation to insect herbivory. *Oecologia* 114:531–540.

Litvak, M. E., F. Loreto, P. C. Harley, T. D. Sharkey, and R. K. Monson. 1996. The response of isoprene emission rate and photosynthetic rate to photon flux and nitrogen supply in aspen and white oak trees. *Plant, Cell and Environment* 19:549–559.

Logan, B. A., R. K. Monson, and M. J. Potosnak. 2000. Biochemistry and physiology of foliar isoprene production. *Trends in Plant Science* 5:477–481.

Long, S. 2001. Genes and signals in the Rhizobium-legume symbiosis. *Plant Physiology.* 125:69–72.

Loreto, F., P. Ciccioli, E. Brancaleoni, R. Valentini, M. De Lillis, O. Csiky, and G. Seufert. 1998. A hypothesis on the evolution of isoprenoid emission by oaks based on the correlation between emission type and Quercus taxonomy. *Oecologia* 115:302–305.

Loreto, F., M. Mannozzi, C. Maris, P. Nascetti, F. Ferranti, and S. Pasqualini. 2001. Ozone quenching properties of isoprene and its antioxidant role in leaves. *Plant Physiology* 126:993–1000.

Marschner, H. 1995. *Mineral nutrition of higher plants.* 2nd ed. London: Academic Press.

Monson, R. K., M. T. Lerdau, T. D. Sharkey, D. S. Schimel, and R. Fall. 1995. Biologi-

cal aspects of constructing volatile organic-compound emission inventories. *Atmospheric Environment* 29:2989–3002.
Peñuelas, J., and J. Llusia. 1998. Seasonal emission of monoterpenes by the Mediterranean tree Quercus ilex in field conditions: Relations with photosynthetic rates, temperature, and volatility. *Physiologia Plantarum* 105:641–647.
Schimel, D. S., J. I. House, K. A. Hibbard, P. Bousquet, P. Ciais, P. Peylin, B. H. Braswell, M. J. Apps, D. Baker, A. Bondeau, J. Canadell, G. Churkina, W. Cramer, A. S. Denning, C. B. Field, P. Friedlingstein, C. Goodale, M. Heimann, R. A. Houghton, J. M. Melillo, B. Moore, D. Murdiyarso, I. Noble, S. W. Pacala, I. C. Prentice, M. R. Raupach, P. J. Rayner, R. J. Scholes, W. L. Steffen, and C. Wirth. 2001. Recent patterns of carbon exchange by terrestrial ecosystems. *Nature* 414:169–172.
Sharkey, T., and S. Yeh. 2001. Isoprene emission from plants. *Annual Review of Plant Physiology and Plant Molecular Biology* 52:407–436.
Sharkey, T. D., F. Loreto, and C. F. Delwiche. 1991. High carbon dioxide and sun/shade effects on isoprene emission from oak and aspen tree leaves. *Plant, Cell and Environment* 14:333–338.
Steinke, M., G. Malin, and P. Liss. 2002. Trophic interactions in the sea: An ecological role for climate relevant volatiles? *Journal of Phycology* 38:630–638.
Trapp, S. C., and R. B. Croteau. 2001. Genomic organization of plant terpene synthases and molecular evolutionary implications. *Genetics* 158:811–832.
Zimmerman, P. 1979. Tampa Bay area photochemical oxidant study: Determination of emission rates of hydrocarbons from indigenous species in the Tampa/St. Petersburg Florida area. United States Environmental Protection Agency, 904/9-77–028, Atlanta, Ga.

10

Element Interactions in Brazilian Landscapes as Influenced by Human Interventions

Luiz Antonio Martinelli

Approximately 4.7 million square kilometers (km^2) of our planet, or 32 percent of the land surface on Earth (Houghton 1994), have been altered by humans, resulting in more rapid changes to global nutrient cycles than could have occurred by natural events (Ojima, Galvin, and Turner 1994). These changes have occurred over millennia but have accelerated during the past century, driven by an increasing world population

Changes in land cover are thought to have occurred first in the temperate zones of the Earth (Ojima, Galvin, and Turner 1994). As a consequence, the terrestrial and aquatic nutrient cycles in these zones have already been under investigation for many years. In contrast, changes in land cover in the tropics have been occurring at an unprecedented scale since the mid-twentieth century, and very little is known about their consequences (Downing et al. 1999; Matson et al. 1999).

The state of Rondônia, located in the Amazon region of Brazil, is a good example of frontier-type production system. Ojima, Galvin, and Turner (1994) associated frontier-type production with developing nations and the coexistence of developed areas (large urban centers, a large number of industries, and high-technology crop-livestock production) and developing areas (with low-technology and frontier-type production systems). The main land use change in this region was forest-to-pasture conversion driven by an increasing population. In 1950 the population of Rondônia was only 37,000 inhabitants (IBGE 2003), and the deforested area was very small (Pedlowski et al. 1997). Thirty years later its population had increased to almost 500,000 inhabitants, and deforestation increased to nearly 30,000 km^2 (12 percent of the state's area) (INPE 2003). Currently,

Rondônia has almost 1.4 million people, and deforestation has almost reached 60,000 km^2 (INPE 2003). The annual population growth (3.5 percent per year from 1950 to 2000) and average deforestation rate (2,500 km^2 per year, 1978–2000) are among the highest in the world, characterizing one of the most impressive land use changes in the past century.

Conversely, the state of São Paulo can be considered a "developed" region of Brazil, as it has more than 90,000 industries and is dominated by sugar cane fields and intensive and extensive livestock systems. But in the late nineteenth century and early twentieth century, this state had a "developing" phase, when coffee plantations replaced the Atlantic forest and Cerrado areas (Brannstrom 2001). As a consequence massive deforestation occurred at the beginning of the twentieth century. Nearly 80,000 km^2 of Atlantic forest (32 percent of the state's area) was cut down between 1907 and 1934—a rate of approximately 2,900 km^2 per year (Brannstrom 2000). Between 1872 and 1920 the number of coffee trees increased from 60,000 to 844 million, and the state population increased from 0.83 to 4.6 million (Brannstrom 2000; IBGE 2003). In the 1960s and 1970s coffee started to be replaced by pasture, sugar cane, or other crops like cotton (Brannstrom and Oliveira 2000), and a strong urbanization-industrialization process of rural areas of the state of São Paulo began. By 2000 the population of the state reached 37 million, and more than 90 percent of the people lived in urban centers (SEADE 2003). As a consequence, only 13 percent of the Atlantic forest is left today, mainly concentrated in coastal mountains (IFLOREST 2003).

Although the deforestation process in Rondônia occurred almost 80 years later than in São Paulo, both states have some common features, such as governmental intervention, use of immigrant labor, and very high initial rates of deforestation (Pedlowski et al. 1997; Brannstrom 2000). The replacement of original vegetation by pasture and sugar cane or coffee plantations appears to be a widespread process in tropical countries of Latin America, though motivated by distinct social and economical reasons (Brannstrom 2001).

Thus a comparison of the changes in the composition of river water caused by Rondônia and São Paulo land use changes could be used to examine alterations in biogeochemical processes and element interactions. Because changes in land use in Rondônia are more recent, the effects of deforestation on ecosystem structure and their functions are better documented, allowing a more comprehensive analysis.

Changes in Element Interactions during Forest-to-Pasture Conversion in Amazônia

When crops replace forests, there is an appreciable loss of biomass and a consequent loss of nutrients (Kauffman et al. 1995; Graça, Fearnside, and Cerri 1999;

Fearnside, Graça, and Rodrigues 2001; Hughes, Kauffman, and Cummings 2002). Moreover, there is a disruption of the biogeochemical cycles and changes in the interactions of elements like carbon and nitrogen (Asner, Seastedt, and Townsend 1997). The reduction of vegetation carbon stock due to burning in the Amazon basin has been found to vary from approximately 20 to 56 percent (Fearnside, Leal Filho, and Fernandes 1993; Carvalho et al. 1995; Kauffman et al. 1995; Carvalho et al. 1998; Guild et al. 1998; Araújo et al. 1999; Graça, Fearnside, and Cerri 1999; Fearnside, Graça, and Rodrigues 2001). For instance, the conversion of forest to pasture in Rondônia decreased biomass from almost 400 tonnes per hectare (t/ha) (primary forest) to approximately 60 t/ha (pasture) (Hughes, Kauffman, and Cummings 2002). Changes in the soil carbon stock due to forest-to-pasture conversion are more controversial. Guo and Gifford (2002), using meta-analysis, concluded that forest-to-pasture conversion led to a gain of soil carbon of 9 percent in relation to the original vegetation. On the other hand, Murty et al. (2002) did not find a clear trend in soil carbon stocks due to forest-to-pasture conversion. In the Amazon region this pattern also persists; studies have shown gains and losses of soil carbon without a clear trend (e.g., Choné et al. 1991; Trumbore et al. 1995; Moraes et al. 1996; Neill et al. 1996; Camargo et al. 1999). Three studies, however, have shown that well-managed pastures in the Amazon region tend to gain carbon in relation to original forests, whereas poorly managed pastures tend to lose carbon (Trumbore et al. 1995; Moraes et al. 1996; Fearnside and Barbosa 1998). Given that well-managed pastures in the Amazon region are the exception and not the rule (Serrão and Toledo 1992; Fearnside and Barbosa 1998), it is reasonable to conclude that forest-to-pasture conversion would imply or result in a loss of soil carbon. Soil carbon losses may be due to a lower input of litterfall in pastures compared with forests. Camargo et al. (1999), working in the western region of the Amazon (Paragominas), found that carbon input from litterfall in the forest was almost 4 times higher in a forest than in a managed pasture and 1.5 times higher in a degraded pasture than in a managed pasture.

The loss of nutrient stocks other than carbon due to forest burning was also variable among sites (Kauffman et al. 1995), but it was higher for elements with a gas phase, like nitrogen (Andreae and Merlet 2001). Losses of N were high, varying from 50 to 70 percent (Kauffman et al. 1995). These heavy losses of N after burning and pasture establishment decreased the N stock in the biomass. Consequently, the soil N stock became higher than the vegetation stock (Kauffman, Cummings, and Ward 1998). On the other hand, the rate of inorganic N cycling in pasture soil became lower than in the forest, especially in old pastures (Piccolo, Neill, and Cerri 1994; Neill et al. 1995, 1997, 1999; Verchot et al. 1999; Garcia-Montiel et al. 2001; Melillo et al. 2001). Tables 10.1 and 10.2 summarize the

Table 10.1. Location, site management, precipitation, and sampling of forest and pastures in the Amazon region

Site[a]	*State*[b]	*Forest*[c]	*Pasture*	*Year of sampling*	*Precipitation (mm/year)*	*Reference*
Nova Vida	RO	1	3	1992	2,200	Piccolo, Neill, and Cerri (1994)
Nova Vida	RO	1	4	1992/1993	2,200	Neill et al. (1995)
Nova Vida	RO	1	2	1993/1194	2,200	Neill et al. (1997)
Nova Vida	RO	1	3	1993	2,200	Neill et al. (1999)
Nova Vida	RO	3	15	1999	2,200	Garcia-Montiel et al. (2001)
Nova Vida	RO	1	9	1992/1993	2,200	Melillo et al. (2001)
Porto Velho	RO	1	1	1993/1994	2,300	Neill et al. (1997)
Jamari	RO	1	2	1993/1994	2,300	Neill et al. (1997)
Cacaolândia	RO	1	1	1993/1994	2,200	Neill et al. (1997)
Ouro Preto	RO	1	3	1993/1994	2,200	Neill et al. (1997)
Vilhena	RO	1	2	1993/1994	2,100	Neill et al. (1997)
Faz. Vitória	PA	2	6	1995/1996	1,850	Verchot et al. (1999)

Note: Most of the data came from the Nova Vida farm, located near the city of Ariquemes in Rondônia, and from the Vitória farm, located near the city of Paragominas in the state of Pará.

[a] A similar sampling and analysis methodology was used among sites.

[b] RO, state of Rondônia; PA, state of Pará.

[c] Number of samples collected in forest and pasture sites. Samples were collected at two soil depths: 0–5 cm and 5–10 cm, except in the Fazenda Vitória, where soil was sampled at one soil depth: 0–10 cm.

Table 10.2. Comparison of N cycling parameters between forests and pastures of the Amazon region

	Forest	*Pasture*
Soil depth, 0–5 cm		
Mineralization (μg - N/g.d)	2.23 ± 1.16 a (n = 18)	0.04 ± 0.54 b (n = 45)
Nitrification (μg - N/g.d)	2.26 ± 1.14 a (n = 18)	0.24 ± 0.32 b (n = 45)
NH_4 (μg - N/g)	6.97 ± 4.98 a (n = 17)	7.68 ± 5.25 a (n = 42)
NO_3 (μg - N/g)	5.93 ± 3.45 a (n = 17)	2.77 ± 3.18 b (n = 42)
$NH_4 + NO_3$ (μg - N/g)	12.90 ± 6.74 a (n = 17)	9.96 ± 6.35 a (n = 42)
NH_4:NO_3	1.62 ± 1.75 a (n = 17)	8.19 ± 11.04 b (n = 42)
N_2O (kg - N/ha.yr)	2.78 ± 2.56 a (n = 8)	1.14b ± 1.38 b (n = 28)
Soil depth, 5–10 cm		
Mineralization (μg - N/g.d)	1.23 ± 0.55 a (n = 17)	0.08 ± 0.56 b (n = 31)
Nitrification (μg - N/g.d)	2.04 ± 1.79 a (n = 18)	0.37 ± 0.44 b (n = 31)
NH_4 (μg - N/g)	8.03 ± 8.25 a (n = 17)	5.05 ± 3.01 a (n = 28)
NO_3 (μg - N/g)	6.36 ± 4.92 a (n = 17)	2.62 ± 2.61 b (n = 28)
$NH_4 + NO_3$ (μg - N/g)	14.39 ± 11.24 a (n = 17)	7.68 ± 3.52 b (n = 28)
NH_4:NO_3	1.58 ± 1.45 a (n = 17)	3.11 ± 2.51 b (n = 28)

Note: Data are presented as averages followed by standard deviations and number of observations. Forest sampling sites from Table 10.1 and soil depths (0–5 cm and 5–10 cm) were pooled together. Pasture sampling sites from different ages and soil depths (0–5 cm and 5–10 cm) were also pooled together. Different letters indicate statistically significant difference at 0.01% level (ANOVA, followed by Tukey Honest Test for unequal number of data).

location and data set used to address N-cycling patterns in forest and pasture soils of the Amazon basin. Figure 10.1 illustrates that mineralization and nitrification rates were lower in pastures than in forests.

Causes for these changes are still speculative, but one of the main causes might be the lower N net availability in pastures (Neill et al. 1999; Verchot et al. 1999). These changes in mineralization and nitrification rates were responsible for changes in the distribution of soil N-inorganic species. The average concentration of soil NO_3^- in forests was higher than in pastures for soil depths of 0–5 centimeters (cm) and 5–10 cm (Table 10.2). Because NH_4^+ content is similar in pasture and forest soils, the ratio NH_4:NO_3 was almost six times and three times higher in forest soils for depths of 0–5 cm and 5–10 cm, respectively, than in pasture soils (Table 10.2). As a consequence of this lower N availability in pasture soils, N_2O emissions were almost two times lower in these soils than in forest soils (Table 10.2), and there was a decrease of these emissions with pasture age (Figure 10.2).

Losses of Ca and P due to the forest burning were much smaller than C and

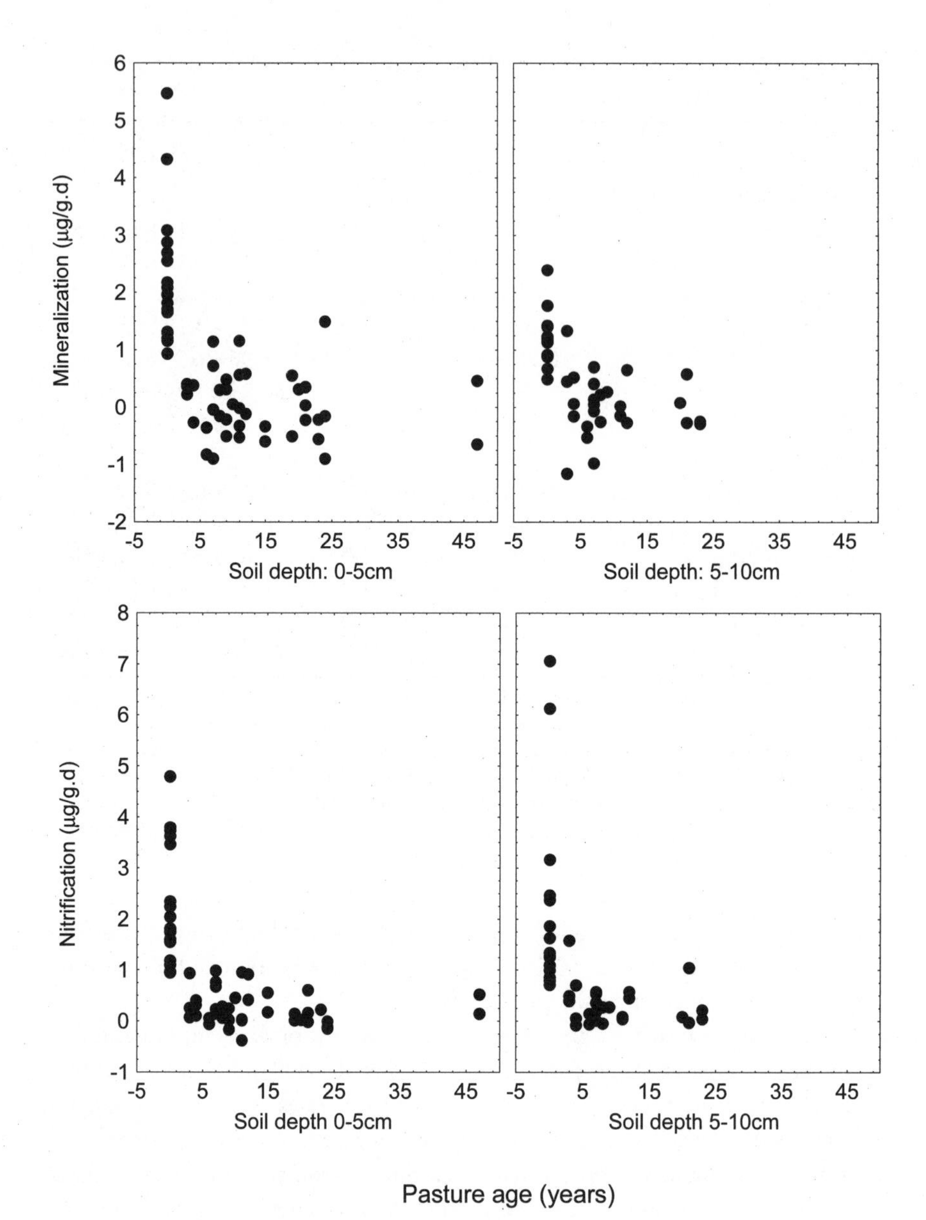

Figure 10.1. Mineralization (upper panel) and nitrification (lower panel) rates in soils of forests (time 0) and pastures of different ages. Data are from Piccolo, Neill, and Cerri (1994), Neill et al. (1995), Neill et al. (1997), Verchot et al. (1999), Garcia-Montiel et al. (2001), and Melillo et al.(2001).

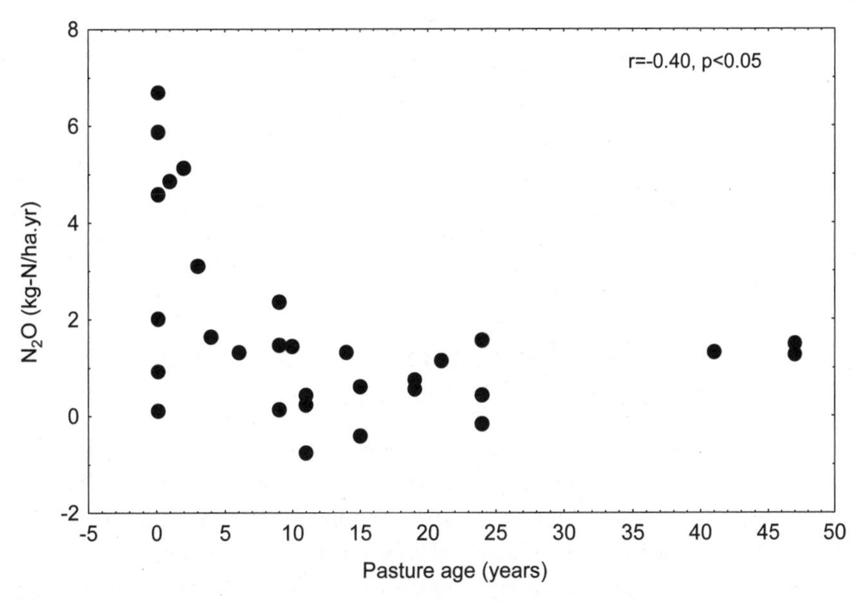

Figure 10.2. Soil emissions of N_2O as a function of the age of pastures; forests are at time 0. Data are from Verchot et al. (1999), Garcia-Montiel et al. (2001), and Melillo et al. (2001).

N, varying from 3 to 11 percent for Ca and from 7 to 32 percent for P, respectively (Kauffman et al. 1995). Stocks of P and Ca became higher in the soil than in the biomass after pasture establishment (Kauffman, Cummings, and Ward 1998), probably owing to the fertilization effect of ashes (Martins et al. 1991; Moraes et al. 1996; Diez et al. 1997). For instance, approximately 50 percent of the Ca and P of the biomass aboveground pool remained in the soil after burning in comparison with N, where only 2–8 percent of N remained in this form (Kauffman et al. 1995). The addition of organic matter via ashes tends to raise the soil pH. This change increases the soil cation exchange capacity, allowing adsorption of the extra Ca input and decreasing P fixation into soil clay particles (Uehara and Gillman 1981; Koutika et al. 1999). Moraes et al. (1996), working in a five-year-old pasture in Rondônia, reported an increase in the soil surface pH from 5.1 to 7.6. This increase was followed by an increase in the base cations content, mainly Ca, from 0.06 kilograms per square meter (kg/m^2) up to 0.24 kg/m^2. The rise in pH and cation contents appears to be only temporary. After initial nutrient input is consumed by crops or leached out from the soil profile, the parent material is not rich enough to replace the soil ions (Buschbacher, Uhl, and Serrão 1986). Nutrient replacement via the atmosphere is possible, but it is a very long process (Chadwick et al. 1999). P is especially problematic in tropical soils, since it is often retained by complexation with Fe and Al and becomes a limiting nutrient to plant growth (Uehara and Gillman 1981). The cutting and burning of tropical forest and the further establishment of pastures completely changed P dynamics (Garcia-Montiel et al. 2000). The control of P dynamics, which before was controlled by efficient P cycling through the forest canopy-litter-soil system, was replaced with P cycling mainly through belowground biomass of pasture grasses (Garcia-Montiel et al. 2000).

Changes in Element Interactions of Streams during the Forest-to-Pasture Conversion in Amazônia

Changes of the magnitude described on the conversion of forest to pasture are likely to be "felt" by streams draining such areas (Howarth et al. 1996). Studies in temperate systems have shown that streams respond to changes in their watersheds by adjusting either their chemical composition or their physical structure (Likens et al. 1970; Jordan, Correll, and Weller 1997). One of the first and most abrupt changes is the increased light intensity on tropical forest and pasture streams. For instance, the mean daylight level in a forest stream in Rondônia varied from 44 to 81 lumens per square meter (lm/m^2) and increased to 1,700 to 3,400 lm/m^2 in the pasture stream (Neill et al. 2001). This increase has the potential to alter the energy and water balance between these areas. The change in water balance will probably affect the pathways through which nutrients move from terrestrial to aquatic environments. For instance, Williams and Melack (1997) compared a small partially deforested catchment (slash-and-burn agriculture) with an intact catchment and concluded that evapotranspiration decreased in the disturbed basin, increasing stream discharge and the export of N, P, and major cations and anions. Differences in light intensity between forest and pasture streams will probably also affect primary productivity. For instance, algal growth did not respond to additions of N and P in a forest stream in Rondônia because of the light limitation in this environment (Neill et al. 2001).

Although the light level was higher in pasture streams of Rondônia, some reaches of these streams were overgrown with *Paspalum* spp., a common grass in the Amazon region (Neill et al. 2001; Thomas et al., in press). This mat of grass provides fuel to heretrophy respiration, increasing the biological oxygen demand (L. F. Charbel, pers. comm.) and decreasing the dissolved oxygen (Neill et al. 2001). Changes in the dissolved oxygen concentration affect the way that organic matter is being decomposed (Ballester et al. 1999; Daniel et al. 2002) and consequently affects the balance between the production of CO_2 and reduced gases like methane. The oxidation status of nitrogen forms could also be affected by changes in the dissolved oxygen concentration. Nitrification was generally inhibited and denitrification increased in streams with low dissolved oxygen concentrations (Christensen et al. 1990; Martinelli et al. 1999). This trend, together with the lower nitrification rates found in pasture soils of Rondônia, may explain the lower NO_3^- concentration in pasture streams compared with forest streams (Neill et al. 2001; Thomas et al. in press). Although the NH_4^+ concentration was the same or similar in forest and pasture streams, the total dissolved inorganic N (DIN) concentration was lower in pasture streams. This led to DIN:DIP (dissolved inorganic P) ratio values lower than 16:1, suggesting an N limitation instead of a P limitation (Neill et al. 2001). Conversely, in the forest stream this ratio was higher than 16:1, suggesting a P limitation to algal growth in addition to a light limitation.

Currently we do not have enough data to estimate how far the changes discussed will propagate along the drainage basin. Results achieved by Biggs et al. (in press), however, suggest that even large rivers may be affected by deforestation. The concentrations of K^+, Na^+, Cl^-, and SO_4^{+2} increased with the deforestation extent in catchments varying from 18 to 12,500 km^2. In addition, Bernardes et al. (in press) found a clear signal of C_4 grasses from Rondônia pastures in the dissolved and particulate organic carbon fractions of the Ji-Paraná River, which has a basin of approximately 75,000 km^2.

Changes in Element Interactions of Streams due to Sewage Inputs in São Paulo State

The major changes in the state of Rondônia are mainly derived from low-technology, frontier-type production systems. The state of São Paulo mirrored the same type of development at the beginning of the century. Nowadays, this state is characterized by a highly developed economy. In addition, water supply and sanitation services are considered adequate by the World Health Organization. Almost 94 percent of the urban households in the state are supplied with clean water, and 80 percent of the urban households have sewer connections. Wastewater treatment in the state, however, is considered far from adequate. Only 17 percent of the wastewater is treated before being discharged into rivers and streams (Martinelli et al. in press). According to the *Global Water Supply and Sanitation Assessment 2000* report (WHO 2003), the lack of proper sewage treatment is a worldwide problem, especially in developing areas of the world. In Africa there is virtually no treatment of wastewater, and in Latin America and the Caribbean region only 14 percent of the sewage is treated.

Besides acute public health problems caused by inadequate treatment of wastewater, a high amount of labile organic matter enters rivers and streams via untreated sewage, disrupting the natural biogeochemical process in several of

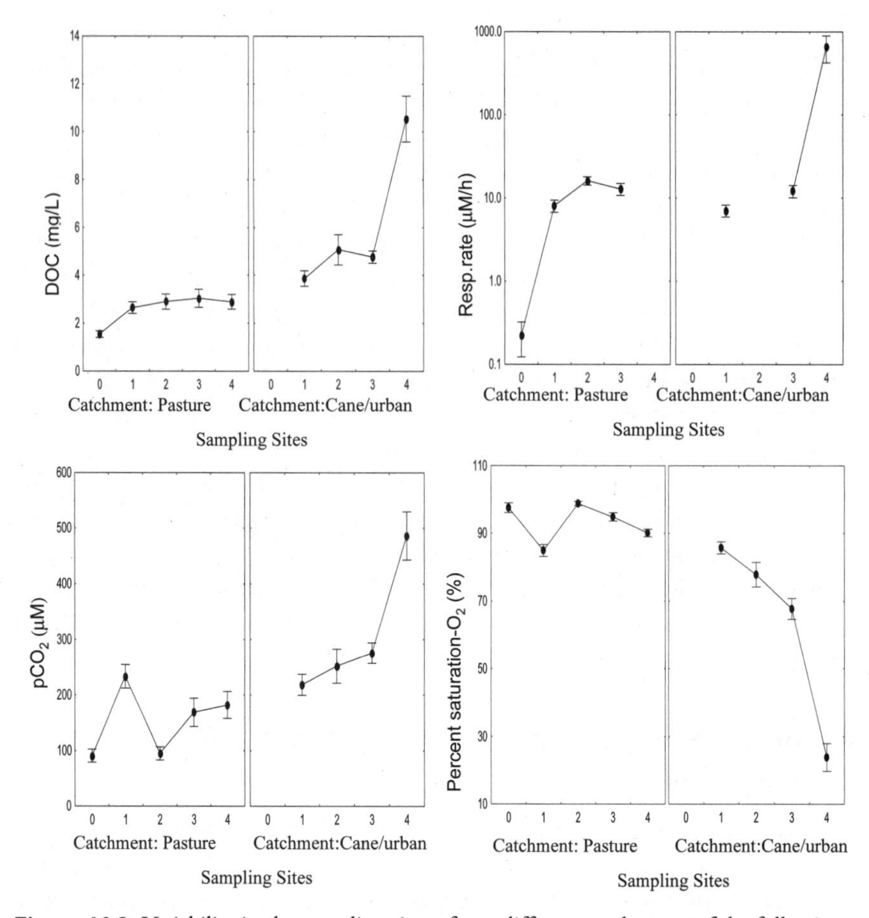

Figure 10.3. Variability in the sampling sites of two different catchments of the following parameters: DOC, respiration rate, pCO_2, and percent saturation of dissolved oxygen

them. In the Piracicaba River basin, one of the most disturbed basins in the state of São Paulo, the input of carbon into rivers and streams is approximately 15 tonnes per hectare per year, and most of the sewage is dumped into small streams instead of in major rivers of the basin (Daniel et al. 2002). It is important to note that small streams are considered to be more important than large rivers in controlling nutrients in watersheds (Peterson et al. 2001). Changes in element distribution caused by urban sewage are illustrated by two small catchments located in the Piracicaba River basin. One is a 50-km^2 catchment dominated by forest (sampling site 0) and by pastures (sampling sites 1 to 3), and the second is a 130-km^2 catchment dominated by sugar cane plantations upstream (sampling sites

1and 2) and by a city downstream (sampling sites 3 and 4) (Figure 10.3; see Ometto et al. 2000 for details).

The dissolved organic carbon (DOC) concentration increased abruptly in the last sampling site of the cane/urban catchment, after most of the urban sewage had already entered the stream (Figure 10.3). As labile carbon enters the systems, the respiration rates increased, especially from sampling site 0 to the following downstream sampling sites in the pasture catchment (Figure 10.3). In the cane/urban catchment the respiration rate increase was especially large in the last sampling site (Figure 10.3). As a by-product of the respiration process, the pCO_2 also sharply increased downstream in the cane/urban catchment, and the percent saturation of dissolved O_2 concomitantly decreased (Figure 10.3).

Such changes in the process of extra labile organic matter also altered the distribution of other elements. The DIN species also had a different distribution among catchments and even among sampling sites of the same catchment. In the pasture catchment NO_3^- increased from 20 μM (micro moles) in the sampling site 0 (forest) up to 70 μM in the sampling site 3 (pasture), decreasing a little in sampling site 4. This is an opposite trend in relation to the higher NO_3^- concentration found in pasture catchments of Rondônia (Neill et al. 2001). The Piracicaba River basin currently receives a total N deposition of approximately 9.0 kilograms per hectare per year owing to atmospheric contamination by industries and biomass burning (Lara et al. 2001). N demand in pasture areas in this region is lower than N demand in sugar cane areas (Filoso et al. in press; Krusche et al. in press). Consequently, the extra NO_3^- added by atmospheric deposition may be leached out of the soil solution in areas of less N demand, ending up in rivers and streams (Matson et al. 1999). Sugar cane plants have a high N demand and generally are underfertilized in this area (Filoso et al. in press). Therefore, it is possible that the smaller NO_3^- concentration in the stream of this catchment reflects the higher uptake of NO_3^- by sugar cane plants (Figure 10.4).

In the last two sampling sites, located in the city, the low dissolved oxygen concentration (Figure 10.3) probably inhibited nitrification and enhanced denitrification, similar to pasture streams of Rondônia (Christensen et al. 1990; Omnes et al. 1996; Nolan 1999; Krusche et al. in press; Thomas et al. in press). This lowered NO_3^- , and together with the high input of NH_4^+ via urban sewage, sharply increased the NH_4^+ concentration in the last sampling site (Figure 10.4).

The same sequence of alterations in C, N, and other elements observed in small streams of the Piracicaba River basin also occurred in the major rivers of the basin (Martinelli et al. 1999). Even in medium to large catchments of temperate zones, the effects of population concentration in urbanized areas have strongly affected the N inputs to riverine systems (Peierls et al. 1991; Howarth et al. 1996; Boyer et al. 2002).

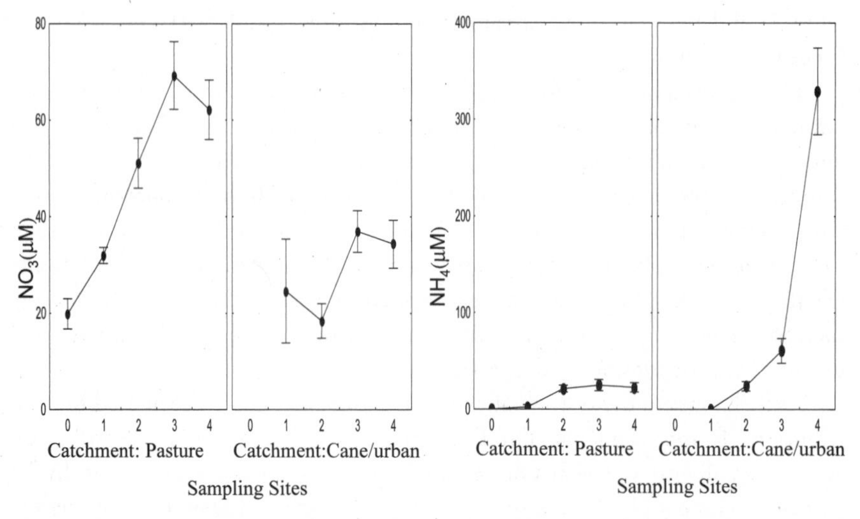

Figure 10.4. Variability in the sampling sites of two different catchments of the following parameters: NO_3^- and NH_4^+

Conclusions

Human interventions in the states of Rondônia and São Paulo represent two ends of a spectrum of changes and may be viewed as typical of several areas of the world. The first phase of such interventions is the replacement of original vegetation by nonintensive agriculture and ranching, followed by urbanization and industrialization of vast areas. The final outcome of these interventions is a mosaic where agricultural zones—characterized by diffusive pollution sources—coexist with urban centers, where point sources of pollution become more and more important. These changes in land use have led to physical, chemical, and biological changes that have in turn affected the biogeochemical cycles of major elements like C, N, and P.

Physical changes include increasing light in pasture streams and altered water balances during forest-to-pasture conversion. Chemical changes include alterations in nutrient stocks in vegetation and soil, nutrient ratios, pH, and dissolved oxygen concentrations in pasture and urban streams. These physical and chemical changes have implications for the biological composition of the ecosystems because they change species distribution (e.g., Ometto et al. 2000). Although interactions among these factors can be complex, several patterns do emerge from the sequence of change that has taken place in Brazil. These include reduced terrestrial N storage and decreasing N cycles during forest-to-pasture conversion

Table 10.3. Sequence of processes that can be altered in rivers and streams by the following land use change: Forest-pasture-urbanization

Process	*Forest*	*Pasture*	*Pasture with paspalum*	*Urban*	*Consequence*
Input of organic matter	Low	Medium	High	Very high	Increase in heterotrophic respiration
Heterotrophic respiration	Low	Medium	High	Very high	Increased consumption of O_2 and increase in DIC, DOC, and conductivity
Redox conditions	Aerobic	Aerobic	Aerobic/ anaerobic	Mostly anaerobic	Changes in the ratio CO_2:CH_4
Photosynthesis limitation	Light and P	N	Light and N	None	Changes in NPP
Nitrification	Medium	High	Low	Low	Increase in NO_3 in pasture streams
Denitrification	Low	Low	Medium	High	Loss of N-NO_3
Eutrophication	Low	Low-medium	Medium	High	Loss of biodiversity

(owing to greater removal of N than of other elements), decreased N flux from soils to the atmosphere, and declining stream NO_3 during initial land cover change followed by increasing NO_3 as urbanization takes place. The long-term consequences of these changes are mostly or largely unknown. Terms like "missing carbon" and "missing nitrogen" are two challenges to our knowledge base and an incentive to work to fully understand these ecosystems.

Acknowledgments

I thank Scott Ollinger, Amy Austin, and Filip Moldan for a careful review of this manuscript.

Literature Cited

Andreae, M. O., and P. Merlet. 2001. Emission of trace gases and aerosols from biomass burning. *Global Biogeochemical Cycles* 15:955–966.

Araújo, T. M., J. A. Carvalho Jr., N. Higuchi, A. C. P. Brasil Jr., and A. L. A. Mesquita. 1999. A tropical rainforest clearing experiment by biomass burning in the state of Pará, Brazil. *Atmospheric Environment* 33:1991–1998.

Asner, G. P., T. R Seastedt, and A. R. Townsend. 1997. The decoupling of terrestrial carbon and nitrogen cycles. *BioScience* 47:226–233.

Ballester, M.V., L. A. Martinelli, A. V. Krusche, R. L. Victoria, M. Bernardes, and P. B. Camargo. 1999. Effects of increasing organic matter loading on the dissolved O_2, free dissolved CO_2 and respiration rates in the Piracicaba River Basin, Southeast Brazil. *Water Research* 33:2119–2129.

Bernardes, M. C., L. A. Martinelli, A. V. Krusche, J. Gudeman, M. Moreira, R. L. Victoria, J. P. H. B. Ometo, M. V. R. Ballester, A. K. Aufdenkampe, J. E. Richey, and J. I. Hedges. Organic matter composition of rivers of the Ji-Paraná Basin as a function of land use changes. *Ecological Applications,* in press.

Biggs, T. W., T. Dunne, T. F. Domingues, and L. A. Martinelli. The relative influence of natural watershed properties and human disturbance on stream concentration in the southwestern Brazilian Amazon basin. *Water Resource Research,* in press.

Boyer, E. W., C. L. Goodale, N. A. Jaworski, and R. W. Howarth. 2002. Anthropogenic nitrogen sources and relationships to riverine nitrogen export in the northeastern U.S.A. *Biogeochemistry* 57/58:137–169.

Brannstrom, C. 2000. Coffee labor regimes and deforestation on a Brazilian frontier, 1915–1965. *Economic Geography* 76:326–246.

———. 2001. Conservation-with-development models in Brazil's agro-pastoral landscapes. *World Development* 29:1345–1359.

Brannstrom, C., and M. S. Oliveira. 2000. Human modification on stream valleys in the western plateau of São Paulo, Brazil: Implication for environmental narratives and management. *Land Degradation and Development* 11:535–548.

Buschbacher, R. J., C. Uhl, and E. A. S. Serrão. Large-scale development in Eastern Amazonia. 1986. In *Amazon rain forests: Ecosystem disturbance and recovery,* edited by C. F. Jordan. New York: Springer-Verlag.

Camargo, P. B., S. E. Trumbore, L. A. Martinelli, E. A. Davidson, D. C. Nepstad, and R. L. Victoria. 1999. Soil carbon dynamics in regrowing forest of eastern Amazonia. *Global Change Biology* 5:693–702.

Carvalho Jr., J. A., J. M. Santos, J. C. Santos, M. M. Leitao, N. Higuchi. 1995. A tropical rainforest clearing experiment by biomass burning in the Manaus region. *Atmospheric Environment* 29:2301–2309.

Carvalho Jr., J. A., N. Higuchi, T. M. Araújo, J. C. Santos. 1998. Combustion completeness in a rainforest clearing experiment in Manaus, Brazil. *Journal of Geophysical Research* 103:13,195–13,199.

Chadwick, O. A., L. A. Derry, P. M. Vitousek, B. J. Huebert, and L. O. Hedin. 1999. Changing sources of nutrients during four million years of ecosystem development. *Nature* 397:491–497.

Choné, T., F. Andreux, J. C. Correa, B. Volkoff, and C. C. Cerri. 1991. Changes in organic matter in an oxisol from the central Amazonian forest during eight years as pasture, determined by ^{13}C composition. In *Diversity of environmental biogeochemistry,* edited by J. Berthelin. New York: Elsevier.

Christensen, P., L. P. Neilson, J. Lorensen, and N. P. Revsbech. 1990. Denitrification in nitrate-rich streams: Diurnal and seasonal variation related to benthic oxygen metabolism. *Limnology and Oceanography* 35:640–651.

Daniel, M. H. B., A. A. Montebelo, M. C. Bernardes, J. P. H. B. Ometto, P. B. de Camargo, A. V. Krusche, M. V. Ballester, R. L. Victoria, and L. A. Martinelli. 2002.

Effects of urban sewage on dissolved oxygen, dissolved inorganic and organic carbon, and electrical conductivity of small streams along a gradient of urbanization in the Piracicaba River Basin. *Water, Air, and Soil Pollution I* 136:189–206.

Diez, J. A., A. Polo, M. A. Diaz-Burgos, C. C. Cerri, B. J. Feigl, and M. C. Piccolo. 1997. Effect of fallow land, cultivated pasture and abandoned pasture on soil fertility in two deforested Amazonian regions. *Scientia Agricola* 54:1–13.

Downing, J. A., M. McClain, R. Twilley, J. M. Melack, J. Elser, N. N. Rabalais, W. M. Lewis, R. E. Turner, J. Corredor, D. Soto, A. Yanez-Arancibia, J. A. Kopaska, and R. W. Howarth. 1999. The impact of accelerating land-use change on the N-cycle of tropical aquatic ecosystems: Current conditions and projected changes. *Biogeochemistry* 46:109–148.

Fearnside, P. M., N. Leal Filho, and F. M. Fernandes. 1993. Rainforest burning and the global carbon budget: Biomass combustion efficiency and charcoal formation in the Brazilian Amazon. *Journal Geophysical Research* 98:16,733–16,743.

Fearnside, P. M., and R. I. Barbosa. 1998. Soil carbon changes conversion of forest to pasture in Brazilian Amazon. *Forest Ecology and Management* 108:147–166.

Fearnside, P. M., P. M. L. de A. Graça, and F. J. A. Rodrigues. 2001. Burning of Amazonian rainforests: Burning efficiency and charcoal formation in forest cleared for cattle pasture near Manaus, Brazil. *Forest Ecology Management* 146:115–128.

Filoso, S., L. A. Martinelli, M. R. Williams, L. B. Lara, A. Krusche, M. V. Ballester, R. L. Victoria, and P. B. de Camargo. Land use and nitrogen export in the Piracicaba River Basin, Southeast Brazil. *Biogeochemistry*, in press.

Garcia-Montiel, D. C., C. Neill, J. Melillo, S. Thomas, P. A. Steudler, and C. C. Cerri. 2000. Soil phosphorus transformations following forest clearing for pasture in the Brazilian Amazon. *Soil Science Society of America Journal* 64:1792–1804.

Garcia-Montiel, D. C., P. A. Steudler, M. C. Piccolo, J. M. Melillo, C. Neill, and C. C. Cerri. 2001. Control on soil nitrogen oxide emissions from forest and pastures in the Brazilian Amazon. *Global Biogeochemical Cycles* 15:1012–1031.

Graça, P. M. L. de A., P. M. Fearnside, and C. C. Cerri. 1999. Burning of Amazonian forest in Ariquemes, Rondônia, Brazil: Biomass, charcoal, formation and burning efficiency. *Forest, Ecology and Management* 120:179–191.

Guild, L. S., J. B. Kauffman, L. J. Ellingston, D. L. Cummings, and E. A. Castro. 1998. Dynamics associated with total above-ground biomass, C, nutrient pools, and biomass burning of primary forest and pasture in Rondônia, Brazil, during SCAR-B. *Journal of Geophysical Research* 103:32,091–32,100.

Guo, L. B., and R. M. Gifford. 2002. Soil carbon stocks and land use change: A meta analysis. *Global Change Biology* 8:345–360.

Houghton, R. A. 1994. The worldwide extent of land-use changes. *BioScience* 44:305–313.

Howarth, R. W., G. Billen, D. Swaney, A. Townsend, N. Jaworski, K. Lajtha, J. A. Downing, R. Elmgren, N. Caraco, T. Jordan, F. Berendse, J. Freney, V. Kudeyarov, P. Murdoch, and Zhu Zhao-Liang. 1996. Regional nitrogen budgets and riverine N and P fluxes for the drainages in the North Atlantic Ocean: Natural and human influences. *Biogeochemistry* 35:75–139.

Hughes, R. F., J. B. Kauffman, and D. L. Cummings. 2002. Dynamics of aboveground and soil carbon and nitrogen stocks and cycling of available nitrogen along a land-use gradient in Rondônia, Brazil. *Ecosystems* 4:244–259.

IBGE (Instituto Brasileiro de Geografia e Estatística). 2003. Brazilian Institute for Geography and Statistics (www.ibge.gov.br) maintained by the government of Brazil.

INPE (Instituto Nacional de Pesquisas Espaciais). 2003. Brazilian National Space Agency data set on Land Set Images (www.inpe.br).

IFLOREST 2003. Forest Reserve Database (www.iflorestsp.br) maintained by the government of Brazil.

Jordan, T. E., D. E. Correll, and D. E. Weller. 1997. Effects of agriculture on discharge of nutrients from coastal plain watersheds of Chesapeake Bay. *Journal of Environmental Quality* 26:836–848.

Kauffman, J. B., D. L. Cummings, D. E. Ward, and R. Babbit. 1995. Fire in the Brazilian Amazon: 1. Biomass, nutrient pools, and losses in slashed primary forests. *Oecologia* 104:397–408.

Kauffman, J. B., D. L. Cummings, and D. E. Ward. 1998. Fire in the Brazilian Amazon: 2. Biomass, nutrient pools, and losses in cattle pastures. *Oecologia* 113:415–427.

Koutika, L.-S., F. Andreux, J. Hassink, T. Choné, and C. C. Cerri. 1999. Characterization of organic matter in topsoils under rain forest and pasture in the eastern Brazilian Amazon basin. *Biology Fertility Soils* 29:309–313.

Krusche, A. V., P. B. Camargo, C. E. Cerri, M. V. Ballester, L. Lara, R. L. Victoria, and L. A. Martinelli. Acid rain and nitrogen deposition in a sub-tropical watershed (Piracicaba): Ecosystems consequences. *Environmental Pollution,* in press.

Lara, L. B. S. L., P. Artaxo, L. A. Martinelli, R. L. Victoria, P. B. Camargo, A. V. Krusche, G. P. Ayers, E. S. B. Ferraz, and M. V. Ballester. 2001. Chemical composition of rain water and anthropogenic influences in the Piracicaba River basin, southeast Brazil. *Atmospheric Environment* 35:4937–4945.

Likens, G. E., F. H. Bormann, N. M. Johnson, D. W. Fisher, and R. S. Pierce. 1970. Effects of forest cutting and herbiced treatment on nutrient budgets in the Hubbard Brook watershed-ecosystem. *Ecological Monographs* 40:23–47.

Martins, P. F. da S., C. C. Cerri, B. Volkoff, F. Andreux, and A. Chauvel. 1991. Consequences of clearing and tillage on the soil of a natural Amazonian ecosytem. *Forest Ecology and Management* 38:273–202.

Martinelli, L. A., A. V. Krusche, R. L. Victoria, P. B. de Camargo, M. C. Bernardes, E. S. Ferraz, J. M. de Moraes, and M. V. Ballester. 1999. Effects of sewage on the chemical composition of Piracicaba River, Brazil. *Water, Air, and Soil Pollution* 110:67–79.

Martinelli, L. A., A. M. da Silva, P. B. de Camargo, L. R. Moretti, A. C. Tomazelli, D. M. L. da Silva, E. G. Fisher, K. C. Sonoda, and M. S. M. B. Salomão. Levantamento das cargas orgânicas lançadas nos rios do Estado de São Paulo. *Biota Neotrópica,* in press.

Matson, P. A., W. H. McDowell, A. R. Townsend, and P. M. Vitousek. 1999. The globalization of N deposition: Ecosystem consequences in tropical environments. *Biogeochemistry* 46:67–83.

Melillo, J. M., P. A. Steudler, B. J. Feigl, C. Neill, D. Garcia, M. C. Piccolo, C. C. Cerri, and H. Tiani. 2001. Nitrous oxide emissions from forests and pastures of various ages in the Brazilian Amazon. *Journal of Geophysical Research* 106:24,179–24,185.

Moraes, J. L., B. Volkoff, C. C. Cerri, and M. Bernoux. 1996. Soil properties under Amazon forest and changes due to pasture installation in Rondônia, Brazil. *Geoderma* 70:63–81.

Murty, D., M. U. F. Kirschbaum, R. E. McMurtie, and H. McGilvray. 2002. Does conversion of forest to agricultural land change soil carbon and nitrogen? A review of the literature. *Global Change Biology* 8:105–124.

Neill, C., M. C. Piccolo, P. A. Steudler, J. M. Melillo, B. J. Feigl, and C. C. Cerri. 1995. Nitrogen dynamics in soils of forests and active pastures in the western Brazilian Amazon Basin. *Soil Biology Biochemistry* 27:1167–1175.

Neill, C., B. Fry, J. M. Mellilo, P. A. Steudler, J. F. L. Moraes, and C. C. Cerri. 1996. Forest- and pasture-derived carbon contributions to carbon stocks and microbial respiration of tropical pasture soils. *Oecologia* 107:113–119.

Neill, C., M. C. Piccolo, C. C. Cerri, P. A. Steudler, J. M. Mellilo, and M. Brito. 1997. Net nitrogen mineralization and net nitrification rates in soils following deforestation for pasture across the southwestern Brazilian Amazon Basin landscape. *Oecologia* 110:243–252.

Neill, C., M. C. Piccolo, J. M. Mellilo, P. A. Steudler, and C. C. Cerri. 1999. Nitrogen dynamics in Amazon forest and pasture soils measured by ^{15}N pool dilution. *Soil Biology and Biochemistry* 31:567–572.

Neill, C., L. A. Deegan, S. M. Thomas, and C. C. Cerri. 2001. Deforestation for pasture alters nitrogen and phosphorus in small Amazonian streams. *Ecological Applications* 11:1817–1828.

Nolan, B. T. 1999. Nitrate behavior in groundwaters of the southeastern USA. *Journal of Environmental Quality* 28:1518–1527.

Ojima, D. S., K. A. Galvin, and B. L. Turner II. 1994. The global impact of land-use change. *BioScience* 44:300–304.

Ometto, J. P. H. B., L. A. Martinelli, M. V. Ballester, A. Gessner, A. Krusche, R. L. Victoria, and M. Williams. 2000. Effects of land use on water chemistry and macroinvertebrate in two streams of the Piracicaba River Basin, Southeast Brazil. *Freshwater Biology* 44:327–337.

Omnes, P., G. Slawyk, N. Garcia, and P. Bonin. 1996. Evidence of denitrification and nitrate ammonification in the River Rhone plume (northwestern Mediterranean Sea). *Marine Ecology – Progress Series* 141:275–281.

Pedlowski, M. A., V. H. Dale, E. A. T. Matricardi, and E. P. da Silva Filho. 1997. Patterns and impacts of deforestation in Rondônia, Brazil. *Landscape and Urban Planning* 38:149–157.

Peierls, B. L., N. F. Caraco, M. L. Pace, and J. J. Cole. 1991. Human influence on river nitrogen. *Nature* 350:386–387.

Peterson, J. P., W. M. Wolheim, P. J. Mulholland, J. R. Webster, J. L. Meyer, J. L. Tank, E. Marti, W. B. Bowden, H. M. Valett, A. E. Hershey, W. H. McDowell, W. K. Dodds, S. K. Hamilton, S. Gregory, and D. D. Morrall. 2001. Control of nitrogen export from watersheds by headwater streams. *Science* 292:86–90.

Piccolo, M. C., C. Neill, and C. C. Cerri. 1994. Net nitrogen mineralization and net nitrification along a tropical forest-to-pasture chronosequence. *Plant and Soil* 162:61–70.

SEADE 2003. Socioeconomic data on São Paulo State maintained on a Brazilian government database (ww.seade.gov.br).

Serrão, E., and J. Toledo. 1992. Sustaining pasture-based production systems for the humid tropics. In *Development or destruction: The conversion of tropical forest to pasture in Latin America,* edited by T. E. Downing, S. Hecht, H. Pearson, and C. Garcia-Downing. Boulder, Colo.: Westview Press.

Thomas, S. M., C. Neill, L. A. Deegans, A. V. Krusche, and R. L. Victoria. Influences of land use and stream size on particulate and dissolved materials in a small Amazonian stream network. *Biogeochemistry,* in press.
Trumbore, S. E., E. A. Davidson, P. B. Camargo, D. C. Nepstad, and L. A. Martinelli. 1995. Below ground cycling of carbon in forests and pastures of eastern Amazonia. *Global Biogeochemical Cycles* 9:515–528.
Uehara, G., and G. Gillman. 1981. *The mineralogy, chemistry, and physics of tropical soils with variable charge clays.* Boulder, Colo.: Westview Press.
Verchot, L. V., E. A. Davidson, J. H. Cattânio, I. L. Ackerman, H. E. Erickson, and M. Keller. 1999. Land use change and biogeochemical controls of nitrogen oxide emissions from soils in eastern Amazonia. *Global Biogeochemical Cycles* 13:31–36.
WHO 2003. Data obtained from the official World Health Organization website (www.who.int) report on water supply and sanitation.
Williams, M. R., and J. M. Melack. 1997. Solute export from forested and partially deforested catchments in the central Amazon. *Biogeochemistry* 38:67–102.

11

Elemental Interactions and Constraints on Root Symbiont Functioning

Mary Scholes and Andrew Woghiren

Depletion of soil fertility, along with the concomitant problems of weeds, pests, and diseases, is a major biophysical cause of low per capita food production in many developing countries (Sanchez 2002). Soil fertility depends on several biological, physical, and chemical processes. Nitrogen and phosphorus limitations are often cited as the major constraints of primary productivity. Even though N is the most abundant element on earth, it is mostly unavailable for plant growth (Vance 2001). Constraints on nitrogen fixation are one of the causes of N limitation of primary production, but hypotheses concerning limitations to N_2 fixation remain largely untested especially in natural terrestrial ecosystems (Uliassi and Ruess 2002). Sustaining agriculture on tropical nutrient-poor soils will require efficient management of the limited nutrient reserves and the appropriate use of fertilizers (Giller and Cadisch 1995). In addition, if ecosystems are to respond positively to elevated carbon dioxide, the necessity for adequate nitrogen and phosphorus will be even greater. The production of sufficient food to feed the world's population has come at the cost of an ever-increasing impact on Earth's sustainability. Crop production is compromising the global future use of N and P as well as water sustainability (Vance 2001).

Symbiotic organisms play an important role in the life of plants, ensuring their nutrition (especially in terms of N and P), defense against pathogens and pests, and adaptation to various environmental stresses. The symbiosis formed by legumes (Fabaceae, in the two subfamilies Mimosoideae and Papillionoidae) and nodule bacteria (Rhizobia) is the best-studied system of any plant-microbe interaction. Rhizobia (Azorhizobium, Bradyrhizobium, Mesorhizobium, Rhizobium, Sinorhizobium) are facultative plant symbionts cultivated easily *ex planta* and accessible to investigation by all modern genetic and molecular methods. The root nodule represents a very suitable model to analyze a range of key plant functions, including signaling and gene expression, cell differentiation and organogenesis, and carbon and nitrogen metabolism. In contrast to the taxonomically restricted N_2-fixing nodules, vesicular arbuscular mycorrhiza (AM) is extremely widespread, associated with 80–90 percent of terrestrial plants. The symbiosis is obligatory for AM fungi (Zyomycetes, Glomales), which can exist *ex planta* only as dormant spores and as such have very limited ability to act as parasites (Provorov, Borison, and Tikhonovich 2002). Ectomycorrhizas are associated with 20–35 percent of terrestrial plants.

The development of N_2-fixing and AM symbioses involves multistep processes that result in the formation of subcellular compartments containing endosymbionts. Different levels of anatomical complexity of the compared symbioses correlate with the specificity of the partner's interactions. From the vast amount of literature on the morphology and genetics of the legume-rhizobial and AM symbioses, it may now be possible to reconstruct their natural histories and conduct comparative and evolutionary research (O'Hara 2001; Provorov, Borison, and Tikhonovich 2002).

AM is a much more ancient symbiosis than the N_2-fixing one. Several lines of evidence suggest that AM originated 400–500 million years ago, together with terrestrial plants. The hypothesis of a coevolutionary colonization of land by plants and fungi is broadly accepted, suggesting that the very origin of terrestrial plants occurred because of their interaction with fungi, which supplied the hosts with water and nutrients. Legume-Rhizobia symbiosis is apparently much younger: it originated 60–70 million years ago when the major angiosperm family diverged (Gianinazzi-Pearson 1996; Doyle and Doyle 1997). The symbioses share a range of common morphological features and genes, which suggests that at least a part of the host genetic system controlling the nodule development originated during coevolution of the ancient plants with AM fungi (Parniske 2000). These ancestral genes apparently pre-adapted plants for the evolution of nodulation. Moreover, the mycorrhizal plant genes might be responsible for the evolution of symbiotic traits in nodule bacteria (Provorov, Borison, and Tikhonovich 2002). The prokaryotic organism, *Trichodesmium*, which is sometimes responsible for massive ocean blooms and contributes significantly to N_2 fixation in the oceans, is also thought to be an ancient ancestor with respect to N fixation ability, dating back to 400–500 million years ago (Karl et al. 2002). An intriguing theory has been proposed to explain the occurrence of nodulation in the *Leguminosae* (McKey 1994). A feature common to all legumes, which tends to differentiate them from most other families of plants, is the high concentration of N in their leaves. Although it has generally been assumed that these N-rich leaves are a consequence of N_2 fixation, this is a feature shared by all caesalpinioid legumes, including those that do not nodulate. McKey (1994) suggested that the ability to fix N_2 in association with bacteria actually evolved in legumes in response to the demands of maintaining high leaf N concentration. A corollary to this theory is that legumes developed N-rich leaves to begin with because the efficiency of photosynthesis increases with tissue N concentrations.

This chapter tries to broaden the debate across organisms and environments. It will review the knowledge on legume based/Rhizobium systems, mycorrhizae, and their interactions, mostly from an agroecological perspective. It will then address natural systems and the most recent findings of an ecological understanding of biological nitrogen fixation. This topic will be broadened from tropical and temperate forest systems to semiarid savanna systems. N and P deposition from aerial sources will be discussed in relation to constraints and saturation. This chapter does not cover other root symbionts, although there is growing evidence that bacteria isolated from the rhizosphere interact in both negative and positive ways with AM fungi and Rhizobia (Gryndler et al. 2002). The chapter will conclude with suggestions for future research.

Root Symbionts

Studies on root symbionts date back at least 100 years. Plants have adopted two strategies that enhance N and P acquisition and use: (1) those directed at improved acquisition (acquiring nutrients from the bulk soil and the rhizosphere) or uptake (movement of these nutrients across the membranes), and (2) those targeted to conserve use (Vance 2001). A great deal of emphasis has been given to understanding the functioning and role, in plant nutrition, of each individual symbiont, with much less work having been done on feedbacks between N and P acquisition and conservation. Although many basic physiological relationships between nutrient availability and plant growth have been addressed, an understanding of the interactions between elements and multiple organisms is still very limited (Aber et al. 2001). Host and symbiont responses to individual and multiple stresses add to the complexity of the understanding. The tropics have the potential to be the most productive cropping environments in the world, yet in many cases the yields are small. The addition of mineral fertilizers would be an immediate solution, but for a variety of social, economic, and political reasons this is generally difficult, especially in Africa. The time is ripe for a re-emergence of the use of legumes; studies need to be initiated that address the functioning of Rhizobium and mycorrhizae in the same study. Progress will be made if root symbionts are studied in more of a systems context, understanding the limitations to fixation and the possibility of multiple stresses and feedbacks.

Rhizobium

Classification

Nodule-forming bacteria (Rhizobia) require inorganic nutrients for metabolic processes to enable their survival and growth as free-living soil saprophytes and for their role as the N-fixing partners in legume symbioses. There are six currently accepted genera and thirty named species. The taxonomy is currently under review, with molecular techniques giving new insights into the relationships of the increasing number of isolates from a wide diversity of root nodules (O'Hara 2001; Thies, Holmes, and Vachot 2001). A good phylogeny combined with functional genomics would improve our understanding of their role in ecosystems.

Nutrient Deficiencies

Mineral nutrient deficiencies affecting root nodule bacteria are agronomically significant because of constraints on symbiotic fixation through effects on the growth, survival, and population diversity of free-living Rhizobia and limitations by specific deficiencies in the development and function of legume root nodules. Nutrient constraints can affect both free-living and symbiotic forms of root nodule bacteria, but whether they do is a function of a complex series of events and interactions. Evidence for nutrient constraints on legume symbiosis dates back to at least the 1930s, and the understanding has progressed slowly. This area was targeted as a research gap in the 1950s, and there has been some progress with a limited number of strains in the past twenty years.

Understandably, most of the work has focused on legumes in agricultural settings and much less on naturally occurring legumes. This imbalance may change as agroforestry practices are extended to larger land areas and savannas are transformed as a function of land use and changed environmental conditions. Rhizobia may be nutrient-limited owing to low absolute supply in the soil, low availability due to soil conditions such as pH, and limited uptake capacity resulting from limited carbon supply.

A number of nutrients essential for the growth of plants or bacteria play specific roles in nodulation and/or nitrogen fixation: Some effects of the deficiency of these elements are (1) reductions in the number and size of nodules formed, and (2) reductions in the amounts of N_2 fixed.

Many reviews focusing on the inorganic nutrition of root nodule bacteria and nitrogen fixation in cropping systems were carried out between 1970 and 2001 (Robson 1978; Giller and Cadisch 1995; Giller 2001). Rhizobia utilize diverse metabolic pathways and can grow on a wide range of substrates. The specific nutritional requirements for symbiotic N fixation reflect a combination of the essential needs of both the Rhizobia and the host plant. The following chemical elements are essential for free-living root nodule bacteria: C, H, O, N, P, S, K, Ca, Mg, Fe, Mn, Cu, Zn, Mo, Ni, Co, and Se. The functions of these nutrients can be considered in four categories: first, some nutrients are essential constituents of cell constituents, macromolecules, and cell structures; second, some are essential constituents of enzymes and cofactors; third, they may have roles in energetic functions; and fourth, a few function as signal molecules or have a role for regulation of gene function. Rhizobia can use a wide range of carbon and nitrogen sources. High exogenous nitrogen sources can lead to nonfunctional nitrogenase activity. A few of these elements will be discussed in more detail later in this chapter. Earlier work in the literature suggested that frequently occurring nutrient deficiencies were the major cause of inefficient nitrogen fixation; however, deficiencies might not be as widespread as initially thought (Giller 2001). The general physiological responses of root nodule bacteria to nutrient stress include a slowing of the growth rate, the induction of high-affinity active uptake systems, and, for some bacteria, the excretion of extracellular metabolites, enzymes, and chelating compounds to increase the availability of the limiting nutrients. Recent molecular work is supplying evidence that deficiencies can be much better defined by understanding plant-rhizobial signaling linked to the nod genes (O'Hara 2001; Vance 2001).

Phosphorus

It is widely believed that limited P availability affects nodule number and functioning. Rhizobia can use inorganic P as well as sources of organophosphates. Phosphodiester compounds can be utilized by Rhizobia as a source of P, especially under phosphate-limiting conditions, linked to the ability to produce phosphatases (acid and alkaline), linked to external environment pH conditions. Signaling systems in bacterial cells play important roles in their interactions with the physical, chemical, and biological components of their environment. The effects of P deficiency in Rhizobia on phosphate-mediated signal systems are not known. At present there is no unequivocal evidence that the phosphate concentrations commonly found in soil solutions limit the growth of free-living root nodule bacteria (O'Hara 2001). This statement may apply only to agroecosystems where P levels are usually significantly higher than in many naturally occurring systems.

Iron and Molybdenum

Iron is an essential component of the heme-containing proteins and the enzyme nitrogenase. The Fe uptake capacity is linked in some cases to the siderophore-based inducible active uptake system; some bacteria do not seem to have the capacity to synthesize siderophores under Fe-limiting conditions, whereas others do (Carson, Meyer, and Dilworth 2000). The role of siderophores in the Fe nutrition of Rhizobia is not well understood in natural ecosystems. Iron deficiency in peanuts did not limit the growth of soil or rhizosphere populations and did not directly affect root infection and nodule initiation, but it did limit nodule development, which led to a failure in nitrogen fixation (O'Hara et al. 1988a). Molybdenum is also an integral element of the nitrogenase enzyme. In some cases the supply of phosphorus to alleviate P deficiency and enhance legume production needs to be used with caution because it may induce Mo deficiency in acid sandy soils (Rebafka, Ndunguru, and Marschner 1993). Molydenum deficiencies in cropping systems can be easily reversed by foliar sprays (Giller 2001).

Calcium

Low calcium availability affects functioning and rhizobial survival. Calcium seems to be more important than previously thought at multiple stages during the infection process. Whether there is a specific Ca-sensitive metabolic function in root hair infection remains to be determined (O'Hara, Boonkerd, and Dilworth 1988b; O'Hara 2001).

Toxicities

Acid mineral soils, associated with high levels of Al^{3+}, Fe, and Mn and a corresponding decrease in P and Mo as well as a lack of Ca, limit formation and survival of nodules. Aluminum sensitivity is through disruption of the rapidly dividing tissue during nodule formation (Clarkson 1965). Sewage sludge contaminated with Zn and Cd is often applied in Europe and England to agricultural fields. Studies show that these metals have an adverse effect on populations of indigenous Rhizobium and the growth of clover. It is suggested that safe limits need to be set for the long-term protection of soils (Chaudri et al. 1993).

It has been frequently asked why there is such enormous variation in the responses of Rhizobia to nutrient deficiencies and toxicities. Will new molecular tools help in the identification of more efficient strains? Will an improved evolutionary understanding of development of the symbiosis be useful in predicting responses? Giller and Cadisch (1995) believe that benefits arising from breeding of legumes for N_2 fixation and rhizobial strain selection have less potential to increase inputs of fixed N than alleviation of environmental stresses or changes in farming systems to include more legumes. The survival and persistence of inoculant strains being used by farmers to increase legume yields and productivity is also an issue. It is clear that an understanding of both field-level trials and molecular approaches are needed to unscramble the responses. The previous discussion clearly illustrates that there is little understanding of elemental interactions on biological nitrogen fixation (BNF). We believe that experimental designs, in future studies, need to include elemental interactions.

Mycorrhiza

Mycorrhizal fungi are associated with 80 percent of terrestrial plants (Marschner and Dell 1994). Ectomycorrhizal trees dominate nitrogen-limited forest ecosytems, and they vary in their nitrogen uptake physiology (Smith and Read 1997). Mycorrhizal associations can improve plant health and growth through several mechanisms, which include increased uptake of nitrogen and phosphorus. The pool of nitrogen and phosphorus is increased through the increase in the absorptive surface area provided by their hyphae, greater uptake efficiency, or ability to access various nitrogen and phosphorus sources that are unavailable to nonmycorrhizal plants. The ability of mycorrhizal plants to assimilate proteins (bovine serum albumin, gliadin) and organic phosphate and nitrogen sources and to transfer these to host plants has been demonstrated extensively (Keller 1996). A large number of studies show that inoculation of plants with mycorrhizae leads to enhanced phosphorus uptake and improved growth, especially on highly weathered, marginal soils (Al-Karaki 2002; Ortas et al. 2002). AM inoculation increases fertilizer efficiency, as much for rock phosphate as for superphosphate, for plants growing in acid, P-fixing soils (Blal et al. 1990).

Mycorrhizal studies have generally been conducted in pots with many fewer studies under field conditions. This approach implies that the variable being tested is usually in constant supply. In a study with lettuce plants, it was shown that the augmentation of AM fungi does not necessarily ameliorate the detrimental effects of temporally variable water availability. Further studies are needed to elucidate when mycorrhiza are beneficial and when they are detrimental. It seems likely that the benefits of mycorrhizal inoculum will tend to be greater under P-limiting conditions but that further studies are needed to resolve the complex effects of mycorrhizae in environments with temporally variable resources. A theoretical framework is needed, which includes multiple organisms together with multiple stresses (Berg, Eaton, and Ayers 2001).

Toxicities

Heavy metal detoxification and enhanced plant nutrient uptake may be brought about by the presence of mycorrhiza—for example, toxic levels of cadmium can be reduced by the introduction of *Pisum* inoculated with various strains of *Glo-*

mus (Rivera-Becerril et al. 2002). The mechanisms for detoxification are diverse and difficult to resolve. They are probably related to binding to extracellular materials and sequestration in the vacuolar compartment, absorption by the hyphal sheath, chelation by fungal exudates, and adsorption onto the external mycelium. Recent reviews by Schutzendubel and Polle (2002) and Hall (2002) present updated information on possible antioxidative systems in mycorrhizal fungi and protection from heavy metal–induced stress. A discussion on the range of mycorrhizal systems involved, such as ectomycorrhizal versus vesicular arbuscular mycorrhizal, is also included. It was concluded that the development of stress-tolerant plant-mycorrhizal associations might be a promising new strategy for phytomediation and soil amelioration measures (Schutzendubel and Polle 2002).

Mycorrhizal Effects on Plant Communities and Diversity

AM was shown to be a major factor contributing to the maintenance of plant biodiversity and ecosystem function. Grime et al. (1987) found that community structure could be significantly altered by mycorrhizal activity in species-rich mixtures of plants in experimental microcosms, depending on the identity and mycorrhizal responsiveness of the dominant plant species. The influence of mycorrhizae in controlling plant diversity may be vitally important in semiarid ecosystems, where communities can be subjected to dramatic seasonal and interannual fluctuations and may rely on high biodiversity to maintain stability (Grime et al. 1987). Examples are available from European calcerous grasslands, which fluctuate in response to changed AM diversity, and North American old pastures, where plant biodiversity significantly increased with increasing fungal diversity (van der Heijden et al. 1998). On the other hand, in a study conducted, both in the field and in a pot experiment, in an overgrazed semiarid herb land in Australia, it was found that there was no change in plant species richness in mycorrhizal-suppressed field plots, but diversity increased owing to an increase in evenness. The treatments had no effects on community productivity, and therefore there was no relationship between diversity and productivity (O'Connor, Smith, and Smith 2002).

Recent studies show that mycorrhizal symbiosis may significantly affect plant competition and community structure through significant increases in phosphorus accumulation in seeds. Improved seed quality may have long-term effects on the structure of plant communities, especially in communities where the seeds are long-lived (Sanders and Koide 1994).

Much research in ecology addresses how the intensity or nature of processes changes along an environmental gradient. There is a statistical phenomenon known as Simpson's paradox, which states that it is impossible to determine the response of the whole based on the response of subgroups. Allison and Goldberg (2002) argue that it is not possible to extrapolate from a response shown by a fixed set of species to the response when the species composition is allowed to change. Hypotheses were tested with a meta-analysis of data on plant responses to arbuscular mycorrhizae. Although species-level response commonly declined as phosphorus availability increased, it was hypothesized that the community-level response could either decline or remain constant. The response averaged across multiple species was negative but not robust, so no clear distinction between the hypotheses could be made. The authors made the point, however, that any experimental approach taken must allow species composition to change. Future studies should use species representative of natural communities at different points along gradients rather than using a subset of species and allowing just the environment to change (Allison and Goldberg 2002).

Effects of Atmospheric N Deposition on Mycorrhizal Species Richness

Atmospheric N deposition has been implicated in the decline of ectomycorrhizal species richness in a gradient study in Alaska. Species richness declined dramatically, from thirty to nine taxa, with increasing N inputs. Organic horizon mineral N and foliar nutrient ratios (N:P, P:Al) were excellent predictors of taxonomic richness. A suggestion is made that the ectomycorrhizal community shifts from taxa specialized for N uptake under low N conditions toward taxa specialized for high overall nutrient availability and finally toward taxa specialized for P uptake under high-N, low-P, acidified conditions (Lilleskov et al. 2002). Further studies of this nature, combining a range of root symbionts, are needed to understand future patterns.

Symbiont Interactions: Use of Plant-Microbe Symbioses for Improving Plant Production

The introduction of target organisms into a system (plants, bacteria, and VAM) can increase the plant available nitrogen and phosphorus levels, as well as increasing soil organic matter and the stability of soil aggregates through a number of interacting soil physical, biological, and chemical processes (Requena et al. 2001). Intercropping of legumes (e.g., pigeon pea) has markedly increased production in subsistence agriculture in Africa and India. Improved germplasm of pigeon pea and appropriate *Rhizobium* sp. strains, especially those tolerant of high salinity levels, has allowed for expansion of cropping agriculture into marginal lands in India (Subbarao et al. 1990). A number of success stories exist for soybean production in Brazil owing to the development of soybean germplasms and rhizobial inoculants for low-fertility, acidic soils of the cerrados. Successes have also been recorded for selecting common bean and lupin germplasm, in association not with mycorrhiza but with cluster roots, which enhance P absorption (Vance 2001). Inoculation with selected AM fungi can improve the establishment and growth of forage legumes in fields that contain ineffective populations of native AM fungi (Medina, Kretschmer, and Sylvia 1990). Further studies have shown the potential of using indigenous AM inoculations in fodder crops, including legumes, grown in marginal soils in India along with *in situ* large-scale production of AM inocula (Gaur and Adholeya 2002).

Ecological Understanding of Root Symbionts

Much of the work conducted by Vitousek and his colleagues on Hawaii elucidated the role of BNF using a chronosequence of soils in which the P availability changed with age. Developing and disturbed ecosystems are frequently dominated by plants with well-developed N_2 fixing symbioses, such as alder and various legumes. With one exception, however, the Hawaiian montane rain forest lacks nitrogen-fixing vascular plant-microbe symbioses. The Hawaiian studies focused on ecosystem dynamics at various successional stages and the BNF potential of the systems as controlled by P availability. The hypothesis emerged stating that early successional stages were limited by N availability and late successional stages by P availability (Vitousek et al. 1993; Herbert and Fownes 1995; Chadwick et al. 1999). Support for this hypothesis was provided by the conceptual model of Walker and Syers (1976) and Crews et al. (1995), which stated that P becomes less available in the long term because of its conversion to recalcitrant organic and adsorbed forms. Unfortunately, many of these studies did not provide details on the presence and role of mycorrhizae across the various successional stages.

Although the studies from Hawaii have made a marked contribution, an understanding of the ecological controls on the rate of biological nitrogen is still sparse. Howarth, Chan, and Marino (1999) suggested that trace-element interactions and zooplankton grazing control the rate of cyanobacterial fixation in lakes versus estuaries. Vitousek and Field (1999) suggested that shade intolerance, P limitation, and grazing on N-rich plant tissues could suppress symbiotic N fixers in late successional forest ecosystems. Nitrogen fixers are almost absent from late successional temperate and boreal forests but present in lowland tropical forests. The study concluded that this pattern could be explained by the premise that N-demanding legumes are able to survive in temperate systems only if they fix nitrogen but that environmental constraints of water availability and temperature limit their distribution, whereas relatively high N availability in lowland tropical forests allows legumes to maintain an N-demanding lifestyle without necessarily paying the costs of fixing N (Vitousek et al. 2002).

In semiarid and arid savanna systems most *Acacias* have the potential for substantial symbiotic N_2 fixation, and this feature makes them attractive tree plantation crops in nitrogen-impoverished soils. In addition, the nitrogen-rich foliage provides the mainstay for the large number of ungulates feeding off *Acacias* in African ecosystems. A study was conducted in 1981 in an effort to better understand the fixation ability of thirteen *Acacia* species, a small subset of a very large genus, in order to determine their Rhizobium requirements. Dreyfus and Dommergues (1981) classified these species into three groups based on effective nodulation with fast- and slow-growing strains of Rhizobium. Further studies have been conducted on additional *Acacia* species, and Galiana et al. (1990) worked on *Acacia mangium* and *Acacia auriculiformis*, two species grown widely in the humid tropics for timber production. The first species was found to be a very specific host, whereas the latter species is a most promiscuous host. A study on *Acacia koa*, in Hawaii, showed that it changed its fixation ability as the stands aged (Pearson and Vitousek 2001). The six-year-old stands had more nodules and therefore fixed more N than the twenty-year-old stands. Thinning increased biomass but not N fixation ability. Again P availability may influence N fixation capacity; however, carbon limitation and mycorrhizal status were not investigated.

N_2 fixation in *Acacias*, in their natural habitat in the semiarid savannas, appears to be most important during early life stages rather than in maturing populations and is often limited by the availability of water and phosphorus. Much of the information is anecdotal and from short-term pot studies on *Acacia* seedlings. A few studies in the field using ^{15}N natural abundance techniques showed that *Acacias* and herbaceous legumes could make a marked contribution to African savanna ecosystems where the losses of N are high owing to fire and herbivory (Aranibar 2002 pers. comm.; Woghiren 2002). Nodulation and nitrogenase activity are more sensitive to ambient nitrate concentrations than those of most herbaceous and many woody legumes, and often substantial N_2 fixation is found only in field situations with very low soil N concentrations. It is hypothesized that the temporal variability in water and phosphorus availability, together with the high takeoff of biomass by herbivory, limits the BNF capacity of *Acacias* growing in semiarid savannas. The interactions of fire and herbivory add to the complexity of understanding ecological constraints on BNF. These factors have not been considered adequately in the current conceptual models proposed by Vitousek et al. (2002) and need to be addressed. There is wonderful scope for expanding the regional and global models that use ecological controls over N fixation to help understand how ecosystems will respond to changing environmental conditions.

Terrestrial biogeochemists traditionally distinguish atmospherically derived from rock-derived elements. Phosphorus, a rock-derived element, is usually considered to be made available through chemical weathering. Recent evidence, however, has shown that phosphorus deposition from dust is crucial to the functioning of two very different ecosystems, the islands of Hawaii and the Okavango Delta system in Botswana (Chadwick et al. 1999; Garstang et al. 1998). Mechanisms that control ecosystem N and P limitation strongly determine the consequences of increased N and P deposition. It is suggested that BNF and mycorrhizal status will be affected in regions where deposition is expected to increase as a result of industrialization. Nitrogen deposition can influence fruit-body formation by ectomycorrhizal fungi, the production and distribution of the extraradical mycelium in the soil, and the formation of ectomycorrhizas (Wallenda and Kottke 1998).

There are many studies on the expected impact of elevated carbon dioxide on BNF in cultivated forage legumes and temperate tree species (Rastetter, Agren, and Shaver 1997; Hungate et al. 1999). Much less work has been conducted on the trees and herbaceous legumes in semiarid tropical environments. A study carried out in Australia showed that *Acacias* grown under elevated CO_2 increased net assimilation rates for all of the species studied, which led to a decrease in plant nitrogen concentrations. In general, the amount of N_2 fixed per unit nodule was unaltered by CO_2 treatment (Schortemeyer et al. 2002). More studies of this nature are required.

Conclusions

The main points that have emerged are (1) the need to investigate legume symbioses and mycorrhizal associations in the same study in order to understand the limitations of fixation and the possibility of multiple stresses and feedbacks; (2) the need to pay attention to experimental designs, including multiple stresses, variability in time and space of nutrients and water, and changing species composition; (3) the need to transfer the knowledge of the functioning of root symbionts gained from agricultural studies to natural ecosystems; and (4) the need to expand the ecological understanding of biological nitrogen fixation and N and P availability that has been proposed for temperate and lowland tropical forests to other ecosystems and to model these controls both regionally and globally.

Literature Cited

Aber, J., R. P. Neilson, S. McNulty, J. M. Lenihan, D. Bachelett, and R. J. Drapek. 2001. Forest processes and global environmental change: Predicting the effects of individual and multiple stressors. *BioScience* 51:735–751.

Al-Karaki, G. N. 2002. Benefit, cost and phosphorus use efficiency of mycorrhizal field-grown garlic at different soil phosphorus levels. *Journal of Plant Nutrition* 25:1175–1184.

Allison, V. J., and D. E. Goldberg. 2002. Species-level versus community-level patterns of mycorrhizal dependence on phosphorus: An example of Simpson's index. *Functional Ecology* 16:346–352.

Berg, E. S., G. K. Eaton, and M. P. Ayers. 2001. Augmentation of AM fungi fails to ameliorate the adverse effects of temporal resource variation on a lettuce crop. *Plant and Soil* 236:251–262.

Blal, B., C. Morel, V. Gianinazzi-Pearson, J. C. Fardau, and S. Gianinazzi. 1990. Influence of vesicular-arbuscular mycorrhizae on phosphate fertilizer efficiency in two tropical acid soils planted with micropropagated oil palm (*Elaeis guineensis* jacq.). *Biology and Fertility of Soils* 9:43–48.

Carson, K. C., J.-M. Meyer, and M. J. Dilworth. 2000. Hydroxamate siderophores of root nodule bacteria. *Soil Biology and Biochemistry* 32:11–21.

Chadwick, O. A., L. A. Derry, P. M. Vitousek, B. J. Huebert, and L. O. Hedin. 1999. Changing sources of nutrients during four million years of ecosystem development. *Nature* 397:491–497.

Chaudri, A. M., S. P. McGrath, K. E. Giller, E. Rietz, and D. R. Sauerbeck. 1993. Enumeration of indigenous *Rhizobium leguminosarum* biovar *trifoli* in soils previously treated with metal-contaminated sewage sludge. *Soil Biology and Biochemistry* 25:301–309.

Clarkson, D. T. 1965. The effect of aluminum and some other trivalent metal ions on cell division in the root apices of *Allium cepa. Annals of Botany* 29:309–315.

Crews, T. E., K. Kityama, J. H. Fownes, R. H. Ripley, D. A. Herbert, D. Mueller-Dombois, and P. M. Vitousek. 1995. Changes in soil phosphorus fractions and ecosystem dynamics across a long chronosequence in Hawaii. *Ecology* 76:1407–1424.

Doyle, J. J., and J. L. Doyle. 1997. Phylogenetic perspectives on the origins and evolution of nodulation in the legumes and allies. Pp. 307–312 in *Biological fixation of nitrogen for ecology and sustainable agriculture*, edited by A. Legocki, H. Bothe, and A. Pfuhler. NATO ASI Series G. Ecological Science 39. Berlin: Springer-Verlag.

Dreyfus, B. L., and Y. R. Dommergues. 1981. Nodulation of *Acacia* species by fast and slow growing tropical strains of *Rhizobium. Applied Environmental Microbiology* 41:97–99.

Galiana, A., J. Chaumont., H. G. Diem, and Y. R. Dommergues. 1990. Nitrogen-fixing potential of *Acacia mangium* and *Acacia auriculiformis* seedlings inoculated with *Bradyrhizobium* and *Rhizobium* spp. *Biology and Fertility of Soils* 9:261–267.

Garstang, M., W. N. Ellery, T. S. McCarthy, M. C. Scholes, R. J. Scholes, R. J. Swap, and P. D. Tyson. 1998. The contribution of aerosol- and water-borne nutrients to the functioning of the Okavango Delta ecosystem, Botswana. *South African Journal of Science* 94:223–229.

Gaur, A., and A. Adholeya. 2002. Arbuscular-mycorrhizal inoculation of five tropical fodder crops and inoculum production in marginal soil amended with organic matter. *Biology and Fertility of Soils* 35:214–218.

Gianinazzi-Pearson, V. 1996. Plant cell responses to arbuscular mycorrhizal fungi: Getting to the roots of the symbiosis. *Plant Cell* 8:1871–1883.

Giller, K. E. 2001. *Nitrogen fixation in tropical cropping systems.* Wallingham, U.K.: CABI.

Giller, K. E., and G. Cadisch. 1995. Future benefits from biological nitrogen fixation: An ecological approach to agriculture. *Plant and Soil* 174:255–277.

Grime, J. P., J. M. Mackey, S. M. Hillier, and D. J. Read. 1987. Floristic diversity in a model system using experimental microcosms. *Nature* 328:420–422.

Gryndler, M., M. Vosatka, H. Hrselova, V. Catska, I. Chvatalova, and J. Jansa. 2002. Effect of dual inoculation with arbuscular mycorrhizal fungi and bacteria on growth and mineral nutrition of strawberry. *Journal of Plant Nutrition* 25:1341–1358.

Hall, J. L. 2002. Cellular mechanisms for heavy metal detoxification and tolerance. *Journal of Experimental Botany* 53:1–11.

Herbert, D. A., and J. H. Fownes. 1995. Phosphorus limitation of forest leaf area and net primary production on highly weather soils. *Biogeochemistry* 29:223–235.

Howarth, R. W., F. Chan, and R. Marino. 1999. Do top-down and bottom-up controls interact to exclude nitrogen-fixing cyanobacteria from the plankton of estuaries? An exploration with a simulation model. *Biogeochemistry* 46:203–231.

Hungate, B. A., P. Dijkstra, D. W. Johnson, C. R. Hinkle, and B. G. Drake. 1999. Elevated CO_2 increases nitrogen fixation and decreases soil nitrogen mineralization in Florida scrub oak. *Global Change Biology* 5:797–806.

Karl, D., A. Michaels, B. Bergman, D. Capone, E. Carpenter, R. Letelier, F. Lipschultz, H. Paerl, D. Sigman, and L. Stal. 2002. Dinitrogen fixation in the world's oceans. *Biogeochemistry* 57/58:47–98.

Keller, G. 1996. Utilization of inorganic and organic nitrogen sources by high-subalpine ectomycorrhizal fungi of *Pinus cembra* in pure culture. *Mycological Research* 100:989–998.

Lilleskov, E. A., T. J. Fahey, T. R. Horton, and G. M. Lovett. 2002. Belowground ectomycorrhizal fungal community change over a nitrogen deposition gradient in Alaska. *Ecology* 83:104–115.

Marschner, H., and B. Dell. 1994. Nutrient uptake in mycorrhizal symbiosis. *Plant and Soil* 159:89–102.

Medina, O. A., A. E. Kretschmer Jr., and D. M. Sylvia. 1990. Growth response of field-grown Sirato (*Macroptilium atropurpureum* Urb.) and *Aeschynomene americana* L. to inoculation with selected vesicular-arbuscular mycorrhizal fungi. *Biology and Fertility of Soils* 9:54–60.

McKey, D. 1994. Legumes and nitrogen: The evolutionary ecology of a nitrogen-demanding lifestyle. Pp. 211–228 in *Advances in legume systematics: Part 5 – The nitrogen factor,* edited by J. L Sprent and D. McKey. Kew, England: Royal Botanic Gardens.

O'Connor, P. J., S. E. Smith, and F. A. Smith. 2002. Arbuscular mycorrhizas influence plant diversity and community structure in a semiarid herbland. *New Phytologist* 154:209–218.

O'Hara, G. W., A. Hartzook, R. W. Bell, and J. F. Loneragan. 1988a. Responses to *Bradyrhizobium* strains of peanut cultivars grown under iron stress. *Journal of Plant Nutrition* 11:843–852.

O'Hara, G. W., N. Boonkerd, and M. J. Dilworth. 1988b. Mineral constraints to nitrogen fixation. *Plant and Soil* 108:115–138.

O'Hara, G. W. 2001. Nutritional constraints on root nodule bacteria affecting symbiotic nitrogen fixation: A review. *Australian Journal of Experimental Agriculture* 41:417–433.

Ortas, I., D. Ortakei, K. Zulkuf, A. Cinar, and N. Onelge. 2002. Mycorrhizal dependency of sour orange in relation to phosphorus and zinc nutrition. *Journal of Plant Nutrition* 25:1263–1279.

Parniske, M. 2000. Intracellular accommodation of microbes by plants: A common developmental program for symbiosis and disease? *Current Opinion in Plant Biology* 3:320–328.

Pearson, H. L., and P. M. Vitousek. 2001. Stand dynamics, nitrogen accumulation, and symbiotic nitrogen fixation in regenerating stands of *Acacia koa*. *Ecological Applications* 11:1381–1394.

Provorov, N. A., A. Y. Borison, and I. A. Tikhonovich. 2002. Developmental genetics and evolution of symbiotic structures in nitrogen-fixing nodules and arbuscular mycorrhiza. *Journal of Theoretical Biology* 214:215–232.

Rastetter, E. B., G. I. Agren, and G. R. Shaver. 1997. Responses of N-limited ecosystems to increased CO_2: A balanced-nutrition, coupled-element-cycles model. *Ecological Applications* 7:444–460.

Rebafka, F.-P., B. J. Ndunguru, and H. Marschner. 1993. Single superphosphate depresses molybdenum uptake and limits yield response to phosphorus in groundnut (*Arachis hypogaea* L.) grown on an acid sandy soil in Niger, West Africa. *Fertilizer Research* 34:233–242.

Requena, N., E. Perez-Solis, C. Azcon-Aguilar, P. Jeffries, and J.-M. Barea. 2001. Management of indigenous plant-microbe symbioses aids restoration of desertified ecosystems. *Applied and Environmental Microbiology* 67:495–498.

Rivera-Becerril, F., C. Calantzis, K. Turnau, J.-P. Caussanel, A. A. Belimov, S. Gianinazzi, R. J. Strasser, and V. Gianinazzi-Pearson. 2002. Cadmium accumulation and buffering of cadmium-induced stress by aruscular mycorrhiza in three *Pisum sativum* L. genotypes. *Journal of Experimental Botany* 53:1177–1185.

Robson, A. D. 1978. Mineral nutrient limiting nitrogen fixation in legumes. Pp. 277–293 in *The mineral nutrition of legumes in tropical and subtropical soils*, edited by C. S. Andrew and E. J. Kamprath. Melbourne, Australia: Commonwealth Scientific and Industrial Research Organisation (CSIRO).

Sanchez, P. A. 2002. Soil fertility and hunger in Africa. *Science* 295:2019–2020.

Sanders, I. R., and R. T. Koide. 1994. Nutrient acquisition and community structure in co-occurring mycotrophic and non-mycotrophic old-field annuals. *Functional Ecology* 8:77–84.

Schortemeyer, M., O. K. Atkin, N. McFarlane, and J. R. Evans. 2002. N_2 fixation by *Acacia* species increase under elevated atmospheric CO_2. *Plant, Cell and Environment* 25:567–579.

Schutzendubel, A., and A. Polle. 2002. Plant responses to abiotic stresses: Heavy metal-induced oxidative stress and protection by mycorrhization. *Journal of Experimental Botany* 53:1351–1365.

Smith, S. E., and D. J. Read. 1997. *Mycorrhizal symbiosis*. 2nd ed. San Diego: Academic Press.

Subbarao, G. V., C. Johansen, J. V. D. K. Kumar Rao, and M. K. Jana. 1990. Response of the pigeonpea-Rhizobium symbiosis to salinity stress: Variation among Rhizobium strains in symbiotic ability. *Biology and Fertility of Soils* 9:49–53.

Thies, J. E., E. M. Holmes, and A. Vachot. 2001. Application of molecular techniques to studies in *Rhizobium* ecology: A review. *Australian Journal of Experimental Agriculture* 41:299–319.

Uliassi, D. D., and R. W. Ruess. 2002. Limitations to symbiotic nitrogen fixation in primary succession on the Tanana river floodplain. *Ecology* 86:88–103.

van der Heijden, M. G. A., J. N. Klironomos, M. Ursic, P. Moutoglis, R. Streitwolf-Engel, T. Boller, A. Wiemken, and I. R. Sanders. 1998. Mycorrhizal fungal diversity determines plant biodiversity, ecosystem variability and productivity. *Nature* 396:69–72.

Vance, C. P. 2001. Symbiotic nitrogen fixation and phosphorus acquisition: Plant nutrition in a world of declining renewable resources. *Plant Physiology* 127:390–397.

Vitousek, P. M., L. R. Walker, L. D. Whitaker, and P. A. Matson. 1993. Nutrient limitation to plant growth during primary succession in Hawaii. *Biogeochemistry* 23:197–215.

Vitousek, P. M., and C. B. Field. 1999. Ecosystem constraints to symbiotic nitrogen fixers: A simple model and its implications. *Biogeochemistry* 46:179–202.

Vitousek, P. M., K. Cassman, C. Cleveland, T. Crews, C. Field, N. Grimm, R. Howart, R. Marino, L. Martinelli, E. Rastetter, and J. Sprent. 2002. Towards an ecological understanding of biological nitrogen fixation. *Biogeochemistry* 57/58:1–45.

Walker, T. W., and J. K. Syers. 1976. The fate of phosphorus during pedogenesis. *Geoderma* 15:1–19.

Wallenda, T., and I. Kottke. 1998. Nitrogen deposition and ectomycorrhizas. *New Phytologist* 139:169–187.

Woghiren, A. J. 2002. Nitrogen characterization of the savanna flux site at Skukuza, Kruger National Park. M.Sc. thesis, University of the Witwatersrand, Johannesburg, South Africa.

12

Multiple Element Interactions and Ecosystem Productivity with Emphasis on Micronutrients in Tropical Agroecosystems

Holm Tiessen and Chao Shang

Micronutrients, such as B, Fe, Mn, Cu, Zn, Mo, and Cl, are required by higher plants in small quantities compared with the macronutrients. Only a small portion of the total soil micronutrient inventory is available to plants. The difference between total and available nutrient amounts is well recognized from soil analyses. With some micronutrients this difference also exists in plants. For instance, total Fe in rice plants does not relate to Fe-deficiency chlorosis, but o-phenanthroline reactive Fe does (Katyal and Sharma 1980). The bioavailability of micronutrients is affected by plant species, and soil conditions such as pH, redox potential, clay content, organic matter level, and mineral composition (Adriano 2001).

Elemental interaction in agricultural ecosystems may be defined as an influence of one element upon another in relation to plant or animal function. This influence can be reciprocal and may result in deficiency symptoms, changed tissue concentration, and yield response (Olsen 1972). Plant responses can show complex patterns of nutrient interactions: in tall fescue (*Festuca arundinacea Schreb.*), P fertilization improved not only P but also Mg and Ca nutrition. Mg fertilization, however, was effective only when P was also applied (Reinbott and Blevins 1997). Such elemental interactions affect not only crop production, but also the health of animals (Miller, Lei, and Ullrey 1991). Grass tetany is induced when ruminants consume forage with a K/(Ca+Mg) ratio greater than 2.2 (Reinbott and Blevins 1997).

Transformations and cycling of C, N, P, S, and micronutrients (e.g., sequestration, mineralization, and redistribution) are interrelated, particularly in tropical ecosystems where the natural fertility relies on the close cycling of nutrients in organic matter. Agricultural management practices such as liming, fertilization, irrigation, and application of fresh plant residues can induce or enhance interactions.

Elemental Interactions in Soils

Ion pairs and complexes between polyvalent trace metals and inorganic ligands such as Cl^-, OH^-, SO_4^{2-}, HPO_4^{2-}, PO_4^{3-}, CO_3^{2-}, F^-, and NO_3^- are common, such as Zn complexes with Cl^-, SO_4^{2-}, PO_4^{3-}, and NO_3^- (Lindsay 1979) or chloride complexes with Hg and Cd (Stumm and Morgan 1981). Equilibrium constants for complexes of Mn, Fe, Cu, and Zn with various anions in water are summarized by Harmsen and Vlek (1985). Formation of ion pairs usually increases the mobility of the cation in soil because its charge is reduced, increasing the availability of trace cations. The formation of $CdCl_2$ may be responsible for the Cl-enhanced Cd uptake by Swiss chard (Smolders and McLaughlin 1996). Sulfate-Cd complexation similarly affects Cd uptake (McLaughlin et al. 1998). The phosphomolybdate complex anion is believed to have high mobility and to be readily absorbed by plants, which may explain the synergistic effects of P and Mo (Barshad 1951).

Ion pair and complex formation also occurs in sorption: the stoichiometric retention of cations and anions together is termed salt sorption. Pearce and Sumner (1997) observed salt sorption on an acid soil involving Ca, Mg, and Na sulfates and Ca and Mg chlorides. A similar process was described by Marcano-Martinez and McBride (1989) for the sorption of Ca and K sulfate and chloride in an Oxisol from the Brazilian cerrado. In such highly weathered, nutrient-poor soils these processes may be critical for nutrient availability since salt adsorption reduces the availability of both ions below that calculated by equilibrium chemistry. Cerrado soils have become agriculturally usable only through massive applications of lime to moderate their low pH. The carbonate anion of the lime will be lost as CO_2 upon contact with the acid soil, leaving the Ca to react with other anions, potentially reducing their availability. On the other hand, ion pair and complex formation may be beneficial, as in the case of aluminum sulfates, which form upon addition of calcium sulfate to high-Al soils, alleviating Al toxicity to plants (Nobel, Sumner, and Alva 1988). Equilibrium constants for complexes of Mn, Fe, Cu, and Zn with various anions in water (summarized by Harmsen and Vlek 1985) show that complexes between divalent and higher ions and particularly carbonate complexes are very stable.

Organic ligands such as organic acids and siderophores form complexes with micronutrient cations and trace metals (Stevenson 1991). The mobility and solubility of metal cations are increased when they are complexed with small organic molecules but reduced when they form complexes with large humic molecules. The latter may be responsible for the Cu deficiency in organic soils. Trivalent Fe is preferentially complexed over the divalent species. Of the divalent cations, binding affinity is generally in the order: Cu > Co > Zn > Fe(II) > Mn (Stevenson 1991). Pb, Ni, and Ca also are active in binding with natural ligands (Bloom 1981). The binding ability of a metal cation may influence the availability of other elements. For example, Fe (III) can effectively replace Mn from Mn-organic complexes leading to Mn deficiency in plants because Mn is oxidized and precipitated as insoluble oxide (Knezek and Greinert 1971; Roomizadeh and Karimian 1996). Complexation of metals with organics also modifies their affinity toward soil particles (Harter and Naidu 1995). Addition of citric and salicylic acid alleviated plant Fe and Zn deficiency (Aly and Soliman 1998). Similar acids are excreted by roots and microbes. Alvarez-Fernandez, Garate, and Lucena (1997) indicated that soil Cu could replace the Fe in commercial chelates applied and affect the Fe availability.

Competition for adsorption sites between micronutrient cations and anions for soil surface sites affects their availability and mobility. Zn competes with Cd, Ca, and Mg for soil exchange sites (Kabata-Pendias and Pendias 2001). The adsorption of $B(OH)_3$ by soils is influenced by oxyanions in the order phosphate > molybdate > sulfate (Goldberg et al. 1996). Silicate ions may also play a role in B adsorption and mobility in soils. Selenite competes with sulfate and phosphate for adsorption sites. Addition of sulfate and phosphate to soil increases the availability of Se by desorption (Rajan and Watkinson 1976; Rajan 1979; Singh, Singh, and Relan 1981). Other examples of ion competition for adsorption sites are molybdate with sulfate, arsenate, and phosphate; arsenate with chloride, nitrate, sulfate, and phosphate (Smith, Brown, and Deuel 1987); and silicate with phosphate (Blair, Freney, and Park 1990).

Coprecipitation, for example, the formation of franklinite ($ZnFe_2O_4$), which converts labile Zn and Fe into an unavailable form, may cause the deficiency of both metals (Pulford 1982). Precipitation of insoluble $FePO_4 \cdot 2H_2O$ is partially responsible for the P-Fe interaction observed in plants (Kabata-Pendias and Pendias 2001). $MnPO_4 \cdot 1.5H_2O$, identified in soils (Boyle and Lindsay 1986), lowers Mn^{2+} activity. Under reduced conditions low-solubility forms of copper minerals, $Cu_2Fe_2O_4$ and Cu_2S, can form, controlling the Cu activity in solution. MoO_4^{2-} is readily coprecipitated with several cations such as Cu^{2+}, Zn^{2+}, Mn^{2+}, Pb^{2+}, and Ca^{2+}.

Redox reactions affect the availability of two micronutrients: As redox potential increases, Fe^{2+} and Mn^{2+} are oxidized to Fe^{3+} and Mn^{4+}, followed by strong hydrolysis forming insoluble oxides, which are very active in immobilizing micronutrients and trace metals. Iron and Mn cations and oxides are also involved in the transformation of As^{3+}/As^{5+}, and Cr^{3+}/Cr^{4+} in soil (Smith, Naidu, and Altson 1998; Adriano 2001).

Interactions in Plants and the Root-Soil Interface

Boron deficiency was once linked to high Ca and Mg concentrations in plant and soil (Adriano 2001), and a Ca/B concentration ratio in plant tissue has been used as a diagnostic for a balanced B nutrition of crops (Gupta 1972a,b). The uptake of B by some crops is decreased at high pH and high Ca supplies; however, the high pH that affects the speciation and adsorption of B may be the true factor behind the antagonistic relationship on alkaline and overlimed acid soils (Gupta and MacLeod 1981). Interactions of B with P, Mg, K, and N are not consistent, and the B-P interaction measured by plant uptake may result from a competitive relationship on soil surfaces. An increased supply of Zn enhanced B uptake by rapeseed under high B supply. A high B supply also increased the Zn uptake under high Zn supply, but this effect varied with genotype (Grewal, Graham, and Stangoulis 1998). Silicon seems to enhance the B uptake by rapeseed under B deficiency but depresses the uptake at normal and high B supplies (Liang and Shen 1994). Swietic (1995) reported that Zn can alleviate B toxicity in sour orange seedlings.

Iron and Mn have similar chemistry, so an antagonistic relation is expected. High Mn supplies reduce the Fe tissue concentration and induce Fe deficiency (Kirsch, Harward, and Petersen 1960; Zaharieva, Kasabov, and Romheld 1988), and the reverse is also true. A suitable Fe/Mn ratio for growth media is 1.5–2.5 (Kabata-Pendias and Pendias 2001). The Fe-Mn interrelationship is complicated by changes in rhizosphere environments resulting from Fe or P deficiencies or applied NH_4^+-N; by soil physical and chemical properties; by applied chelates and other nutrients; and by microorganisms associated with the plant roots (Reisenauer 1994). Fe and Mn oxides may form root plaques that may serve as a temporary Fe and Mn source/sink around roots. At low Fe or a high supply of Mo, Mo depressed the Fe uptake by plants, but the mechanisms are not clear (Olsen 1972). Liu, Reid, and Smith (2000) reported that the Fe concentration in mung bean leaves was reduced by 80 percent in the presence of 5 millimol (mM) Co in culture solution, whereas the Fe uptake in roots was not affected, suggesting an inhibition of Fe transport from root to shoots. The authors showed that the toxic effects of Co are additive to those of Co-induced Fe deficiency, and high Co concentrations might disrupt a range of metabolic processes contributing to the symptoms.

Manganese deficiency has been induced or Mn toxicity reduced by Fe chelates

(Knezek and Greinert 1971; Foy, Chaney, and White 1978; Heenan and Campbell 1983). The decrease in Mn concentration and uptake caused by Fe application to soybean in calcareous soils was attributed to the restriction of Mn translocation from soil to root and/or from root to tops (Roomizadeh and Karimian 1996). Iron deficiency leads to rhizosphere acidification and increased reductant (e.g., oxalic and other organic acid) production, enhancing Mn uptake from soils (Reisenauer 1994). High Mo levels accentuate the Mn-induced Fe deficiency, probably because Mo depletes the already-low Fe. Mn-P interactions were also observed but not consistent. Manganese nutrition of beets benefits from the application of Cl^- (Elmer 1997).

Copper (II) has a high affinity for organic materials. Calcium can stabilize polygalacturonate that is used as a model compound for the natural root mucilage forming a fibrillar structure, a porous network with large free space (Deiana et al. 2001). This free space, which serves as a pathway for the transport of nutrients to the root surface, can be collapsed by Cu^{2+} replacing Ca^{2+} (Deiana et al. 2001). Pb^{2+}, Zn^{2+}, and Cd^{2+} also compete for the same sites, with Pb^{2+} being preferred. This might account for the interactions of Cu with these elements in plants (Kabata-Pendias and Pendias 2001). The Cu-Fe antagonism results in Fe chlorosis in crops at excess Cu (Olsen 1972; Adriano 2001). Phosphorus-Cu and Cu-Mn interactions are also reported (Adriano 2001). The Cu-Mo interaction is antagonistic and is greater in the presence of NO_3, probably because Cu interferes with Mo in the enzymatic reduction of NO_3^- (Olsen 1972). Nitrogen can induce Cu deficiency, possibly because protein-Cu complexes in plant roots retard the translocation of Cu to the shoots.

Zinc-P interaction—that is, P-induced Zn deficiency—is the most systematically studied among elemental interactions in soil-plant systems, as reviewed by Olsen (1972). Bean and corn are subject to P-induced Zn deficiency, particularly at marginal soil levels of Zn (Shang and Bates 1987). Five hypotheses on the cause of P-Zn interaction have been proposed (Olsen 1972; Adriano 2001): (1) P-Zn interaction in soil, (2) a dilution effect on Zn concentration in plants due to growth response to P, (3) a slower rate of translocation of Zn from the roots to shoots compared with P, (4) a metabolic disorder within plants related to an imbalance between P and Zn, and (5) the impact of P on mycorrhizal infection. Formation of sparingly soluble P-Zn precipitates in plant parts may be related to the accumulation of P in leaves, which can be interpreted either as P-induced Zn requirement or enhanced P translocation due to Zn deficiency (Webb and Loneragan 1990; Adriano 2001). This mechanism is specific for Zn and did not occur with Fe, Mn, or Cu. Zinc has an antagonistic relationship with Cu, Cd, and Pb, possibly because these cations compete for the same absorption sites on root surfaces (Deiana et al. 2001). Pb also may affect the translocation of Zn from roots to shoots.

Molybdenum uptake in plants is enhanced by P, probably due to P effects on the absorption and internal transport of Mo (Olsen 1972). Formation of phosphomolybdate complex may favor the absorption and translocation of Mo in plants. Molybdenum was found, together with P, to enhance the N fixation by Azolla, used as green manure in paddy soils (Adriano 2001). Since both sulfate and phosphate compete with molybdate for soil sorption sites (Smith, Brown, and Deuel 1987), increased availability and uptake of Mo would be expected upon application of S or P fertilizers. The Mo-S relationship in plants, however, is antagonistic, possibly owing to sulfate interference with Mo absorption, transport, or metabolism in plants (Smith, Brown, and Deuel 1987). Molybdenum-Cu antagonism is strongly related to N metabolism.

Management Effects

Liming normally decreases the availability of metal cations as pH rises, whereas Mo availability increases (Adam 1984). Flooding (irrigation in paddy soils) and/or addition of fresh organic matter can significantly decrease the redox potential in soils so that the availability of Mn and Fe increase and interactions of Fe-Mn, P-Mn, and Fe-Mo may become evident. Acidity caused by fertilization and acid deposition decreases the absorption of B, Fe, Mn, Ca, and P by inhibiting root elongation, and Al toxicity in acid soils induces B absorption (Poschenrieder, Llugany, and Barcelo 1995).

Plant genotype showed a significant interaction with the effects of Cd treatment on nutrient concentrations in both shoot and root (Zhang, Fukami, and Sekimoto 2000). Grewal, Graham, and Stangoulis (1998) reported that B supply was improved only by improving Zn availability in one of several cultivars of rapeseed tested. Such genotypic effects on element interactions in plants are common, and it may be easier to select or breed for varieties that tolerate microelement deficiencies than try to alleviate the deficiency in the soil. For instance, susceptibility to Fe-chlorosis may be low in plant varieties that efficiently release acids or siderophores that aid Fe acquisition (Jolley et al. 1996). In deep, sandy, infertile soils of Australia, cereal varieties that are more effective in Zn uptake are required to promote root development in lower soil layers, since it is impractical to fertilize the subsoil. Deep rooting is needed for moisture efficiency. Graham, Ascher, and Hynes (1993) report that the Zn efficiency traits of cereals for sandy and clay soils differ genetically. Kramer and Chardonnens (2001) suggest that the genetic control on trace element uptake by plants could be used to create varieties that are efficient in bioremediation of contaminated sites. Even apparently soil-based elemental interactions may have genetic modifiers. Shim and Vose (1965) reported that Fe deficiency in plants was augmented by Ca in only one of the rice varieties they tested.

Tropical Environments

The high biological potential of the humid tropics is reflected not only in large biomass production and organic matter turnover, but also in very intensive land use. Up to four crops are planted and harvested per year on a single plot. As a result, elemental cycles and stocks are governed to a greater extent by biological processes than in most temperate environments. Highly weathered soils with mostly 1:1 minerals and low cation exchange capacities predominate on many stable land surfaces in the tropics. Because of the small amounts of 2:1 clays and limited mineral-associated cation exchange capacity, a greater proportion of the exchange is linked to organic matter. The low capacity of weathered soils to hold nutrient elements further increases the importance of biological and organic nutrient reservoirs. This situation creates a close linkage between organic matter (largely C, N, P, and S) and other elements, especially metals. The strong tie-in of elements to organic ecosystem components has the consequence that biological transformations play an important role and elemental linkages occur through the composition and transformations of organic matter, plants, and other organisms.

Carbon and Macronutrients

In many dystrophic tropical forests the largest reservoir of nutrient elements is in the plant biomass. Nykvist (1997) showed that in a Malaysian rainforest with low soil Ca, 20 percent of the ecosystem's Ca was held in the boles of trees. A very close link thus exists between C (of the plant mass) and cations. The total accumulation of nutrients in the aboveground biomass of a seasonal tropical forest in China was 2,100 kilograms (kg) Ca, 2,000 kg N, 1,000 kg Mg, 800 kg K, and 200 kg P (Shanmughavel et al. 2001), whereas in a Colombian tropical rainforest the total nutrients stored in the 300 Mg per hectare of biomass were 1,100 kg Ca, 2,000 kg N, 500 kg Mg, 700 kg K, and 40 kg P (Rodriguez-Jimenez 1988). It is apparent that the Colombian forest, for a similar amount of organic matter (N), is greatly impoverished in cations and P relative to the Chinese forest, reflecting plant community adaptation to geochemical constraints. Rodriguez-Jimenez (1988) estimated primary productivity at 8 Mg per hectare per year, giving a turnover time of the biomass of approximately forty years. A major portion of the nutrient element flow in dystrophic and oligotrophic tropical rainforest is thus tied into the biotic cycle of plant biomass and soil organic matter.

A biotic reservoir of elements reduces soil nutrient stocks. Nutrient sequestration in tree biomass, and particularly in boles, of a mixed species plantation reduced soil nutrient stocks measurably within five years after planting, and 50–60 percent of the biomass Ca, Mg, K, and P may be contained in boles and be lost from the ecosystem upon harvest (Montagnini 2000). Internal conservation and recycling of nutrients in tropical trees is important for productivity and nutrient budgets of ecosystems (Singh and Singh 1991; Herbon and Congdon 1998). Lal et al. (2001) estimated that the N demands for leaf production of several dry forest species in India were satisfied to 46–80 percent by retranslocation. Corresponding figures for other nutrients were: 20–91 percent for P, 20–57 percent for K, 1–54 percent for Na, and 0–30 percent for Ca. Nearly half the major nutrients in this ecosystem are thus internally recycled and closely tied to biomass and C accumulation. Retranslocation is not only related to the tree species but may also reflect other biological linkages such as mycorrhizal associations. Chuyong, Newberry, and Songwe (2000) showed twofold greater retranslocation of N and P in non-ectomycorrhizal trees than in mycorrhizal trees.

In addition to the biological cycling of forest elements, rain and the resulting through-fall mobilize elements in characteristic ratios. Oliviera-Leite, Valle, and Oliviera-Leite (1990) measured 40, 20, 10, and 1 kg per hectare per year of N, Ca, Mg, and P respectively in the rainfall on a shaded cocoa plantation. The through-fall was slightly enriched in Mg and Ca, twice as high in K and ten times enriched in P. These ratios were substantially modified under unshaded cocoa.

Plant community composition and biodiversity can modify element cycles and linkages (Wang et al. 1991; Parrotta 1999). Lata et al. (1999) measured 240-fold variations in the nitrification potential at Lamto (Ivory Coast), and *Hyparrhenia diplandra* plantlets taken from microsites characterized by widely differing nitrification showed nitrate reductase levels reflecting the field variation. Variability in the plant community thus relates to an uncoupling between C and N cycling and retention.

Elemental links are closely dependent on climate and landscape (Silver et al. 1994). Tiessen, Chacon, and Cuevas (1994) showed high P availability despite low total P levels in the soil of wet Amazon Caatinga sites relative to a drier Tierra Firme forest and attributed this to the frequent reducing conditions. Similarly, Silver et al. (1994) concluded that organic matter was closely linked to N, P, and K at upper slope positions, whereas hydromorphic conditions at lower slopes created a more variable environment.

When land use changes and organic matter stocks fluctuate under changing inputs and outputs of carbon, other elements will follow these transformations. Mailly, Chritanty, and Kimmins (1997) measured the element budget throughout a whole cultivation-bamboo fallow cycle in Java. Along the six-year cycle some 790 kg N, 690 kg K, 250 kg Mg, 220 kg Ca, and 130 kg P accumulated in plants, half of which were removed in harvested materials. Fertilization replenished outputs between 60 percent for P and 22 percent for Mg. This imbalance illustrates the typical situation for most agricultural settings: Management inputs are not balanced relative to crop outputs. For some elements, such as N and possibly K, surpluses will be leached; for others, such as P, surplus will be adsorbed by soil. The fallow phase is likely to reestablish a partial nutrient balance, but with continuing imbalance of inputs and outputs and shortened fallows, declining yields will eventually force a fine-tuning of fertilizer combinations (Spaargarden 1989). Perennial vegetation can accommodate unbalanced nutrient supplies through increased dry matter production or luxury consumption, but even relatively minor disturbances such as grazing will increase net losses (Castilla, Ayarza, and Sanchez 1995).

Minimum input management as part of subsistence farming typically relies on local or regional recycling of nutrient elements in plant residues or animal manures. Hoffmann et al. (2001) estimated recycling of nutrients under Haussa-managed subsistence farming to be 120 kg K, 87 kg N, and 33 kg P per hectare per year and concluded that management had greater impact on P and K status of soils than on organic matter or N. This situation might result in a drift of elemental ratios under low input farming, but the common inclusion of legumes in subsistence agriculture of the West African savanna may contribute a compensatory belowground N input. Dalal et al. (1997) provide a review of C and nutrient depletion rates under cropping.

Micronutrients

One of the problems associated with evaluating amounts, availabilities, and interactions of micronutrients in tropical soils is the wide range encountered for any one element. In a summary for Asian soils, Katyal and Vlek (1985) reported ranges for total element contents of soils of B (5–80), Cu (4–140), Mn (20–4,000), Mo (0.4–14), and Zn (5–250) (all μg [micrograms] per gram). Analyzing a large survey by the Food and Agriculture Organization of the United Nations of micronutrients in tropical and subtropical regions (Sillanpää and Vlek 1985; Sillanpää 1982) showed that regional means and medians for micronutrient amounts are often far apart, so that "average" recommendations for micronutrient management are not useful. Values obtained in the micronutrient analyses were then fitted to plant response, and regressions were corrected for covariances. Best results were obtained when B was corrected for cation exchange capacity (although B is anionic), Cu was corrected for organic C, and Mn and Mo were corrected for soil pH. These corrections indicate that the chemical extractions used are not indicators of bioavailability. They also indicate that there are universal interactions between micronutrients and organic matter or other elements in the soil-plant system.

Silver at al. (2000) suggested that P limitation can not only reduce ecosystem productivity but also limit organic matter decomposition, thereby throttling the release of other nutrients for recycling in an ecosystem (Lisanework and Michelsen 1994). Davies (1997) reviewed micronutrient problems in tropical soils.

Kanwar and Youngdahl (1985) have tabulated average micronutrient removals by crops: Maize and soybeans are identified as removing large amounts of Zn and Cu, rice and soybeans remove much Mn, and groundnuts and maize remove much Fe. Root crops with their much higher yields remove large amounts of most nutrients. All of these exports will increase proportionally when projected optimum yields are realized, which are some three to five times greater than current typical production levels in the tropics. Shifting cultivation usually recycles enough micronutrients in organic matter associated with slash and burn to avoid deficiencies (León, López, and Vlek 1985).

Elemental Interactions in Tropical Crops

Antagonistic relationships between P and Zn were observed in coffee and rice plants (Blair, Freney, and Park 1990). Zinc deficiency in rice is common and might be a major nutritional constraint to yield (Randhawa and Katyal 1982). Foliar application of Zn, Mn, and B increased yield and N-P-K uptake of sorghum (Raja and Reddy 1987). A synergistic effect of N and Zn was also observed in mango, and a foliar application of 0.4 percent Zn combined with 1 percent urea produced the highest fruit yield among treatments (Banik, Sen, and Bose 1997). A high Mn supply decreased the Zn uptake by banana (Turner and Barkus 1983).

Application of Mn, Zn, Cu, and Fe increased not only the uptake of these nutrients, but also the plant and seed yields of peanut, although some antagonistic relationships were observed (Moussa, Dahdoh, and Shehata 1996). Shanjida, Ullah, and Sarwar (2001), however, reported that Cu (and Cr) had antagonistic effects on the N-P-K uptake by rice because the critical concentration of Cu in culture medium above which a decrease in N-P-K uptake was observed was lower than the Cu concentration at which dry matter production started decreasing. The Cu uptake by banana was decreased by an increased K supply but was highest at a moderate Mg rate (Turner and Barkus 1983; Jaime et al. 1992; Daliparthy, Barker, and Mondal 1994). Copper had an antagonistic relationship with Mo at a Mo concentration above 2.5 parts per million (Iorio et al. 1998). Molybdenum accumulation in plants determines N fixation efficiency. Brodrick and Giller (1991) reported that two genotypes of *Phaseolus vulgaris* L. differed in uptake, seed storage, and distribution of Mo, and the nodule Mo content and N fixation rate of one genotype was three times greater than that of the other.

The interaction between B and Zn is antagonistic. Swietik (1995) showed that sour orange seedlings with low Zn but normal B supply developed Zn deficiency symptoms, and a high B but normal Zn supply resulted in B toxicity, which could be alleviated by a high rate of Zn application. The B uptake by plants could also be decreased by a high K application (Jaime et al. 1992; Daliparthy, Barker, and Mondal 1994). Boron is an essential micronutrient for legume-Rhizobium systems (Bolanos, Brewin, and Bonilla 1996; Redondo Nieto et al. 2001).

Chlorine is important to plant nutrition, and the interactions of this element with other nutrients in food crops and palms were reviewed by Halstead et al. (1992). Von Uexkull (1990) showed that a low Cl^- could result in slow growth and low N uptake in coconut and oil palm.

Liming caused a decrease in Mn and Zn concentrations in banana (Johns and Vimpany 1999) and P, Zn, Cu, Mn, and Fe uptake by rice, wheat, common bean, and corn (Fageria, Zimmermann, and Baligar 1995), probably because of the reduced availability of these nutrients under increased pH (Raju and Deshpande 1986b). The decreased K uptake upon liming may be due to the antagonistic effects with associated Ca and Mg (Fageria, Zimmermann, and Baligar 1995).

Daliparthy, Barker, and Mondal (1994) reviewed the interaction of K with other elements in tropical soils, distinguishing levels of interaction in the soil and plant. Cation interactions in the soils affect the availability of Ca and Mg but also of N, because of the importance of NH_4 fixation in tropical soils. Increased K supply increased the uptake rate of K and P but decreased that of N, Ca, and Mg by banana (Turner and Barkus 1983). Other effects of K were the lowering of toxic Mn or Fe levels in plants and an effect of the Na:K ratio on the performance of root crops.

Tropical soils are sensitive to trace element additions caused by management such as fertilizer use because of their low ion exchange and buffering capacities. Raven and Loeppert (1997) analyzed thirty-five elements in various soil amendments and reported that the level of contamination decreased in the order rock phosphate > sewage sludge > phosphorus fertilizers > organic amendments and liming materials > K fertilizers > N fertilizers. The move toward using regional rock phosphates for managing fertility of P-depleted tropical soils increases the risk of building up heavy metal levels in these soils.

Information on trace-element interactions in the tropics is limited because of several systematic problems. There are relatively few studies on micronutrients, most of which focus on deficiency identification and correction by blanket treatments (Takkar 1993; Davies 1997). Many studies cannot be compared because sample preparation, extractions, and analytical procedures varied. There are some fifteen different extraction procedures in use to characterize availability. Laboratory contamination may be a significant problem (Moraghan 1985).

Literature Cited

Adam, F. 1984. Crop response to lime in the Southern United States. Pp. 212–265 in *Soil acidity and liming,* edited by F. Adam. 2nd ed. Madison, Wisc.: American Society of Agronomy (ASA), Crop Science Society of America (CSSA), and Soil Science Society of America (SSSA).

Adriano, D. C. 2001. *Trace elements in terrestrial environments.* 2nd ed. New York: Springer-Verlag.

Aly, S. S. M., and S. M. Soliman. 1998. Impact of some organic acids on correcting iron chlorosis in two soybean genotypes grown in calcareous soils. *Nutrient Cycling in Agroecosystems* 51:185–191.

Alvarez-Fernandez, A., A. Garate, and J. J. Lucena. 1997. Interaction of iron chelates with several soil materials and with a soil standard. *Journal of Plant Nutrition* 20:559–572.

Banik, B. C., S. K. Sen, T. K. Bose. 1997. Effect of zinc, iron, and boron in combination with urea on growth, flowering, fruiting, and fruit quality of mango cv Fazli. *Environment Ecology* 15:122–125.

Barshad, I. 1951. Factors affecting the molybdenum content of pasture plants. 2. Effect of soluble phosphates, available nitrogen, and soluble sulfates. *Soil Science* 71:387–398.

Blair, G. J., J. R. Freney, and J. K. Park. 1990. Effect of sulfur, silicon, and trace metal interactions in determining the dynamics of phosphorus in agricultural systems. Pp. 269–280 in *Phosphorus requirements for sustainable agriculture in Asia and Oceania,* Proceedings of a symposium, 6–10 March 1989. Manila, Philippines: International Rice Research Institute.

Bloom, P. R. 1981. Metal-organic matter interactions in soil. Pp. 129–150 in *Chemistry soil environment.*. ASA Special Publication 40. Madison, Wisc.: American Society of Agronomy (ASA), Crop Science Society of America (CSSA), and Soil Science Society of America (SSSA).

Bolanos, L., N. J. Brewin, and I. Bonilla. 1996. Effects of boron on Rhizobium-legume cell-surface interactions and nodule development. *Plant Physiology* 110:1249–1256.

Boyle, F. W., Jr., and W. L. Lindsay. 1986. Manganese phosphate equilibrium relationships in soils. *Soil Science Society of America Journal* 50:588–593.

Brodrick, S. J., and K. E. Giller. 1991. Genotypic differences in molybdenum accumulation affects N_2-fixation in tropical Phaseolus vulgaris L. *Journal of Experimental Botany* 42:1339–1343.

Castilla, C. E., M. A. Ayarza, and P. A. Sanchez, 1995. Carbon and potassium dynamics in grass/legume grazing systems in the Amazon. Pp. 191–210 in *Livestock and sustainable nutrient cycling in mixed farming systems of sub-Saharan Africa,* Vol. 2. Technical Papers. Addis Ababa, Ethiopia: International Livestock Centre for Africa (ILCA).

Chuyong, G. B., D. M. Newberry, and N. C. Songwe. 2000. Litter nutrients and retranslocation in central African rain forest dominated by ectomycorrhizal trees. *New Phytologist* 148:493–510.

Dalal, R. C., M. E. Probert, A. L. Clarke, and P. B. Wylie. 1997. Soil nutrient depletion. Pp. 42–63 in *Sustainable crop production in the sub-tropics: An Australian perspective.* Brisbane, Australia: Queensland Department of Primary Industries, Information Centre.

Daliparthy, J., A. V. Barker, and S. S. Mondal. 1994. Potassium fractions with other nutrients in crops: A review focusing on the tropics. *Journal of Plant Nutrition* 17:1859–1886.

Davies, B. E. 1997. Deficiencies and toxicities of trace elements and micronutrients in tropical soils: Limitations of knowledge and future research needs. *Environmental Toxicology and Chemistry* 16:75–83.

Deiana, S., B. Manunza, A. Palma, A. Premoli, and C. Gessa. 2001. Interactions and mobilization of metal ions at the soil-root interface. In *Trace elements in the rhizosphere*, edited by G. R. Gobran, W. W. Wenzel, and E. Lombi. Boca Raton, Fl.: CRC Press.

Elmer, W. H. 1997. Influence of chloride and nitrogen form on Rhizoctonia root and crown rot of table beets. *Plant Disease* 81:635–640.

Fageria, N. K., F. J. P. Zimmermann, and V. C. Baligar. 1995. Lime and phosphorus interactions on growth and nutrient uptake by upland rice, wheat, common bean, and corn in an oxisol. *Journal of Plant Nutrition* 18:2519–2532.

Foy, C. D., R. L. Chaney, and M. C. White. 1978. The physiology of metal toxicity in plants. *Annual Review of Plant Physiology* 29:511–566.

Goldberg, S., H. S. Forster, S. M. Lesch, and E. L. Heich. 1996. Influence of anion competition on boron adsorption by clays and soil. *Soil Science* 161:99–103.

Graham, R. D., J. S. Ascher, and S. C. Hynes. 1993. Selecting zinc efficient cereal genotypes for soils of low zinc status. *Developments in Plant and Soil Science* 50:349–358.

Grewal, H. S., R. D. Graham, and J. Stangoulis. 1998. Zinc-boron interaction effects in oilseed rape. *Journal of Plant Nutrition* 21:2231–2243.

Gupta, U. C. 1972a. Interaction effects of boron and lime on barley. *Soil Science Society of America Proceedings* 36:332–334.

———. 1972b. Effects of boron and line on boron concentration and growth of forage legumes under greenhouse conditions. *Communications in Soil Science and Plant Analysis* 3:355–365.

Gupta, U. C., and J. A. MacLeod. 1981. Plant and soil boron as influenced by soil pH and calcium sources on Podzol soils in the Atlantic region of Canada, barley, pea. *Soil Science* 131:20–25.

Halstead, E. H., J. D. Beaton, J. C. W. Keng, and M. Y. Wang. 1992. Chloride: New understanding of this important plant nutrient. Pp. 314–325 in *International symposium on the role of sulphur, magnesium, and micronutrients in balanced plant nutrition (1991), Szechwan Province, China,* edited by S. Portch. Saskatoon: Potash and Phosphate Institute of Canada.

Harmsen, K., and P. L. G. Vlek. 1985. The chemistry of micronutrients in soil. *Fertilizer Research* 7:1–42.

Harter, R. D., and R. Naidu. 1995. Role of metal-organic complexation in metal sorption by soils. *Advances in Agro*nomy 35:219–263.

Heenan, D. P., and L. C. Campbell. 1983. Manganese and iron interactions on their uptake and distribution in soybean (*Glycine max* L. Merr.). *Plant Soil* 70:317–326.

Herbon, J. L., and R. A. Congdon. 1998. Ecosystem dynamics at disturbed and undisturbed sites in North Queensland wet tropical rain forest. III. Nutrient returns to the forest floor through litterfall. *Journal of Tropical Ecology* 14:217–229.

Hoffmann, I., D. Gerling, U. B. Kyiogwom, and A. Mane-Bielfeldt. 2001. Farmers' management strategies to maintain soil fertility in a remote area of northwest Nigeria. *Agriculture Ecosystems and Environment* 86:263–275.

Iorio, A. F. de., A. Rendina, M. J. Barros, M. Bargiela, A. Garcia, and A. Iriarte. 1998. Copper and molybdenum interaction in Elytrigia elongata growing on a halomorphic soil. *Journal of Plant Nutrition* 21:937–947.

Jaime, S., M. J. Subires, J. T. Soria, and A. Aguilar. 1992. Interaction K-B in avocado (*Persea americana* Mill.) culture. *Acta Horticulturae* 296:75–80.

Johns, G. G., and I. A. Vimpany. 1999. Interaction of pH amendment and potassium fertiliser on soil chemistry and banana plant growth. *Australian Journal of Agricultural Research* 50:199–210.

Jolley, V. D., K. A. Cook, N. C. Hansen, and W. B. Stevens. 1996. Plant physiological responses for genotypic evaluation in iron efficiency in strategy I and strategy II plants: A review. *Journal of Plant Nutrition* 19:1241–1255.

Kabata-Pendias, A., and H. Pendias. 2001. *Trace elements in soils and plants.* 3rd ed. Boca Raton, Fl.: CRC Press.

Kanwar, J. S., and L. J. Youngdahl. 1985. Micronutrient needs for tropical food crops. *Fertilizer Research* 7:43–67.

Katyal, J. C., and B. D. Sharma. 1980. A new technique of plant analysis to resolve iron chlorosis. *Plant and Soil* 55:105–108.

Katyal, J. C., and P. L. G. Vlek. 1985. Micronutrient problems in tropical Asia. *Fertilizer Research* 7:69–94.

Kirsch, R. K., M. E. Harward, and R. G. Petersen. 1960. Interrelationships among iron, manganese, and molybdenum in the growth and nutrition of tomatoes grown in culture solutions. *Plant and Soil* 12:259–267.

Knezek, B. D., and H. Greinert. 1971. Influence of soil iron and manganese chelate interactions upon the iron and manganese nutrition of bean plants (*Phaseolus vulgaris* L.). *Agronomy Journal* 63:617–619.

Kramer, U., and A. N. Chardonnens. 2001. The use of transgenic plants in the bioremediation of soils contaminated with trace elements. *Applied Microbiology and Biotechnology* 55:661–672.

Lal, C. B., C. Annarurna, A. S. Raghubanshi, and J. S. Singh. 2001. Foliar demand and resource economy of nutrients in dry tropical forest species. *Journal of Vegetation Science* 12:5–14.

Lata, J. C., J. Durand, R. Lensi, and L. Abbadie. 1999. Stable coexistence of contrasted nitrification statuses in a wet tropical savanna ecosystem. *Functional Ecology* 13:762–768.

León, L. A., A. S. López, and P. L. G. Vlek. 1985. Micronutrient problems in tropical Latin America. *Fertilizer Research* 7:95–130.

Liang, Y., and Z. Shen. 1994. Interaction of silicon and boron in oilseed rape plants. *Journal of Plant Nutrition* 17:415–425.

Lindsey, W. L. 1979. *Chemical equilibria in soils.* New York: John Wiley Interscience.

Lisanework, N., and A. Michelsen. 1994. Litterfall and nutrient release by decomposition in three plantations compared with a natural forest in the Ethiopian highland. *Forest Ecology and Management* 65:148–164.

Liu, J., R. J. Reid, and F. A. Smith. 2000. The mechanism of cobalt toxicity in mung beans. *Physiologia Plantarum* 110:104–110.

Mailly, D., L. Chritanty, and J. P. Kimmins. 1997. Without bamboo, the land dies:

Nutrient cycling and biogeochemistry of a Javanese bamboo talun-kebun system. *Forest Ecology and Management* 91:155–173.

Marcano-Martinez, E., and M. B. McBride. 1989. Calcium sulfate retention by two oxisols of the Brazilian cerrado. *Soil Science Society of America Journal* 53:63–69.

McLaughlin, M. J., S. J. Andrew, M. K. Smart, and E. Smolders. 1998. Effects of sulfate on cadmium uptake by Swiss chard. I. Effects of complexation and calcium competition in nutrient solutions. *Plant and Soil* 202:211–216.

Miller, E. R., X. Lei, and D. E. Ullrey. 1991. Trace elements in animal nutrition. Pp. 593–662 in *Micronutrients in agriculture*, edited by J. J. Mortvedt et al. 2nd ed. Madison, Wisc.: Soil Science Society of America.

Montagnini, F. 2000. Accumulation in above ground biomass and soil storage of mineral nutrients in pure and mixed plantations in a humid tropical lowland. *Forest Ecology and Management* 134:257–270.

Moraghan, J. T. 1985. Plant tissue analysis for micronutrient deficiencies and toxicities. *Fertilizer Research* 7:201–220.

Moussa, B. I. M., M. S. A. Dahdoh, and H. M. Shehata. 1996. Interaction effect of some micronutrients on yield, elemental composition, and oil content of peanut. *Communications in Soil Science and Plant Analysis* 27:1995–2004.

Nobel, A. D., M. E. Sumner, and A. K. Alva. 1988. The pH dependence of aluminum phytotoxicity alleviation by calcium sulfate. *Soil Science Society of America Journal* 52:1398–1402.

Nykvist, N. 1997. Total distribution of plant nutrients in a tropical rainforest ecosystem, Sabah, Malaysia. *Ambio* 26:152–157.

Oliviera-Leite, J. de, R. R. Valle, and R. de Oliviera-Leite. 1990. Nutrient cycling in the cacao ecosystem: Rain and throughfall as nutrient sources for the soil and the cacao tree. *Agriculture Ecosystems and Environment* 32:143–154.

Olsen, S. R. 1972. Micronutrient interactions. In *Micronutrients in agriculture*, edited by J. J. Mortvedt et al. Madison, Wisc.: Soil Science Society of America.

Parrotta, J. A. 1999. Productivity, nutrient cycling, and succession in single- and mixed species plantations of casuarina equisetifolia, Eucalyptus robusta, and leucaena leucocephala in Puerto Rico. *Forest Ecology and Management* 124:45–77.

Pearce, R. C., and M. E. Sumner. 1997. Apparent salt sorption reactions in an unfertilized acid subsoil. *Soil Science Society of America Journal* 61:765–772.

Poschenrieder, C., M. Llugany, and J. Barcelo. 1995. Short-term effects of pH and aluminum on mineral nutrition in maize varieties differing in proton and aluminum tolerance. *Journal of Plant Nutrition* 18:1495–1507.

Pulford, I. D. 1982. Controls on the solubility of trace metals in soils. Pp. 486–491 in Proceedings of the Ninth International Plant Nutrition Colloquium. Farnham Royal, U.K.: CAB.

Raja, V., and K. R. Reddy. 1987. Grain and straw yields and the uptake of N, P and K in sorghum (CSH-5) as affected by micronutrients under varying levels of nitrogen. *International Journal of Tropical Agriculture* 5:102–109.

Rajan, S. S. S. 1979. Adsorption of selenite, phosphate, and sulfate on hydrous alumina. *Journal of Soil Science* 34:709–718.

Rajan, S. S. S., and J. H. Watkinson. 1976. Adsorption of selenite and phosphate on an allophane clay. *Soil Science Society of America Journal* 40:51–54.

Raju, T., and P. B. Desphande. 1986a. Phosphorus and zinc interactions in coffee (*Coffea arabica* L.) seedlings. *Journal of Coffee Research* 16:1–13.
———. 1986b. Lime-zinc interactions in coffee (*Coffea arabica* L.) seedlings. *Journal of Coffee Research* 16:79–88.
Randhawa, N. S., and J. C. Katyal. 1982. Micronutrients management for submerged rice soils. Vertisols and rice soils of the tropics. Symposia papers II. Pp. 192–211 in *Transactions of the 12th International Congress of Soil Science.* City?: International Soil Science Society.
Raven, K. P., and R. H. Loeppert. 1997. Trace element composition of fertilizers and soil amendments. *Journal of Envionmental Quality* 26:551–557.
Redondo Nieto, M., R. Rivilla, A. El Hamdaoui, I. Bonilla, and L. Bolanos. 2001. Boron deficiency affects early infection events in the pea-Rhizobium symbiotic interaction. *Australian Journal of Plant Physiology* 28:819–823.
Reinbott, T. M., and D. G. Blevins. 1997. Phosphorus and magnesium fertilization interaction with soil phosphorus level: Tall fescue yield and mineral element content. *Journal of Production Agriculture* 10:260–265.
Reisenauer, H. M. 1994. The interactions of manganese and iron. Pp. 147–164 in *Biochemistry of metal micronutrients in the rhizosphere,* edited by J. A. Manthey, D. E. Crowley, and D. G. Luster. Boca Raton, Fl.: CRC Press.
Rodriguez-Jimenez, L. V. A. 1988. Consideraciones sobre la biomasa, composicion quimica y dinamica del bosque pluvial tropical de colinas bajas, Bajo Calima, Buenaventura, Colombia. In Serie Documentacion, Corporacion Nacional de Investigacion y Fomento Forestal n. 16. Bogota, Colombia.
Roomizadeh, S., and N. Karimian. 1996. Manganese-iron relationship in soybean grown in calcareous soils. *Journal of Plant Nutrition* 19:397–406.
Shang, C., and T. E. Bates. 1987. Comparison of zinc soil tests adjusted for soil and fertilizer phosphorus. *Fertilizer Research* 11:209–220.
Shanjida, K., S. M. Ullah, and K. S. Sarwar. 2001. Interaction of chromium and copper with nutrient elements in rice (Oryza sativa cv. BR-11). *Bulletin of the Institute of Tropical Agriculture,* Kyushu University 23:35–39.
Shanmughavel, P., L. Q. Sha, Z. Zheng, and M. Cao. 2001. Nutrient cycling in a tropical seasonal rainforest of Xishuangbanna, Southwest China. Part 1: Trees species, nutrient distribution and uptake. *Bioresource Technology* 80:163–170.
Shim, S. C., and P. B. Vose. 1965. Varietal differences in the kinetics of iron uptake by excised rice roots. *Journal of Experimental Botany* 16:216–232.
Sillanpää, M. 1982. Micronutrients and the nutrient status of soils: A global study. *FAO Soils Bulletin* 264. Rome: Food and Agriculture Organization of the United Nations.
Sillanpää, M., and P. L. G. Vlek. 1985. Micronutrient problems in tropical Asia. *Fertilizer Research* 7:151–168.
Silver, W. L., F. N. Scatena, A. H. Johnson, T. G. Siccama, and M. J. Sanchez. 1994. Nutrient availability in a montane wet tropical forest: Spatial patterns and methodological considerations. *Plant and Soil* 164:129–145.
Silver, W. L., J. Neff, M. McGroddy, E. Veldkamp, M. Keller, and R. Cosme. 2000. Effects of soil texture on belowground carbon and nutrient storage in a lowland Amazon forest ecosystem. *Ecosystems* 3:193–209.
Singh, L., and J. S. Singh. 1991. Storage and flux of nutrients in a dry tropical forest in India. *Annals of Botany* 68:275–284.

Singh, M., N. Singh, and P. S. Relan. 1981. Adsorption and desorption of selenite and selenate selenium on different soils: Normal, calcareous, high organic carbon saline, and alkali. *Soil Science* 132:134–141.

Smith, C., K. W. Brown, and L. E. Deuel Jr. 1987. Plant availability and uptake of molybdenum as influenced by soil type and competing ions. *Journal of Environmental Quality* 16:377–382.

Smith, E., R. Naidu, and A. M. Altson. 1998. Arsenic in the soil environment: A review. *Advances in Agronomy* 64:149–195.

Smolders, E., and M. J. McLaughlin. 1996. Chloride increases cadmium uptake in Swiss chard in a resin-buffered nutrient solution. *Soil Science Society of America Journal* 60:1443–1447.

Spaargarden, O. 1989. Management of acid soils in Africa. Pp. 32–35 in *Management of acid soils in the humid tropics of Asia*, edited by E. T. Craswell and E. Pushparajah. ACIAR Monograph No. 13 (IBSRAM Monograph No.1). Bangkok: International Soil Management Board.

Stevenson, F. J. 1991. Organic matter-micronutrient reactions in soil. Pp. 145–186 in *Micronutrients in agriculture*, edited by J. J. Mortvedt et al. 2nd ed. Madison, Wisc.: Soil Science Society of America.

Stumm, W., and J. J. Morgan. 1981. *Aquatic chemistry*. 2nd ed. New York: John Wiley and Sons.

Swietik, D. 1995. Interaction between zinc deficiency and boron toxicity on growth and mineral nutrition of sour orange seedlings. *Journal of Plant Nutrition* 18:1191–1207.

Takkar, P. N. 1993. Requirement and response of crop cultivars to micronutrients in India: A review. *Developments in Plant and Soil Science.* 50:341–348.

Tiessen, H., P. Chacon, and E. Cuevas. 1994. Phosphorus and nitrogen status in soils and vegetation along a toposequence of dystrophic rainforests on the upper Rio Negro. *Oecologia* 99:145–150.

Turner, D. W., and B. Barkus. 1983. Long-term nutrient absorption rates and competition between ions in banana in relation to supply of K, Mg and Mn. *Fertilizer Research* 4:127–134.

Von Uexkull, H. R. 1990. Chloride in the nutrition of coconut and oil palm. Pp. 134–139 in *Transactions of the 14th International Congress of Soil Science, Kyoto, Japan, 12–18 August 1990.* Vol. 4. City?: International Soil Science Society.

Wang, D., F. H. Bormann, A. E. Lugo, and R. D. Bowden. 1991. Comparison of nutrient efficiency and biomass production in five tropical tree taxa. *Forest Ecology and Management* 46:1–21.

Webb, M. J., and J. F. Loneragan. 1990. Zinc translocation to wheat roots and its implications for a phosphorus/zinc interaction in wheat plant. *Journal of Plant Nutrition* 13:1498–1512.

Zaharieva, T., D. Kasabov, and V. Romheld. 1988. Responses of peanut to iron-manganese interaction in calcareous soil. *Journal of Plant Nutrition* 11:6–11.

Zhang, G., M. Fukami, and H. Sekimoto. 2000. Genotypic differences in effects of cadmium on growth and nutrient compositions in wheat. *Journal of Plant Nutrition* 23:1337–1350.

PART IV
Atmosphere

13

Element Interactions and Trace Gas Exchange

Torben R. Christensen and Michael Keller

The climate of the Earth is controlled by a series of complex interactions between incoming solar radiation, reflectance in the atmosphere, surface albedo, and the absorption by greenhouse gases (most notably H_2O and the biogenic trace gases CO_2, CH_4, and N_2O) of infrared radiation coming back out from the Earth's surface. The latter process is heating the lower atmosphere, making the climate of the Earth unique compared with neighboring planets in the solar system and making life possible on Earth. Understanding the role the trace gases CO_2, CH_4, and N_2O play in the climate system is important for predictions of future climate. This point is accentuated because these gases are the ones most dramatically affected by human activities, with annual concentration increases of 0.25 percent per year for N_2O and 0.4 percent per year for CO_2 (IPCC 2001). The growth rate of CH_4 in the atmosphere was about 1 percent per year during the 1980s but has recently taken some so far unexplained detours from this rate of increase (IPCC 2001).

As of 1998 the mean atmospheric concentrations of CO_2, CH_4, and N_2O were 365 parts per million (ppm), 1.75 ppm, and 0.314 ppm respectively (IPCC 2001). The relative atmospheric concentrations of the trace gases are, however, of little use for evaluating the importance of these gases as their individual global warming potentials (GWPs) vary. For CH_4 and N_2O the GWPs are 23 and 296 over a 100-year time horizon. These numbers increase or decrease depending on the time horizon used (IPCC 2001).

Although the trace gases are involved in this complex interaction with the physical climate system, they also have important interactions among themselves and in their exchanges between ecosystems and the atmosphere. Important element interactions cause ecosystem CO_2 exchange, often phrased as the "carbon balance," to affect, in certain ecosystems, the exchange of both CH_4 and N_2O.

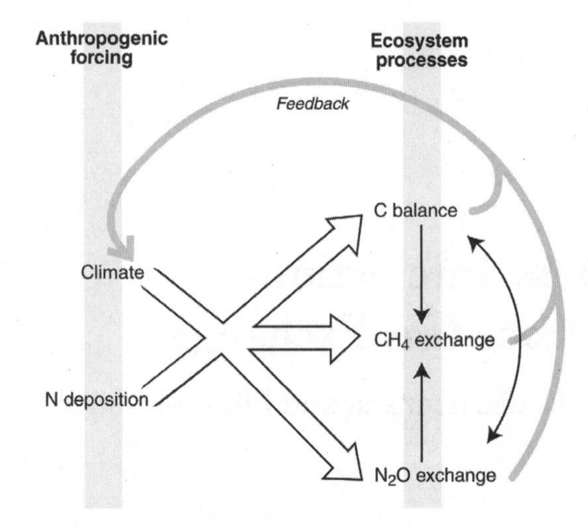

Figure 13.1. Graphic illustration of how anthropogenic forcing and ecosystem processes interact. In this chapter the focus is on the interactions between climate and the trace gas exchanges. The linkages, however, go beyond these and also include the effects of other anthropogenic forcing such as N deposition as discussed in Austin et al. (Chapter 2) and Holland and Carroll (Chapter 15; both this volume).

N_2O and CH_4 are in turn strongly regulated by nutrient cycling (especially nitrogen cycling), a strong control on processes controlling the net ecosystem CO_2 uptake. It is important to consider all major trace gases and the interactions between them when evaluating the effect of ecosystem trace gas exchange on the greenhouse forcing of the Earth's atmosphere (Figure 13.1). The integrated effect of the trace gases may then in turn be evaluated in relation to the energy exchange (albedo) and other physical processes associated with the Earth surface (Betts 2000). This chapter will briefly outline a number of processes and situations where the element interactions in relation to trace gas exchanges are of pivotal importance. The emphasis will be on terrestrial ecosystem exchange of the three major biogenic trace gases: CO_2, CH_4, and N_2O.

Scaling Integrated Effects

The exchanges of trace gases between ecosystems and the atmosphere are most easily and frequently quantified with techniques covering a rather small spatial scale (typically less than one square meter). Major developments in micrometeorological techniques using masts and aircrafts (e.g., Lindroth, Grelle, and Moré 1998; Oechel et al. 2000; Vourlitis et al. 2001) during recent decades have made it possible to measure trace gas exchanges at larger scales (typically around one hectare for masts and thousands of hectares for aircraft) and with a high time resolution. This advance has allowed huge improvements in our ability to estimate annual budgets for landscape-scale carbon exchange. The use of remote sensing has been applied to scaling even further up to regional and global scales (Ranson et al. 1998; Soegaard et al. 2000). Although the micrometeorological techniques have been introduced as powerful tools for measuring also CH_4 and N_2O fluxes (Christensen et al. 1996; Suyker et al. 1996; Friborg et al. 2000), most scaling exercises have so far been conducted with attention to net CO_2 exchange and carbon cycling (Desjardins et al. 1998; Oechel et al. 2000; Soegaard et al. 2000).

Relatively few studies have integrated the landscape to regional scale interactions between multiple trace gases in their radiative forcing of the atmosphere. Robertson, Paul, and Harwood (2000) studied intensive agriculture in the Midwest of the United States for the contributions of individual gases to the greenhouse forcing. They found that nitrous oxide was the single biggest source of forcing, superior to both CO_2 and CH_4. Similar calculations for other ecosystem types are hindered largely because of the scarcity of fully comparable annual flux estimates for all three gases. In this chapter we will, however, make an attempt at synthesizing the relative contribution of the individual gases in different natural, seminatural, and agricultural ecosystems based on generalized flux numbers from different studies.

Soil C and Interactions with Trace Gas Exchanges

Aboveground biomass cannot increase indefinitely. Ultimately the most important storage place for organic carbon is the soil, and it holds the largest potential for acting as a long-term sink (or source) for atmospheric CO_2. The global soil carbon pools are detailed in Table 13.1. The total amount of carbon in soils globally is about twice the atmospheric load of CO_2. Clearly, the potential for the soils to affect atmospheric CO_2 is huge. Changes in soil carbon storage, however, will not occur in isolation from changes in other trace gas exchanges as indicated in Table 13.1.

Methane is produced through anaerobic decomposition in soils and sediments. Slow decomposition rates generally cause nontidal wetlands to accumulate carbon as dead soil organic matter and/or peat; in other words, these systems act as an atmospheric CO_2 sink. At the same time, wetlands and rice paddies emit about 175 teragrams (Tg) CH_4 annually (Table 13.1). This source is highly sensitive to temperature (Christensen et al. 2003) so that changes in the global climate toward warmer conditions could, depending on the moisture changes, lead to significant increases in CH_4 emissions and a positive feedback causing further

Table 13.1. Global ecosystem soil carbon storage and associated atmospheric exchanges of the trace gases CH_4 and N_2O

Ecosystem type	*Soil C (10^{15} g C)*	*CH_4 exchanges (10^{12} g CH_4/year)*	*N_2O emissions (10^{12} g N/year)*
Tropical forest	255		3
Temperate and boreal forest	321	-10	1
Woodland and shrubland	59	-5	
Savanna	56		1
Temperate grassland	173	-10	1
Tundra and wetland	310	115	
Desert scrub	101	-5	
Cultivated land	178	60	3.5
Total ecosystem associated	1,453	Source 175[a], sink -30	9.5[b]

Note: Most processes that would change the C storage are also likely to alter the trace gas exchanges (extracted from Schlesinger 1997; Prather et al. 1995).
[a]Out of total global sources of 535×10^{12} g CH_4/yr.
[b]Out of total global identified sources of 15.7×10^{12} g N/yr.

warming (Christensen and Cox 1995). A significant drying of current wetland areas, however, would lead to diminished CH_4 emissions and most likely to a net release of CO_2 to the atmosphere.

In rice paddies the annual net emissions are strongly dependent on the management regimes. Amendments of rice straw, a C source to paddy soils, consistently increases CH_4 production (Rennenberg et al. 1992; Wassmann, Papen, and Rennenberg 1993). The additions of N fertilizers give varied responses in CH_4 emissions. For example, in an early study, $(NH_4)_2SO_4$ additions caused a fivefold increase in methane emissions compared with unfertilized fields in California (Cicerone and Shetter 1981). Possibly, increased productivity led to increases in the carbon available to methanogens. Schütz et al. (1989) found, however, that the same fertilizer composition, when incorporated into soils of Italian rice paddies, led to a 50 percent reduction in CH_4 emission. The addition of N to rice paddies has important implications with respect to trace gases when the paddies are drained either within the cropping season or during the fallow period. Drainage leads to reductions in CH_4 emissions that may be accompanied by increased fluxes of N_2O when N availability is high (as will be discussed). Processes controlling trace gas exchanges are, hence, linked, and they carry significant potentials for feedback mechanisms on climate change, whether directly in response to climate forcing or driven by human-induced land management schemes.

Process Linkages

Carbon cycling and CH_4 emissions from wetlands and rice paddies are intimately linked at the process level. In some cases vascular plants act as conduits for transport of supersaturated CH_4 in pore water to the atmosphere (Schimel 1995; Kelker and Chanton 1997). Because they transport oxygen to their roots, some species may actively oxidize the immediate area surrounding their roots (Frenzel and Rudolph 1998); this oxidation inhibits the methane formation and facilitates methane oxidation. These functional aspects of the vascular plants vary from species to species so the plant composition (i.e., the diversity of plants present) represents an important controlling factor for the net CH_4 emissions. More important, vascular and nonvascular plants also provide substrate for the methanogenesis as a function of their production (net CO_2 exchange) (Joabsson, Christensen, and Wallén 1999). The multiple ways in which CO_2 cycling and productivity in vascular plants may affect CH_4 exchange in wetland ecosystems are illustrated in Colorplate 2.

Emissions of N_2O from soils can follow from both nitrification and denitrification (Umarov 1990), but it is difficult in practice to separate these sources (Arah 1997). With losses in the order of 3.5 Tg N per year as N_2O emissions, agriculture is by far the most significant anthropogenic source of atmospheric N_2O (Kroeze, Mosier, and Bouwman 1999). These emissions are closely associated with agricultural management practices, particularly fertilization, which may change in response to local and/or national policies (Matthews 1994). Management strategies may increase or even initiate N_2O emissions in order to achieve other goals such as increased plant productivity.

Linkages between productivity, CH_4 emissions, and N_2O releases are illustrated in the rice paddies where increased N_2O fluxes has been observed following draining. Bronson et al. (1997) presented an especially convincing case study made using high temporal resolution automatic chambers. They studied the effects of four amendment regimes ($(NH_4)_2SO_4$, urea, green manure + urea, and rice straw + urea) with and without midseason drainage of the paddy on research plots in the Philippines. Regardless of the amendment, midseason drainage led to a strong reduction of CH_4 emission. The drained fields, however, particularly when fertilized with nitrogen, emitted far more N_2O during the growing season. Bronson et al. (1997) calculated the net greenhouse gas balance comparing plots with and without midseason drainage for both urea and straw + urea treatment. These results are recalculated in Table 13.2 using present-day values for the 100-year GWP. In the urea fertilization case the excess N_2O produced in the drained paddies nearly completely offsets the advantage of the reduction in CH_4 emission. In contrast, in the rice straw–amended paddies, where CH_4 emissions are greater,

Table 13.2. The net greenhouse gas balance of rice paddies studied by Bronson et al. (1997)

Treatment	*N_2O emisssion (mg m^{-2})*	*CH_4 emission (mg m^{-2})*	*Net CO_2 equivalent (g m^{-2})*
Drained urea	100.3	667	45
Undrained urea	40.7	1,733	52
Drained - undrained urea	59.6	-1,067	-7
Drained straw	25.9	21,200	495
Undrained straw	4.4	37,067	854
Drained - undrained straw	21.5	-15,867	-359

Note: GWPs of 296 and 23 were used for N_2O and CH_4, respectively, using 100-year integrations to arrive at the net CO_2 equivalent.

the drainage results in a net greenhouse gas balance that is equivalent to a carbon uptake of 359 g C per square meter.

As with CH_4 emissions from wetlands, N_2O emissions from tropical forests have also been shown to be strongly affected by plant species composition. Erickson, Keller, and Davidson (2001) and Erickson, Davidson, and Keller (2002) found that species determination of litter C:N ratios significantly influences N oxide fluxes across a wide range of tropical forest environments. Vascular plant species composition in grasslands also strongly influences element interactions including trace gas fluxes (Craine, Wedin, and Reich 2001). Plant composition and biodiversity is therefore intimately coupled to the element interactions (Austin et al., Chapter 2, this volume), in this context the trace gas exchanges.

An associated example of element interactions in relation to trace gas exchange is the case of N_2O exchanges and forest productivity. Fertilization may increase the CO_2 sink capacity of forests by increasing productivity (Bergh et al. 1999) and also humus formation in the long term (Berg and Meentemeyer 2002). Although the emissions of N_2O are highly spatially variable (Ambus and Christensen 1994), they are generally strongly increased by fertilization (Smith and Dobbie 2001). The theory of ecosystem nitrogen saturation suggests that there is a threshold following which N_2O emissions and N-leaching may increase significantly (Mosier et al. 1991; Aber et al. 1989, 1998; Moldan et al., Chapter 5, this volume). This threshold has already been surpassed in tropical forests where N additions are easily lost (Hall and Matson 1999; Matson, Lohse, and Hall 2002).

Fertilization may, therefore, very well lead to increases in the C sinks, but it is likely that this may be outweighed by an associated increase in N_2O emissions. An estimated quantification of this increase can be found in Austin et al. (Chapter 2, this volume) and also briefly as discussed here.

Trace Gas Forcing Interactions

Following from the last process example mentioned in the previous section, let us make a simple calculation quantifying the consequences. If the normal practice of fertilization in agriculture is applied to forests and the soils respond similarly, this could in due course (following threshold effects, etc.) induce an emission in the order of 5 kg N_2O-N per hectare per year (Groffman et al. 2000), which equals 7.86 kg N_2O per hectare per year. Using a GWP for N_2O of 296, this equals 2.3 tonnes CO_2 per hectare per year or 0.6 tonnes C per hectare per year. The average northern hemisphere forest C sink activity is estimated at 0.7 tonnes C per hectare per year (Goodale et al. 2002) so unless C uptake was more than doubled, fertilizer induced N_2O emissions has the potential to cancel out the CO_2 sink activity on a GWP basis. A reduction of net CO_2 fluxes by 0.6 tonnes C per hectare per year would potentially turn almost half of the twenty-six measurement-years in the EUROFLUX project (Valentini et al. 2000) from negative (if only CO_2 is considered) to positive forcing of greenhouse warming.

A further aggravating aspect, in the sense already mentioned, is that addition of ammonia-based fertilizer to aerobic soils generally leads to a decrease in the atmospheric CH_4 uptake that usually takes place (Steudler et al. 1989; Mosier et al. 1991). Dry ecosystem CH_4 uptake, although minor compared with the wet ecosystem emissions, is still a significant number in the global methane budget. Widespread fertilizer additions hence have the effect of not only inducing or increasing N_2O emissions, but also generally reducing atmospheric CH_4 uptake, with both processes representing increasing radiative forcing.

Wetlands provide further good examples when illustrating the importance of element interactions for the greenhouse forcing of the Earth. Many measurements show that typical wetland habitats, including the huge west Siberian wetlands, are significant net sinks for atmospheric CO_2 (Friborg et al., submitted). In the long term this sink has been estimated at around 20 grams C per square meter per year (Gorham 1991; Turunen et al. 2001), or the equivalent of 73 grams CO_2 per square meter per year. The same wetlands, however, typically emit CH_4 in the range of at least 10–20 grams CH_4 per square meter per year (e.g., Panikov and Dedysh 2000). Given a 100-year GWP of CH_4 of 23, this is equivalent to emissions of 230–460 grams CO_2 per square meter per year. The CH_4 emissions are, in other words, having a significantly greater and opposing impact on the atmosphere compared with the CO_2 sink. Therefore, under natural conditions, depending somewhat on the timeframe used in the GWP calculation (Figure 13.2), these

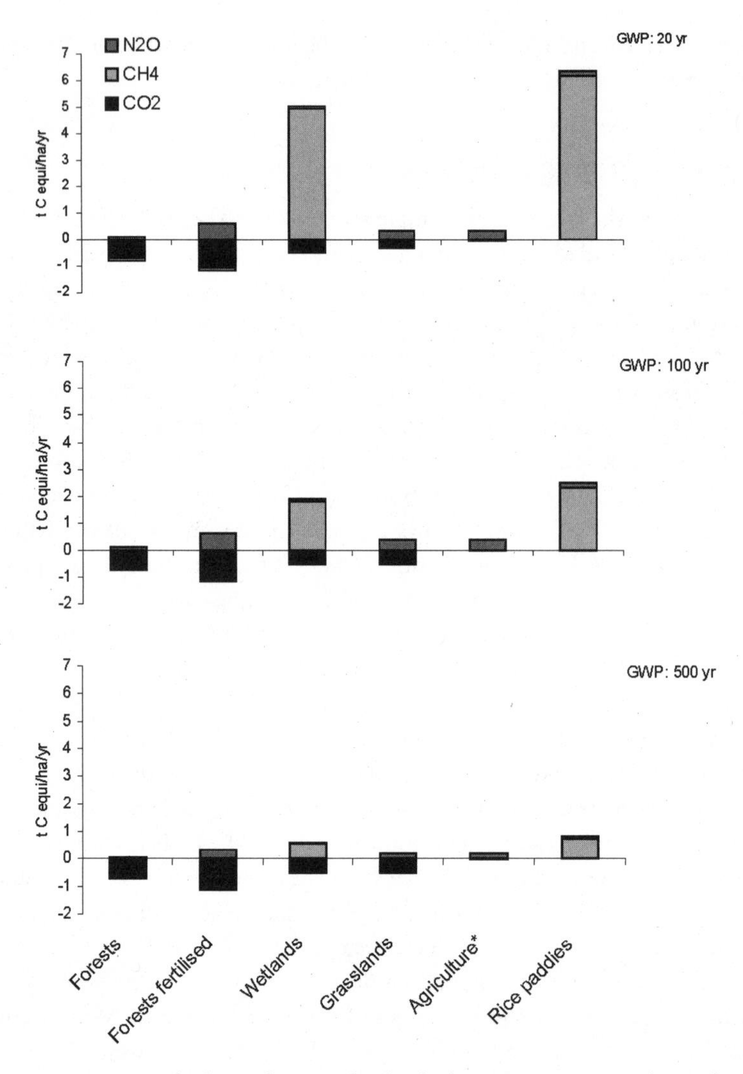

Figure 13.2. Estimated relative forcing of individual trace gases expressed as carbon equivalents. A positive number means a net source of greenhouse forcing. The GWPs used for CH_4 and N_2O relative to CO_2 for the respective time horizons are according to IPCC (2001): 20 years—CH_4: 62, N_2O: 275; 100 years—CH_4: 23, N_2O: 296; 500 years—CH_4: 7, N_2O: 156. Flux numbers are extracted from Goodale et al. (2002), Whalen, Reeburgh, and Kizer (1991), Schultze et al. (2001), Bergh et al. (1999), Steudler et al. (1989), and Panikov and Dedysh (2000). Asterisk (*) indicates agriculture excluding rice paddies.

wetlands are generally carbon sinks while at the same time sources of greenhouse warming because of the counteracting CH_4 emissions. From a climatic perspective it is important to evaluate the whole range of trace gas exchanges for an ecosystem and how they interact.

In Figure 13.2 we have compiled general numbers on annual fluxes of the trace gases CO_2, CH_4, and N_2O between different vegetation types and the atmosphere. The numbers are calculated as C-equivalents based on the global warming potentials of a 20-, 100-, and 500-year time horizon (IPCC 2001), respectively. From these general numbers it can be seen that when all individual trace gases are taken into account few ecosystem types have a net negative effect in terms of radiative forcing of the Earth. In the shorter time perspectives the methane-emitting wetlands and rice paddies demonstrate forceful impacts on the climate. Because of the relatively short lifetime of CH_4 in the atmosphere (about 12 years) the temporal perspective becomes important, as illustrated by the marked difference in wetland and rice paddy forcing comparing 20- and 500-year GWP calculations. According to this very crude analysis unfertilized forest is the only potential candidate for a small but consistent long-term sink for atmospheric radiative forcing. Forests considered in this analysis, however, are mainly aggrading secondary forests. The same may not be true for mature forests.

Literature Cited

Aber, J. D., K. J. Nadelhoffer, P. Steudler, and J. M. Melillo. 1989. Nitrogen saturation in northern forest ecosystems. *BioScience* 39:378–386.

Aber, J., W. McDowell, K. Nadelhoffer, A. Magill, G. Berntson, M. Kamakea, S. McNulty, W. Currie, L. Rustard, and I. Fernandez. 1998. Nitrogen saturation in temperate forest ecosystems: Hypothesis revisited. *BioScience* 48:921–934.

Ambus, P., and S. Christensen. 1994. Measurement of N_2O emission from a fertilized grassland: An analysis of spatial variability. *Journal of Geophysical Research* 99:16549–16555.

Arah, J. R. M. 1997. Apportioning nitrous oxide fluxes between nitrification and denitrification using gas-phase mass spectrometry. *Soil Biology and Biochemistry* 29:1295–1299.

Berg, B., and V. Meentemeyer. 2002. Litter quality in a north European transect versus carbon storage potential. *Plant and Soil* 242:83–92.

Bergh, J., S. Linder, T. Lundmark, and B. Elfving. 1999. The effect of water and nutrient availability on the productivity of Norway spruce in northern and southern Sweden. *Forest Ecology and Management* 119:51–62.

Betts, R. A. 2000. Offset of the potential carbon sink from boreal forestation by decreases in surface albedo. *Nature* 408:187–189.

Bronson, K. F., U. Singh, H.-U. Neue, and E. B. Abao Jr. 1997. Automated chamber measurements of methane and nitrous oxide flux in a flooded rice soil: I. Residue, nitrogen, and water management. *Soil Science Society of America Journal* 61:981–987.

Christensen, T. R., A. Joabsson, L. Ström, N. Panikov, M. Mastepanov, M. Öquist, B. H. Svensson, H. Nykänen, P. Martikainen, and H. Oskarsson. Factors controlling large scale variations in methane emissions from wetlands. *Geophysical Research Letters*, in press.

Christensen, S., P. Ambus, R. M. Arah, H. Clayton, B. Galle, D. W. Griffith, K. J. Hargreaves, L. Klemedtsson, A.-M. Lind, M. Maag, A. Scott, U. Skiba, K. A. Smith, M. Welling, and F. G. Wienhold. 1996. Nitrous oxide emission from an agricultural field: Comparison between measurements by flux chamber and micrometerological techniques. *Atmospheric Environment* 30:4183–4190.

Christensen, T. R., and P. Cox. 1995. Response of methane emission from Arctic tundra to climatic change: Results from a model simulation. *Tellus* 47B:301–310.

Cicerone, R. J., and J. D. Shetter. 1981. Sources of atmospheric methane: measurements in rice paddies and a discussion. *Journal of Geophysical Research* 86:7203–7209.

Craine, J. M., D. A. Wedin, and P. B. Reich. 2001. The response of soil CO_2 flux to changes in atmospheric CO_2, nitrogen supply and plant diversity. *Global Change Biology* 7:947–953.

Desjardins, R. L., J. L. MacPherson, L. Mahrt, P. Schuepp, E. Pattey, H. Neumann, D. Baldocchi, S. Wofsy, D. Fitzjarrald, H. McCaughey, and D. W. Joiner. 1998. Scaling up flux measurements for the boreal forest using aircraft-tower combinations. *Journal of Geophysical Research* 102:29125–29134.

Erickson, H., E. A. Davidson, and M. Keller. 2002. Former land-use and tree species affect nitrogen oxide emissions from a tropical dry forest. *Oecologia* 130:297–308.

Erickson, H. E., M. Keller, and E. A. Davidson. 2001. Nitrogen oxide fluxes and ecosystem nitrogen cycling during postagricultural succession and forest fertilization in the humid tropics. *Ecosystems* 4:67–84.

Frenzel, P., and J. Rudolph. 1998. Methane emission from a wetland plant: The role of CH_4 oxidation in Eriophorum. *Plant and Soil* 202:27–32.

Friborg, T., H. Soegaard, T. R. Christensen, C. R. Lloyd, and N. S. Panikov. Siberian wetlands: Source or sink of greenhouse warming? *Geophysical Research Letters*, submitted.

Friborg, T., T. R. Christensen, B. U. Hansen, C. Nordstrøm, and H. Søgaard. 2000. Trace gas exchange in a high arctic valley 2: Landscape CH_4 fluxes measured and modelled using eddy correlation data. *Global Biogeochemical Cycles* 14:715–724.

Goodale, C. L., M. J. Apps, R. A. Birdsey, C. B. Field, L. S. Heath, R. A. Houghton, J. C. Jenkins, G. H. Kohlmaier, W. Kurz, S. Liu, G.-J. Nabuurs, S. Nilsson, and A. Z. Shvidenko. 2002. Forest carbon sinks in the northern hemisphere. *Ecological Applications* 12:891–899.

Gorham, E. 1991. Northern peatlands: Role in the carbon cycle and probable responses to climatic warming. *Ecological Applications* 1:182–195.

Groffman, P. M., R. Brumme, K. Butterbach-Bahl, K. E. Dobbie, A. R. Mosier, D. Ojima, H. Papen, W. J. Parton, K. A. Smith, and C. Wagner-Riddle. 2000. Evaluating annual nitrous oxide fluxes at the ecosystem scale. *Global Biogeochemical Cycles* 14:1061–1070.

Hall, S. J., and P. A. Matson. 1999. Nitrogen oxide emissions after nitrogen additions in tropical forests. *Nature* 400:152–155.

IPCC (Intergovernmental Panel on Climate Change). 2001. *Climate change 2001: The scientific basis. Contribution of Working Group I to the Third Assessment Report of the Intergovernmental Panel on Climate Change,* edited by J. T. Houghton, Y. Ding, D. J. Griggs, M. Noguer, P. J. van der Linden, X. Dai, K. Maskell, and C. A. Johnson. Cambridge: Cambridge University Press.

Joabsson, A., and T. R. Christensen. 2002. Wetlands and methane emission. In *Encyclopedia of Soil Science,* edited by R. Lal. New York: Marcel Dekker.

Joabsson, A., T. R. Christensen, and B. Wallén. 1999. Vascular plant controls on methane emissions from northern peatforming wetlands. *Trends in Ecology and Evolution* 14:385–388.

Kelker, D., and J. Chanton. 1997. The effect of clipping on methane emissions from Carex. *Biogeochemistry* 39:37–44.

Kroeze, C., A. Mosier, and L. Bouwman. 1999. Closing the global N_2O budget: A retrospective analysis 1500–1994. *Global Biogeochemical Cycles* 13:1–8.

Lindroth, A., A. Grelle, and A.-S. Moré. 1998. Long-term measurements of boreal forest carbon balance reveal large temperature sensitivity. *Global Change Biology* 4:443–450.

Matson, P., K. A. Lohse, and S. J. Hall. 2002. The globalization of nitrogen deposition: Consequences for terrestrial ecosystems. *Ambio* 2002:113–119.

Matthews, E. 1994. Nitrogenous fertilizers: Global distribution of consumption and associated emissions of nitrous oxide and ammonia. *Global Biogeochemical Cycles* 8:411–439.

Mosier, A. R., D. S. Schimel, D. W. Valentine, K. Bronson, and W. J. Parton. 1991. Methane and nitrous oxide fluxes in native, fertilized, and cultivated grasslands. *Nature* 350:330–332.

Oechel, W. C., G. Vourlitis, J. Verfaillie, T. Crawford, S. Brooks, E. Dumas, A. Hope, D. Stow, B. Boynton, V. Nosov, and R. Zulueta. 2000. A scaling approach for quantifying the net CO_2 flux of the Kuparuk River Basin, Alaska. *Global Change Biology* 6:160–173.

Panikov, N. S., and S. N. Dedysh. 2000. Cold season CH_4 and CO_2 emission from boreal peat bogs (West Siberia): Winter fluxes and thaw activation dynamics. *Global Biogeochemical Cycles* 14:1071–1080.

Prather, M. J., R. Derwent, D. Ehhalt, P. Fraser, E. Sanhueza, and X. Zhou. 1995. Other trace gases and atmospheric chemistry. In *Climate change 1994,* edited by J. T. Houghton et al. Cambridge: Cambridge University Press.

Ranson, K. J., G. Sun, R. H. Lang, N. S. Chauhan, R. J. Cacciola, and O. Kilic. 1998. Mapping of boreal forest biomass from spaceborne synthetic aperture radar. *Journal of Geophysical Research* D102:29599–29610.

Rennenberg, H., R. Wassmann, H. Papen, and W. Seiler. 1992. Trace gas exchange in rice cultivation. *Ecological Bulletins* 42:164–173.

Robertson, G. P., E. A. Paul, and R. R. Harwood. 2000. Greenhouse gases in intensive agriculture: Contributions of individual gases to the radiative forcing of the atmosphere. *Science* 289:1922–1925.

Schimel, J. P. 1995. Plant transport and methane production as controls on methane flux from arctic wet meadow tundra. *Biogeochemistry* 28:183–200.

Schlesinger, W. H. 1997. *Biogeochemistry: An analysis of global change.* San Diego: Academic Press.

Schultze, E. D., A. J. Dolman, P. Jarvis, R. Valentini, P. Smith, P. Ciais, J. Grace, S. Linder, and C. Brüning, eds. 2001. *The carbon sink: Absorption capacity of the European terrestrial biosphere.* Brussels: European Commission.

Schütz, H., A. Holzapfel Pschorn, R. Conrad, H. Rennenberg, and W. Seiler. 1989. A three year continuous record on the influence of daytime, season, and fertilizer treatment on methane emission rates from an Italian rice paddy field. *Journal of Geophysical Research* 94:16405–16416.

Smith, K. A., and K. E. Dobbie. 2001. The impact of sampling frequency and sampling times on chamber-based measurements of N_2O emissions from fertilized soils. *Global Change Biology* 8:933–946.

Soegaard, H., T. Friborg, B. U. Hansen, C. Nordstrøm, and T. R. Christensen. 2000. Trace gas exchange in a high arctic valley 3: Integrating and scaling CO_2 fluxes from canopy to landscape using flux data, footprint modelling, and remote sensing. *Global Biogeochemical Cycles* 14:725–744.

Steudler, P. A., R. D. Bowden, J. M. Mellillo, and J. D. Aber. 1989. Influence of nitrogen fertilization on methane uptake in temperate forest soils. *Nature* 341:314–316.

Suyker, A. E., S. B. Verma, R. J. Clement, and D. P. Billesback. 1996. Methane flux in a boreal fen: Season-long measurements by eddy correlation. *Journal of Geophysical Research* 101:28637–28647.

Turunen, J., A. Pitkänen, T. Tahvanainen, and K. Tolonen. 2001. Carbon accumulation in West Siberian mires, Russia. *Global Biogeochemical Cycles* 15:285–296.

Umarov, M. M. 1990. Biotic sources of nitrous oxide (N_2O) in the context of the global budget of nitrous oxide. Pp. 263–268 in *Soils and the greenhouse effect,* edited by A. F. Bouwman. Chichester, U.K.: John Wiley and Sons.

Valentini, R., G. Matteucci, A. J. Dolman, E.-D. Schulze, C. Rebmann, E. J. Moors, A. Granier, P. Gross, N. O. Jensen, K. Pilegaard, A. Lindroth, A. Grelle, Ch. Bernhofer, T. Grünwald, M. Aubinet, R. Ceulemans, A. S. Kowalski, T. Vesala, Ü. Rannik, P. Berbigier, D. Loustau, J. Gumundsson, H. Thorgeirsson, A. Ibrom, K. Morgenstern, R. Clement, J. Moncrieff, L. Montagnani, S. Minerbi, and P. G. Jarvis. 2000. Respiration as the main determinant of European forests carbon balance. *Nature* 404:861–865.

Vourlitis, G. L., N. Priante Filho, M. M. S. Hayashi, J. S. Nogueira, F. T. Caseiro, and J. H. Campelo. 2001. Seasonal variations in the net ecosystem CO_2 exchange of a mature Amazonian transitional tropical forest (cerradão). *Functional Ecology* 15:388–395.

Wassmann, R., H. Papen, and H. Rennenberg. 1993. Methane emission from rice paddies and possible mitigation strategies. *Chemosphere* 26:201–217.

Whalen, S. C., W. S. Reeburgh, and K. S. Kizer. 1991. Methane consumption and emission from taiga. *Global Biogeochemical Cycles* 5:261–273.

14

Acid Deposition: S and N Cascades and Elemental Interactions

James N. Galloway

Most S and N atoms on the planet are stored in long-term reservoirs and have little interaction with other elements. About 90 percent of the S is stored as sulfates and sulfides in sedimentary rocks; about 80 percent of the N is stored as N_2 in the atmosphere, with sedimentary rocks containing most of the remainder (Mackenzie 1998). Although the amount of S and N emitted to the atmosphere (and thus available for biogeochemical interactions) is low relative to the amount in long-term storage, the impact of S and N emissions on biogeochemical processes can be significant. In the natural world N and S emissions contribute to the atmosphere's reactivity and radiation balance, as well as providing nutrients (i.e., N) to ecosystems. These effects are magnified owing to the increased emissions of S and N species due to human activities. One of these effects is the increasing acidity of atmospheric deposition.

Over the past twenty-five years large national and international programs have investigated the phenomena and impacts of acid deposition on soils, forests, streams, lakes, and structures. The state of knowledge has been summarized every five years at an international conference on acid deposition (Galloway 2001). Starting in 1975 in Columbus, Ohio, U.S.A., the following questions were being addressed: What were the acids? Where did they come from? Did acid rain cause lake and stream acidification? Was there a link between acid rain and fish loss? (Dochinger and Seliga 1976). At the subsequent conferences (1980—Sandefjord, Norway [Drablos and Tollan 1980]; 1985—Muskoka, Canada [Martin 1986]; 1990—Glasgow, Scotland [Last and Watling 1991]; 1995—Goteborg, Sweden [Grennfelt et al. 1995]; 2000—Tsukuba, Japan [Satake 2001]), more complex systems were investigated, national programs presented, interactions with other environmental changes noted, and investigations of acid deposition in other regions of the world presented.

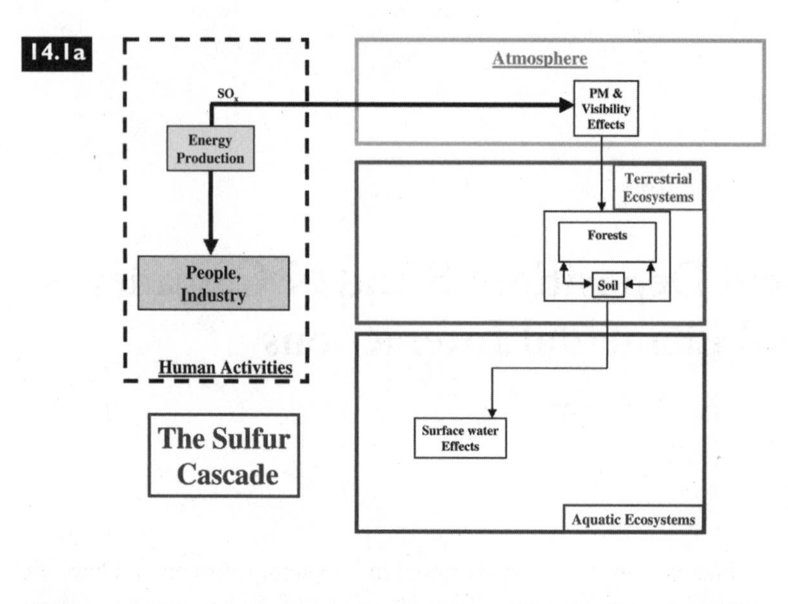

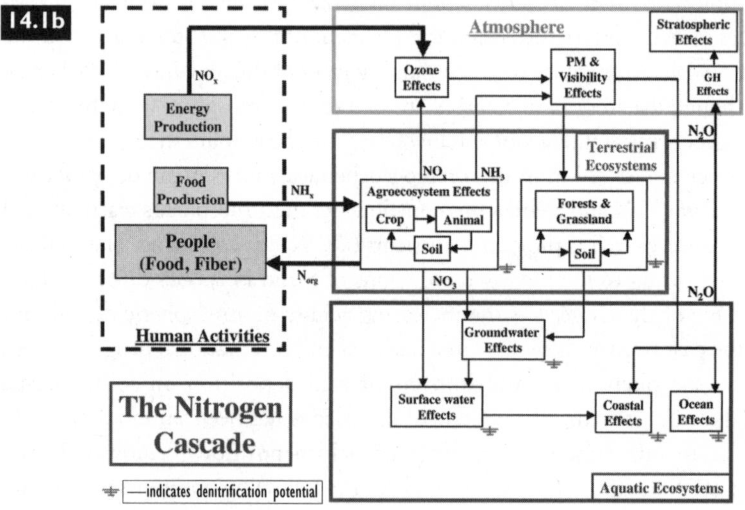

Figure 14.1. S and N deposited into acid-sensitive watersheds can cascade to downstream and downwind systems and can impact those systems directly or indirectly through element interactions. This figure illustrates the sequential effects that a single atom of (a) S or (b) N can have in various reservoirs after it has been mobilized in the environment. *Note:* PM = particulate matter; GH Effects = greenhouse effects.

This chapter builds on this history and expands the focus of acid deposition assessment. Where much of the previous research has focused on the direct effects of increased S and N emissions, this chapter examines how increased S and N emissions have resulted in interactions with the biogeochemical cycles of other elements. The chapter begins with the atmosphere and then follows S and N as they enter successive reservoirs (forested ecosystems, headwaters, rivers, and coastal waters).

The interactions of S and N with other elements can be characterized in three ways: (1) those that result in a net gain of elements to the forested ecosystem (e.g., increased CO_2 uptake due to N fertilization); (2) those that result in a redistribution of elements within the system (transfer of P to biomass from soil due to N fertilization); and (3) those that result in a loss of elements from the ecosystem (e.g., loss of base cations due to sulfate and nitrate discharges to streams). An important aspect of these three interactions is the effective time scale—how long will the interactions last? For example, biomass additions will remain in place until the forest experiences a disturbance that results in a net reduction of standing crop. Base cation losses from the soil can be permanent, especially for those soils with low cation exchange capacity (CEC) and percent base saturation (%BS).

Another important issue is the extent to which S and N deposited into forested ecosystems cascade to systems downstream and downwind. As will be seen, the effects of S, while regionally extensive, are relatively local to the point of deposition—primarily low-order streams and headwater lakes and their associated watersheds (Figure 14.1a). In contrast, N deposited to terrestrial ecosystems is lost via the atmosphere or the hydrosphere, and there is the potential for biogeochemical interactions in most downstream and downwind systems on a regional and global basis (Figure 14.1b) (Galloway et al. 2002).

Atmosphere

The short residence times (~days) of SO_2, NO_x, and NH_3 and their reaction products limit their interactions with other elements in the atmosphere. For example, NH_3 and SO_2 either are removed directly by dry deposition or react with an acid (e.g., H_2SO_4, HNO_3, HCl) in the case of NH_3 or a base (e.g., NH_3) in the case of SO_2 to form salts (e.g., $(NH_4)_2SO_4$, NH_4NO_3, NH_4Cl). These processes create hydroscopic aerosols that can act as cloud condensation nuclei. Before removal, by virtue of their small size the aerosols also decrease atmospheric visibility, alter the earth's radiation balance, and contribute to human respiratory disease (Figure 14.2a).

NO_x is quickly oxidized to HNO_3 in the atmosphere and likewise is either deposited or converted to an aerosol (e.g., NH_4NO_3). Between NO_x emission and

14.2a

Sulfur Biogeochemical Interactions

Atmosphere to Forests to headwaters to Rivers to Coastal Waters

	Transfers	*Interactions*	*Consequences*
Atmosphere	SO_x	Reacts with NH_3 to form PM2.5 aerosols	increases regional haze, changes radiation balance, and impacts human health
		Reacts with Ca^+ & Mg^{++} to form PM10 aerosols	increases atmospheric removal rates
		Reacts with NaCl in coastal atmosphere	increases atmospheric removal rates
Forests	SO_x	Adsorbs to soil surfaces	stores sulfate; delays surface water acidification
		removes cations from exchange surfaces	H^+ & Al^{n+} acidify soil and decrease forest productivity
		increases Ca^{++}, Mg^{++}, K^+, H^+, Al^{n+} in soil solution	Ca^{++}, Mg^{++}, K^+ losses decrease forest productivity
Headwaters	$SO_4^=$	Transfer of H^+, Al^{n+} to surface waters	Increase in acidity of surface waters; loss of alkalinity and biodiversity
Rivers	$SO_4^=$		
Coastal waters	$SO_4^=$		

Nitrogen Biogeochemical Interactions

14.2b

Atmosphere to Forests to headwaters to Rivers to Coastal Waters

	Transfers	*Interactions*	*Consequences*
Atmosphere	NH_x NH_y	NH_3 reacts w/H_2SO_4 & HNO_3 to form PM2.5 aerosols	increases regional haze, changes radiation balance, and impacts human health
		NO reacts w/sunlight to form OH which increases O_3	increases GWP and decreases forest productivity
		NO reacts w/sunlight to form OH which decreases CO, CH_4	decreases GWP
		reacts w/NaCl in coastal atmosphere	increases Nr inputs to coastal ecosystems
Forests	Nr	biota take up Nr	Increases forest productivity; transfers P to biomass from soil; sequesters C in biomass
		microbes convert NH_4^+ to NO_3^-	Soil acidifies
		microbes produce N_2O and NO	Increases in tropospheric O_3 & GWP; decreases in stratospheric O_3
		NO_3^- increases in soil solution; cations also	H^+ & Al^{n+} acidify soil and decrease forest productivity
		increases transfer to cations to headwaters	Ca^{++}, Mg^{++}, K^+ losses decrease forest productivity
Headwaters	Nr	transfer of H^+, Al^{n+} to surface waters	increase in acidity of surface waters; loss of alkalinity and biodiversity
		microbes produce N_2O and NO	Increase in tropospheric O_3 & GWP; decreases in stratospheric O_3
Rivers	Nr	microbes produce N_2O and NO	Increases in tropospheric O_3 & GWP; decreases in stratospheric O_3
Coastal waters	Nr	increase in productivity, hypoxia, and euthrophication	increase in HAB, loss of biodiversity, etc.
		microbes produce N_2O and NO	Increases in tropospheric O_3 & GWP; decreases in stratospheric O_3

Figure 14.2. The biochemical interactions of (a) S and (b) N and the associated consequences. NO_y refers to the sum of all combinations of N and O except N_2O (e.g., NO_x, HNO_3, NO_3^-). NH_x is defined as $NH_3 + NH_4^+$. GHP is greenhouse warming potential, GWP is global warming potential. HAB are harmful algal blooms.

HNO_3 formation, however, there is a critical interaction of NO_x with other elements that has a profound impact on the atmosphere. Not only does NO_x catalyze tropospheric O_3 formation, but in the process, NO enhances tropospheric OH abundance and thus reduces the atmospheric burdens of CO, CH_4, and HFCs (Prather et al. 2001). Thus emissions of NO both increase (O_3) and decrease (CH_4) concentrations of greenhouse gases, and in general increase the reactivity of the troposphere (Figure 14.2b). It is interesting to note that since NO_x acts as a catalyst, it is not destroyed but rather is regenerated and eventually converted to HNO_3, which is deposited to the Earth's surface.

In marine atmospheres SO_2 and HNO_3 have a more complicated chemistry that involves interactions with other elements. Sea salt is basic and thus will react with acidic gases and aerosols. In the marine environment it is thus an important sink for SO_2 and NO_3. This interaction with sea salt promotes more rapid removal of SO_2 and NO_x from the marine atmosphere because sea salt is generally a larger particle and thus will be removed faster. More rapid removal means less transport of anthropogenic S and N over the ocean and greater inputs to coastal systems. The former means decreased impact of sulfate aerosol on visibility and atmospheric radiation; the latter means greater inputs of nutrient N into the coastal zone and potential impacts on productivity.

In summary, SO_2, NO_x, and NH_3 emissions to the atmosphere are transferred to the Earth's surface generally within seven to ten days. While these emitted materials are in the atmosphere, their interactions with other species result in (1) aerosol formation, (2) increases in tropospheric O_3, (3) decreases in tropospheric CO and CH_4, and (4) interactions with sea salt resulting in more rapid removal of N and S from the atmosphere in the coastal zone. These interactions have a number of effects, including changes in the Earth's radiation balance, atmospheric oxidation capacity, and both human and ecosystem health impacts.

Forested Ecosystems

Enhanced S and N deposition to terrestrial ecosystems can be (1) passively stored, (2) actively recycled within the system, (3) lost via atmospheric emissions, or (4) lost via streamwater discharge. As will be seen later in this chapter, S is either stored or lost via discharge. N is more dynamic and has the potential to be stored, recycled, lost to the atmosphere, or lost via drainage waters. As such it has a more extensive interaction with other elements over much large scales of time and space. The following expands on these points in the context of forested ecosystems. For a related analysis on changes in interelement interactions as land use changes between forest and grassland, refer to Martinelli (Chapter 10, this volume).

Sulfur

There is limited recycling of S because inputs are high relative to the system nutrient requirements. In addition, there are limited gaseous S losses. Therefore most S deposition is either stored in the soil or lost via streamwater.

Sulfate Storage

Storage occurs by sulfate adsorption, which occurs in older, acidic soils. These soils are characterized by low organic matter content and high abundance of iron and aluminum oxy-oxides exchange surfaces (Johnson et al. 1980). As dictated by soil characteristics, the sulfate adsorption capacity (SAC) is regionally dependent. Regions with young soils (i.e., soils developed on material from the last glaciation, 10,000–14,000 years ago) have a low SAC (e.g., southeastern Canada, northeastern United States, Scandinavia). Regions with older soils have higher SACs (e.g., southeastern United States) (Johnson et al. 1980; Harrison, Johnson, and Todd 1989; Galloway 1996). There are limited interactions with other elements owing to sulfate adsorption. The process of adsorption generally results in a release of OH^- that has the potential to increase the pH of the soil solution, unless the incoming $SO_4^=$ is accompanied by H^+ or NH_4^+. In some cases, instead of a release of OH^-, co-uptake of cations can occur, a process that would result in no pH change but a potential increase in the soil CEC (Mitchell, David, and Harrison 1992). In both cases the change in pH or CEC would be small. An important aspect of the SAC of a soil is that it is finite and can reach saturation in time scales of years to decades. As the adsorption sites become saturated with sulfate, soil solution levels of sulfate and cations increase and a larger and larger portion of the deposited sulfur experiences the other fate—discharge to surface waters. The cations that increase in concentration with the increasing sulfate will be a mixture of base cations, H^+ and Al^{n+}. The exact composition of the mixture is dependent on the soil %BS. Generally, soils associated with acid-sensitive waters are acidic, and thus as sulfate concentrations increase in soil water, base cations H^+ and Al^{n+} increase in soil solution, resulting in further acidification of the soil. In addition, the increased Al^{n+} concentrations in the soil solution can interfere with divalent cation uptake or root growth (Cronan and Grigal 1995). These changes in soil characteristics induced by S deposition result in losses of biodiversity at many levels of the ecosystem (Eviner and Chapin, Chapter 8, this volume).

Sulfate Discharge via Streamwater

The loss of sulfate generally has no direct impact on forested ecosystems. The loss of the associated cations (for charge balance), however, has impacts on both the terrestrial system and the aquatic system. Base cations are required nutrients for forests. In acid soils their availability is low, and additional base cation losses may result in decreased forest growth (e.g., Lawrence and Huntington 1999) until base cation losses are reduced and mineral weathering (Galloway, Norton, and Church 1983) and atmospheric deposition (Kennedy, Hedin, and Derry 2002) can replenish base cations on ion exchange sites. There are also biogeochemical consequences in the receiving surface waters. In waters draining acid soils with elevated sulfate levels, both Al^{n+} and H^+ will be enhanced. There is a well-developed literature on surface water acidification documenting loss of biodiversity due to the stream acidification that accompanies S deposition to forested ecosystems (Baker and Christensen 1991; Bulger et al. 1999; Driscoll et al. 2001; Overrein, Seip, and Tollan 1980; Havas and Rosseland 1995) (Figure 14.2a).

In summary, the input of S into forested ecosystems will result in internal interactions with other elements as a consequence of the adsorption process. As the SAC becomes saturated, increasing amounts of S will be discharged from the ecosystem to surface waters. Interactions (as a consequence of charge balance) will result in an equivalent loss of cations from the forested ecosystem. The exact mixture of base cations and acid cations (Al, H^+) will be dependent on soil %BS. The former has the potential to decrease forest productivity; the latter will increase surface water acidity and decrease biodiversity. Once the surface waters become higher-order systems, however, there are generally no impacts of the increased sulfur and its associated cations because of deeper soils and the influence of bedrock with greater availability of base cations. Important exceptions to this statement are the salmon rivers of southern Norway and southwestern Sweden, which are large high-order systems and have decreasing salmon populations because of increased S and N deposition (Sandøy and Langåker 2001).

Nitrogen

The potential for elemental interactions with N inputs is larger than for S. In addition to being passively stored and discharged in surface water, N can also be actively cycled within the system and emitted as reactive gases from the system (e.g., Holland and Carroll, Chapter 15, this volume).

N Impacts on Forest Productivity

Most temperate terrestrial ecosystems are N-limited, and as such forested ecosystems will retain most of the N deposited until some other element becomes limiting. The retention process is complicated, and there is substantial cycling within the soil-biota system (Tamm 1991; Aber et al. 1998; Gundersen et al. 1998; Tietema et al. 1998; Nadelhoffer 2001; Aber et al. 2002; Driscoll et al. 2002). Increased N availability will increase both the cycling of material within the ecosystem (e.g., P and base cations) as well as result in the addition of new material to the ecosystem (e.g., C; Kauppi et al. 1997; Nadelhoffer et al. 1999).

In addition to the increase in productivity of forested ecosystems due to N deposition, there is also an indirect impact on productivity in the opposite direction. The increase in atmospheric O_3 due to increased NO_x emissions has the potential to reduce productivity as a result of ozone damage to the vegetation (Chappelka and Samuelson 1998; Felzer et al. 2002; Ollinger et al. 2002; Figure 14.2b). And as with S, N deposition to terrestrial ecosystems results in changes in biodiversity (Eviner and Chapin, Chapter 8, this volume).

N Saturation

N saturation is used to describe the process of the response of forested soils to increased deposition of N (Aber et al. 1995). Most temperate forests are N--limited (Tamm 1991). Increased N deposition has the potential to change that status, however, and as forest soils accumulate N by biotic and abiotic processes (Davidson, Chorover, and Dail 2002), N losses via surface discharge and gaseous losses increase. The length of time before a forest soil begins to lose N is dependent on land-use history and forest type and can range from years to centuries (Emmett et al. 1998; Goodale, Aber, and McDowell 2000; Aber et al. 2002). Two significant events occur at the onset of N saturation. The first is that the rate of N inputs exceed the ability of forests to retain N and thus N losses increase. In some cases something else becomes limiting. When this occurs, the basic biogeochemical nature of the forest changes. Schulze (1989) demonstrated that following N saturation of a European forest, Mg^{++} became the limiting nutrient. In the northeastern United States, climate (length of growing season, lack of sunshine) may become the next limiting factor, but there is concern about Ca^{++} limitations in certain forests (e.g., DeHayes et al. 1999; Lawrence and Huntington 1999). The second is that N is transferred from the forested ecosystem to downstream and downwind systems.

N Losses

Once N inputs are larger than the uptake in the terrestrial ecosystem, NO_3^- concentrations will increase in the soilwater (NH_4^+ is largely nitrified to NO_3^-) and there will be increased N losses to the atmosphere and to surface waters (Aber et al. 1995). Atmospheric losses include emissions of NO, N_2O, and N_2 formed during nitrification and/or denitrification. N losses to surface waters are primarily as NO_3^-.

NO_3^- losses to surface waters will be accompanied by equivalent losses of cations (Ca^{++}, Mg^{++}, Al^{n+}, H^+), as was the case with sulfate (Figure 14.2b). The exact composition of the cation mixture depends on the soil characteristics (e.g.,

CEC and %BS). Again, as with sulfate, acidic soils will become impoverished in Ca^{++} and Mg^{++} and aquatic ecosystems will become acidified by Al^{n+} and H^+ with the attendant biogeochemical consequences on the terrestrial and aquatic ecosystems. The systems most affected by this double consequence of NO_3^- as a carrier ion are high-elevation watersheds with shallow soils and low-order streams. Once soil depth and stream-order increase, these consequences of increased NO_3^- discharge to surface waters diminish and eventually become negligible, except as noted earlier for large rivers (e.g., southern Norway, southwestern Sweden). The same is not the case for N as a nutrient. As discussed in following sections, discharge of N from terrestrial systems to the stream-river-estuary continuum has significant impacts on the atmosphere and on coastal ecosystems. The continuum is a source of N_2O to the atmosphere (Seitzinger et al. 2002) and a source of a limiting nutrient to coastal water systems (Rabalais 2002).

In summary, the input of N into forested ecosystems will result in internal interactions with other elements as N is incorporated into soil organic matter and biomass. N-stimulated growth will redistribute P, base cations, and other nutrients into biomass and will result in an increase in biomass C. When N saturation is reached, increasing amounts of N will be lost to the atmosphere. In addition nitrate will be discharged to surface waters. Interactions (as a consequence of charge balance) will result in an equivalent loss of cations (a mixture of base cations and acid cations Al and H^+) from the forested ecosystem. Base cation losses have the potential to decrease forest productivity; acid cations will increase surface water acidity and decrease biodiversity in headwater streams and lakes.

Headwaters, Rivers, and Coastal Ecosystems

Acidification

In acid soils where base cations are available, the primary interaction of increased S and N levels (as $SO_4^=$ and NO_3^-) is an increase in base cation H^+ and Al^{n+} concentrations in the soilwater and streamwater. The latter two will result in acidification of the water and a loss of biodiversity and decrease in living biomass in the streamwater. This change in turn could reduce the amount of P and C held in biomass in these headwaters.

Denitrification

Increased inputs of reactive N (e.g., nitrate, ammonium, organic N, N_2O) (Nr) from forested catchments can increase rates of denitrification along the entire aquatic continuum (Seitzinger et al. 2002). In the sediments denitrification can be increased through a number of mechanisms. In wetlands increased nitrate in groundwater and surface water runoff can increase the supply of nitrate for denitrification in the wetland soils. In streams, lakes, estuaries, and continental shelf environments increased nitrate inputs can increase the nitrate concentration in the water column and thus increase the diffusive supply of nitrate to the sediments. Increased inputs of Nr can enhance primary production, particularly in estuarine and continental shelf waters, thus increasing organic matter deposition to the sediments, and subsequently sediment nitrification and denitrification. Both processes will result in increased emissions of N_2O and NO to the atmosphere.

Fertilization

Given that freshwaters are generally P-limited (Vollenweider 1976), the increased concentrations of N will not impact biomass. Primary production in most coastal rivers, bays, and seas of the temperate zone, however, is limited by nitrogen supplies (Vitousek and Howarth 1991; Nixon et al. 1996; NRC 2000). As a consequence, increasing Nr inputs leads to increased growth of algae. In moderation, this growth can be viewed as beneficial, as it can lead to increased production of harvestable fish (Nixon 1988; Jorgensen and Richardson 1996). High Nr inputs can also lead, however, to excessive algal growth, or eutrophication. In the United States the increased Nr flux is now viewed as the most serious pollution problem in coastal waters (Howarth et al. 2000; NRC 2000; Rabalais 2002). One-third of the nation's coastal rivers and bays are severely degraded, and another one-third are moderately degraded from nutrient overenrichment (Bricker et al. 1999). The situation is probably equally severe in other regions of the globe where human activity is leading to high Nr inputs to the coast (e.g., Baltic Sea, North Sea).

These impacts have significant consequences on the biogeochemical cycling of other elements. The direct consequences are the demands for other nutrients (e.g., C, P, Si, Ca, Fe) as the N supplies increase. There are also indirect effects. Eutrophication, caused by increased N inputs, can induce hypoxia in large regions of the coastal environment (e.g., Gulf of Mexico; Rabalais 2002). This loss of O_2 will change the basic biogeochemical characteristics of the coastal system from oxic to near-anoxic. There are a number of biogeochemical processes that will decrease in magnitude (e.g., respiration) and others that will increase (e.g., denitrification).

Conclusions

S and N emitted to the atmosphere by human activities can be transported over long distances and deposited into forested ecosystems that are more than 1,000 kilometers from point of emission. Once deposited, S and N species can alter a host of biogeochemical characteristics (e.g., acid-base status; nutrient status) of receiving systems, as well as those that are downwind or downstream. In addition to ecosystem acidification, S and N deposition, through element interactions, can promote loss of base cations from terrestrial ecosystems, change nutrient status of terrestrial and aquatic systems, and increase hypoxia of coastal ecosystems. In general, N is more reactive with the environment than is S. It affects more processes, and following deposition, its effects have the potential to be noted on the global scale. S, however, primarily affects the acid-base status of systems with a low buffering capacity (e.g., high-elevation watersheds and first-order water bodies).

Acknowledgments

This chapter is a contribution to the Shenandoah Watershed Study, supported by the U.S. National Park Service, the U.S. Forest Service, and the U.S. Environmental Protection Agency. I am appreciative to Jack Cosby, Rick Webb, and Dick Wright for helpful comments on this chapter. I thank Mary Ann Seifert for putting the chapter into the correct format and to Sue Donovan for assistance in figure preparation. I am grateful to the Ecosystems Center of the Marine Biological Laboratory, and the Woods Hole Oceanographic Institution for providing a sabbatical home to write this paper and to the University of Virginia for the Sesquicentennial Fellowship.

Literature Cited

Aber, J. D., C. L. Goodale, S. V. Ollinger, M.-L. Smith, A. H. Magill, M. E. Martin, R. A. Hallett, J. L. Stoddard, and NERC Workshop Participants. 2002. Is nitrogen deposition altering the nitrogen status of Northeastern forests? *BioScience*, submitted.

Aber, J. D., A. Magill, S. G. McNulty, R. D. Boone, K. J. Nadelhoffer, M. Downs, and R. Hallett. 1995. Forest biogeochemistry and primary production altered by nitrogen saturation. *Water, Air, and Soil Pollution* 85:1665–1670.

Aber, J. D., W. H. McDowell, K. J. Nadelhoffer, A. Magill, G. Berntson, M. Kamakea, S. G. McNulty, W. Currie, L. Rustad, and I. Fernandez. 1998. Nitrogen saturation in temperate forest ecosystems: Hypotheses revisited. *BioScience* 48:921–934.

Baker, J. P., and S. W. Christensen. 1991. Effects of acidification on biological communities. Pp. 83–106 in *Acidic deposition and aquatic ecosystems,* edited by D. F. Charles. New York: Springer-Verlag.

Bricker, S. B., C. G. Clement, D. E. Pirhalla, S. P. Orland, and D. G. G. Farrow. 1999. *National estuarine eutrophication assessment: A summary of conditions, historical trends, and future outlook.* Silver Spring, Md.: National Ocean Service, National Oceanic and Atmospheric Administration.

Bulger, A. B., B. J. Cosby, C. A. Dolloff, K. N. Eshleman, J. N. Galloway, and J. R. Webb. 1999. The Shenandoah National Park Fish in Sensitive Habitats (SNP FISH) Project Final Report: An Integrated Assessment of Fish Community Response to Stream Acidification. National Park Service.

Chappelka, A. H., and L. J. Samuelson. 1998. Ambient ozone effects on forest trees of the eastern United States: A review. *New Phytologist* 139:91–108.

Cronan, C. S., and D. F. Grigal. 1995. Use of Ca/Al ratios as indicators of stress in forest ecosystems. *Journal of Environmental Quality* 24:209–226.

Davidson, E. A., J. Chorover, D. B. Dail. 2002. A mechanism of abiotic immobilization of nitrate in forest ecosystems: The ferrous wheel hypothesis. *Global Change Biology*, in press.

DeHayes, D. H., P. G. Schaberg, G. J. Hawley, and G. R. Strimbeck. 1999. Acid rain impact on calcium nutrition and forest health. *BioScience* 49:789–800.

Dochinger, L. S., and T. A. Seliga. 1976. *Proceedings of the First International Acid Precipitation Conference and The Forest Ecosystem.* U.S. Department of Agriculture Forest Service General Technical Report NE-23.

Drablos, D., and A. Tollan, eds. 1980. *Ecological impact of acid precipitation: Proceedings of an international conference, Sandefjord, Norway, March 11–14, 1980.* SNSF Project. The Norwegian Interdisciplinary Research Programme, Acid Precipitation-Effects on Forest and Fish, Oslo.

Driscoll, C. T., G. B. Lawrence, A. J. Bulger, T. J. Butler, C. S. Cronan, C. Eagar, K. F. Lambert, G. E. Likens, J. L. Stoddard, and K. C. Weathers. 2001. Acidic deposition in the Northeastern U.S.: Sources and inputs, ecosystem effects, and management strategies. *BioScience* 51:180–198.

Driscoll, C., D. Whitall, J. Aber, E. Boyer, M. Castro, C. Cronan, C. Goodale, P. Groffman, C. Hopkinson, K. Lambert, G. Lawrence, S. Ollinger. 2002. Nitrogen pollution in the Northeastern United States: Sources, effects, and management options. *BioScience*, submitted.

Emmett, B. A., A. D. Boxman, M. Bredemeier, F. Moldan, P. Gundersen, O. J. Kjønaas, P. Schleppi, A. Tietema, and R. F. Wright. 1998. Predicting the effects of atmospheric nitrogen deposition in conifer stands: Evidence from the NITREX project. *Ecosystems* 1:352–360.

Felzer, B., D. Kicklighter, J. Melillo, C. Wang, and Q. Zhuang. 2002. Surface ozone effects on productivity using a biogeochemistry model, EOS. Trans. AGU, 83(19), Spring Meet. Suppl., Abstract GC31A-02.

Galloway, J. N. 1996. Anthropogenic mobilization of sulfur and nitrogen: Immediate and delayed consequences. *Annual Reviews of Energy and Environment* 21:261–292.

Galloway, J. N. 2001. Acidification of the world: Natural and anthropogenic. *Water, Air and Soil Pollution* 130:17–24.

Galloway, J. N., J. D. Aber, J. W. Erisman, S. P. Seitzinger, R. H. Howarth, E. B. Cowling, and B. J. Cosby. 2002. The nitrogen cascade. *BioScience*, in press.

Galloway, J. N., S. A. Norton, and M. R. Church. 1983. Freshwater acidification from atmospheric deposition of sulfuric acid: A conceptual model. *Environmental Science and Technology* 17:541A.

Goodale, C. L., J. D. Aber, and W. H. McDowell. 2000. Long-term effects of disturbance on organic and inorganic nitrogen export in the White Mountains, New Hampshire. *Ecosystems* 3:433–450.

Grennfelt, P., H. Rodhe, E. Thornelof, and J. Wisniewski. 1995. Acid reign '95? Proceedings from the 5th International Conference on Acidic Deposition: Science and policy, Goteborg, Sweden, 26–30 June. *Water, Air and Soil Pollution* 85:1–4.

Gundersen, P., B. A. Emmett, O. J. Kjønaas, C. J. Koopmans, and A. Tietema. 1998. Impact of nitrogen deposition on nitrogen cycling in forests: A synthesis of NITREX data. *Forest Ecology and Management* 101:37–56.

Harrison, R. B., D. W. Johnson, and D. E. Todd. 1989. Sulfate adsorption and desorption reversibility in a variety of forest soils. *Journal of Environmental Quality* 18:419–426.

Havas, M., and B. O. Rosseland. 1995. Response of zooplankton, benthos, and fish to acidification: An overview. *Water, Air and Soil Pollution* 85:51–62.

Howarth, R.W., D. Anderson, J. Cloern, C. Elfring, C. Hopkinson, B. Lapointe, T. Malone, N. Marcus, K. McGlathery, A. Sharpley, and D. Walker. 2000. Nutrient pollution of coastal rivers, bays, and seas. *Issues in Ecology* (Ecological Society of America) 7:15.

Johnson, D. W., J. W. Hornbeck, J. M. Kelly, W. T. Swank, and D. E. Todd. 1980. Regional patterns of soil sulfate accumulation: Relevance to ecosystem budgets. Pp. 507–519 in *Atmospheric sulfur deposition, environmental impact, and health effects,* edited by D. S. Shriner, C. R. Richmond, and S. E. Lindberg. Ann Arbor, Mich.: Ann Arbor Science.

Jorgensen, B. B., and K. Richardson. 1996. *Eutrophication in coastal marine systems.* Washington, D.C.: American Geophysical Union.

Kauppi, P. E., M. Posch, P. Hänninen, H. M. Henttonen, A. Ihalainen, E. Lappalainen, M. Starr, and P. Tamminen. 1997. Carbon reservoirs in peatlands and forests in the Boreal regions of Finland. *Silva Fennica* 31:13–25.

Kennedy, M. J., L. O. Hedin, and L. A. Derry. 2002. Decoupling of unpolluted temperate forests from rock nutrient sources revealed by natural 87Sr/86Sr and 84Sr tracer addition. *Proceedings of the National Academy of Sciences* 99:9639–9644.

Last, F. T., and R. Watling. 1991. *Acidic deposition: Its nature and impacts.* Proceedings of the Royal Society of Edinburgh, Section B (Biological Sciences), Volume 97. Edinburgh: Royal Society of Edinburgh.

Lawrence, G. B., and T. G. Huntington. 1999. Soil-calcium depletion linked to acid rain and forest growth in the eastern United States. USGS WRIR 98-4267. Washington, D.C.: U.S. Geological Survey.

Mackenzie, F. T. 1998. *Our changing planet: An introduction to earth system science and global environmental change.* 2nd ed. Upper Saddle River, N.J.: Prentice Hall.

Martin, H. C. 1986. Acidic precipitation, Proceedings of the International Symposium on Acidic Precipitation, Muskoka, Ontario, September 15–20, 1985. *Water, Air and Soil Pollution* 30.

Mitchell, M. J., M. B. David, and R. B. Harrison. 1992. Sulfur dynamics of forest ecosystems. Pp. 215–254 in *Sulphur cycling on the continents: Wetlands, terrestrial ecosystems, and associated water bodies,* edited by R. W. Howarth, J. W. B. Stewart, and M. V. Ivanov. SCOPE 48. Chichester, U.K.: John Wiley and Sons.

Nadelhoffer, K. J. 2001. The impacts of nitrogen deposition on forest ecosystems. Pp. 311–331 in *Nitrogen in the environment: Sources, problems, and management,* edited by R. F. Follett and J. L. Hatfield. Amsterdam: Elsevier Science.

Nadelhoffer, K. J., B. A. Emmett, P. Gundersen, O. J. Kjønaas, C. J. Koopmans, P.

Schleppi, A. Tietema, and R. F. Wright. 1999. Nitrogen deposition makes a minor contribution to carbon sequestration in temperate forests. *Nature* 398:145–148.

NRC (National Research Council). 2000. *Clean coastal waters: Understanding and reducing the effects of nutrient pollution.* Washington, D.C.: National Academy Press.

Nixon, S. W. 1988. Physical energy inputs and the comparative ecology of lake and marine ecosystems. *Limnology Oceanography* 33:1005–1025.

Nixon, S. W., J. W. Ammerman, P. Atkinson, V. M. Berounsky, G. Billen, W. C. Boicourt, W. R. Boynton, T. M. Church, D. M. Ditoro, R. Elmgren, J. H. Garber, A. E. Giblin, R. A. Jahnke, N. J. P. Owens, M. E. Q. Pilson, and S. P. Seitzinger. 1996. The fate of nitrogen and phosphorus at the land-sea margin of the North Atlantic Ocean. *Biogeochemistry* 35:141–180.

Ollinger S. V., J. D. Aber, P. B. Reich, and R. J. Freuder. 2002. Interactive effects of nitrogen deposition, tropospheric ozone, elevated CO_2, and land use history on the carbon dynamics of northern hardwood forests. *Global Change Biology* 8:1–18.

Overrein, L., H. M. Seip, and A. Tollan. 1980. Acid precipitation: Effects on forest and fish. Final report of the SNSF Project, 1972–1980. Fagrapport FR 19-80, Oslo-Ås, Norway. Oslo: Government of Norway.

Prather, M., D. Ehhalt, F. Dentener, R. Derwent, E. Dlugokencky, E. Holland, I. Isaksen, J. Katima, V. Kirchhoff, P. Matson, P. Midgley, and M. Wang. 2001. Atmospheric chemistry and greenhouse gases. In *Climate change 2001: The scientific basis,* edited by J. T. Houghton, D. J. Griggs, M. Noguer, P. J. van der Linden, X. Dai, K. Maskell, and C. A. Johnson. Contributions of Working Group I to the Third Assessment Report of the Intergovernmental Panel on Climate Change. Cambridge: Cambridge University Press.

Rabalais, N. 2002. Nitrogen in aquatic ecosystems. *Ambio* 31:102–112.

Sandøy, S., and R. Langåker. 2001. Atlantic salmon and acidification in southern Norway: A disaster in the 20th century, but a hope for the future? *Water, Air and Soil Pollution* 130:1343–1348.

Satake, K., ed. 2001. *Acid rain 2000.* Proceedings from the 6th International Conference on Acidic Deposition: Looking Back to the Past and Thinking of the Future, Tsukuba, Japan, 10–16 December 2000. Dordrecht, the Netherlands: Kluwer.

Schulze, E. D. 1989. Air pollution and forest decline in a spruce (picea abies) forest. *Science* 244:776–783.

Seitzinger, S. P, R. V. Styles, E. Boyer, R. B. Alexander, G. Billen, R. Howarth, B. Mayer, and N. van Breemen. 2002. Nitrogen retention in rivers: Model development and application to watersheds in the eastern U.S. *Biogeochemistry* 57/58:199–237.

Tamm, C. O. 1991. *Nitrogen in terrestrial ecosystems.* Ecological Studies 81. Berlin: Springer-Verlag.

Tietema, A., A. D. Boxman, M. Bredemeier, B. A. Emmett, F. Moldan, P. Gundersen, P. Schleppi, and R. F. Wright. 1998. Nitrogen saturation experiments (NITREX) in coniferous forest ecosystems in Europe: A summary of results. *Environmental Pollution* 102 S1:433–437.

Vitousek, P. M., and R. W. Howarth. 1991. Nitrogen limitation on land and in the sea. How can it occur? *Biogeochemistry* 13:87–115.

Vollenweider, R. A. 1976. Advances in defining critical loading levels for phosphorus in lake eutrophication. *Memorie dell'Istituto Italiano di Idrobiologia* 33:53–83.

15

Atmospheric Chemistry and the Bioatmospheric Carbon and Nitrogen Cycles

Elisabeth A. Holland and Mary Anne Carroll

Progress in atmospheric chemistry includes information on important new mechanisms for interactions among the biosphere, atmosphere, and global biogeochemical cycles. Important links between the global biogeochemical cycles of carbon and nitrogen were documented decades ago (Bolin and Cook 1983). The nitrogen cycle, together with water and other nutrients, regulates primary productivity, soil fertility, and agricultural productivity. Over the past century anthropogenic influence on key biogeochemical cycles has changed through fossil fuel combustion, use of N fertilizer, agriculture intensification, and other land management (Mosier et al. 2002). Feedbacks operating through atmospheric chemistry are key determinants of the terrestrial N cycle and its interactions with the carbon cycle. The aim of this chapter is to review the present state of knowledge regarding the bioatmospheric C and N cycles with an emphasis on atmospheric chemistry feedbacks, modeling approaches, and interfaces.

Human perturbations to the global nitrogen cycle over the past century are profound and are only likely to grow in the future (Howarth 1996; Vitousek et al. 1997; Townsend 1999; AGU 2000; IPCC 2001; Boyer and Howarth 2002). This fact has significant consequences for a host of Earth system properties and services—such as ecosystem health and productivity, tropospheric oxidizing capacity, the carbon cycle, biogeochemistry, and climate—many with direct and important societal implications (Figures 15.1 and 15.2; Tables 15.1 and 15.2). Currently many processes contribute to the production of reactive N such as lightning, microbes, and, to a growing degree, humans through fossil fuel combustion and fertilizer use. These processes, together with urbanization, industri-

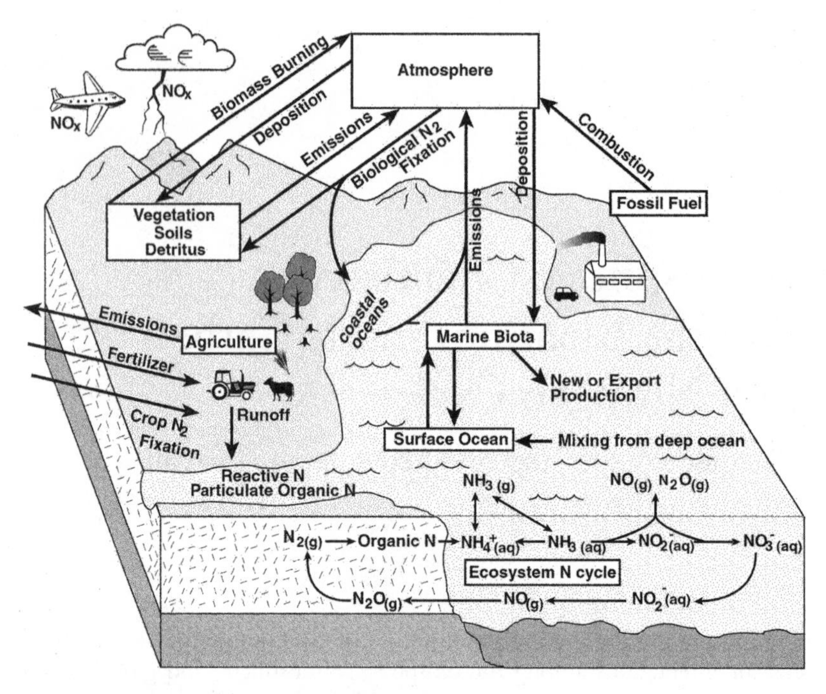

Figure 15.1. The global nitrogen cycle including the internal transformations of nitrogen, mineralization, nitrification, and denitrification, which occur in soils and water, the external inputs of nitrogen through nitrogen deposition, nitrogen fixation, and fertilizer application. The transformation of the modern global nitrogen cycle compared with the preindustrial nitrogen cycle has been greater than for any other biogeochemical cycle (Vitousek et al. 1997).

alization, land management, and the expansion and intensification of agriculture, have over the past century greatly accelerated release of N_2O (nitrous oxide), NO_x (NO+ NO_2), NH_x (NH_3+ NH_4^+), and organic nitrogen to the atmosphere, resulting in geologically unprecedented changes to the global nitrogen cycle (Figure 15.1; Husar 1986; Mayewski, Lyons, and Spencer 1990; Vitousek 1994; Galloway et al. 1995; Holland et al. 1999). The current behavior and potential future trajectories of nitrogenous products are difficult to quantify because reactive nitrogen occurs in a range of redox states and chemical forms; is modulated by a host of biological and chemical reactions; has significant mobile land, ocean, and atmospheric reservoirs; and undergoes substantial large-scale horizontal transport via surface emissions, atmospheric circulation, and deposition as well as rivers and ocean circulation. Further, many of the biological and chemical processes governing nitrogen dynamics are, in turn, sensitive to nitrogen perturbations either directly or indirectly, through modifications to the biotic, chemical, or climate

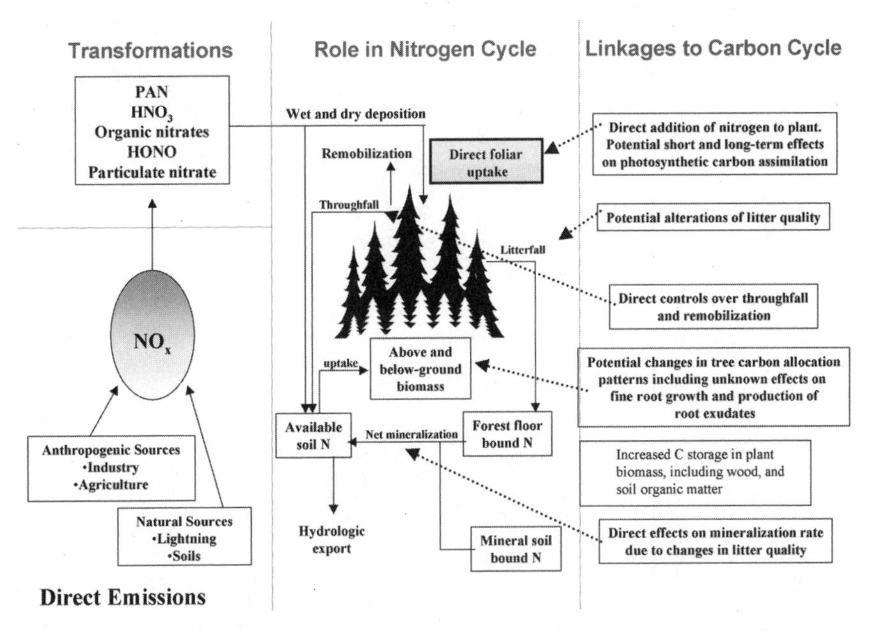

Figure 15.2. Overview of NO_x, carbon cycle interactions.

state. This sensitivity leads to complicated feedback-response relationships and non-linearities (IGAC 2002). Unraveling these complex interactions is a major scientific challenge.

Evidence has accumulated that illustrates subtle, yet important, coupled system interactions and feedbacks among biology, atmospheric chemistry, and climate. Dickinson et al. (2002), for example, have shown that incorporating nitrogen feedbacks on stomatal conductance (through Rubisco) into the land biosphere-atmosphere transfer scheme and including this mechanism in the NCAR CCM3 (National Center for Atmospheric Research, Community Climate Model, version 3) general circulation model produced a more realistic simulation of seasonal and diurnal temperature ranges over a seventeen-year run. Climate simulations using a more complete constellation of greenhouse gases underscored the importance of CO_2, N_2O, O_3, and feedbacks operating through air pollution (Hansen et al. 1998; Hansen 2002). Although it is difficult to quantify the magnitude of the impacts of nitrogen deposition on the carbon cycle through N fertilization and saturation (Townsend et al. 1996; Holland et al. 1997; Aber et al. 1998; Nadelhoffer et al. 1999a,b; Nadelhoffer, Downs, and Fry 1999), the global balance of di-nitrogen fixation and denitrification has been suggested as a contributor to glacial-interglacial atmospheric CO_2 variations on longer time scales (Broecker and Henderson 1998). It is also well established that the documented

Table 15.1. Summary of the key N exchanges and their role in the Earth system

Role	*NO_x emissions*	*NO_y deposition*	*NH_x emissions/ deposition*	*N_2O emissions*
Anthropogenic forcing	Strong	Strong	Strong	Strong
Terrestrial biological interactions	Strong	Strong	Strong	Strong
Ocean biological interactions	None	Strong	Intermediate	Intermediate
Tropospheric photochemistry	Strong	Strong	Intermediate/weak	None
Carbon cycle interactions	Weak	Strong	Strong	Weak/intermediate
Tropospheric ozone	Strong	Strong	Weak	Weak
Radiative forcing	Indirect	Indirect	Indirect	Strong
Other indirect climate feedbacks	Weak	Intermediate/strong	Intermediate/strong	None
Aerosol formation	Through VOC interactions	Intermediate	Strong	None

Table 15.2. Contrast between preindustrial and contemporary CO_2, N_2O, and NO_x atmospheric concentrations, global NO_x emissions, and global tropospheric ozone abundance

Chemical species	*1750*	*1998*	*Atmospheric lifetime (years)*
CO_2 (ppm)	280	365	5–200
N_2O (ppb)	270	314	120
NO_x (ppb)	<1	0.002–200	<0.01–0.03
Fossil fuel NO_x emissions (Tg N)	0.2	25	NA
Surface O_3 (ppb)	5–8	8–200	NA
Tropospheric O_3 (DU)[a]	25	30	0.01–0.05

Source: Olivier et al. (1999), Albritton et al. (2001), and Prather et al. (2001).
Note: NA indicates not applicable.
[a]1 ppb of tropospheric O_3 = 0.65 Dobson units (DU).

changes in tropospheric ozone concentrations over the past 150 years and their projected rise into the future are the product of increased NO_x, CO, and VOC emissions (Prather et al. 2001). Current day surface ozone concentrations are high enough to impact crop productivity and carbon dioxide uptake in natural ecosystems (Reich 1987; Reich et al. 1990; Chameides et al. 1994; Ollinger, Aber, and Reich 1997; NARSTO 2000). Interactions between biogenic organic carbon compounds and reactive N can be important determinants of the fate of nitrogen in terrestrial ecosystems (Jacob, Logan, and Murti 1999; Jacob 2000). Our ability to calculate global warming potentials for NO, NO_2, N_2O, and ozone, however, is compromised by inadequate representations of the biotic and chemical feedbacks operating through the nitrogen cycle (Prather et al. 2001; Tables 15.1 and 15.2). These system interactions argue strongly for considering coupled carbon and nitrogen cycles for both radiatively active gases—that is, CO_2 and N_2O—and the reactive carbon and nitrogen cycles, in light of feedbacks to air quality and climate.

Overview of the Bioatmospheric N Cycle

The bioatmospheric N cycle (Figure 15.1) depicts a balance between N_2 fixation, N gaseous emissions, and deposition. Gaseous emissions occur in various chemical forms: N_2, NO_x ($NO+NO_2$), N_2O, and NH_3 through microbial production, fossil fuel combustion, biomass burning, and lightning production. Nitrogen deposition fluxes are a point of key interaction with the carbon cycle and are most easily split into wet and dry components for description (NADP/NTN 1995; Whelpdale and Kaiser 1997; U.S Environmental Protection Agency 2003; Holland et al. submitted). The spatial pattern and magnitude of N deposition is strongly influenced by patterns of precipitation interacting with emissions and chemical processing (Galloway, Hiram, and Kasibhatla 1994; Lamb and Bowersox 2000). Wet deposition of nitrogen includes deposition of NH_4^+, NO_3^-, HNO_3, and a range of organic reduced and oxidized compounds in precipitation (NADP/NTN 1995; Neff et al. 2002a; Holland et al. submitted). The timing of nitrogen deposition events is driven by the episodic nature of precipitation and is thought to be important for coastal and open ocean ecosystems (Paerl 1997), but the influence of the timing of deposition onto land is less clear. Dry deposition of nitrogen is the removal of N containing gases by deposition onto a surface, such as leaves, soil, and open water. The N compounds exchanged via dry deposition include NH_3 (g), NO_2 (g), HNO_3 (g), particulate NH_4^+ and NO_3^-, and organic compounds including PAN (peroxyacetyl nitrate), organic acids, nitrogen containing aerosols, and simple amino acids. A detailed comprehensive study quantifying all of the different N compounds found in wet and dry deposition of N over a single site, marine or terrestrial, has not been carried out. Much of our understanding of the spatial patterns and magnitude of N deposition is restricted to deposition network sites in the United States and Europe, where for the most part data are obtained using buckets placed in open areas (Edgerton, Lavery, and Boksleitner 1992; EMEP 1995; NDAP/NTN 1995; Lamb and Bowersox 2000; Holland et al. submitted). Both the measurements and models are incomplete. The uncertainty associated with wet deposition measurements is much less than the substantial uncertainty associated with measurements/estimates of dry deposition (Weseley and Hicks 1999), but there are important questions regarding the remobilization of N compounds during the relatively long sampling period.

These fundamental uncertainties in understanding and estimating deposition limit our ability to successfully model the interactions of terrestrial nitrogen and carbon cycles. Furthermore, most models do not include interactions between reactive oxidized N species and volatile organic carbon compounds or the deposition of organic nitrates, even though these mechanisms could represent important additional (and non-acidic) forms of nitrogen that may have considerably different uptake rates for terrestrial ecosystems. These organic nitrates are more likely to be assimilated into terrestrial ecosystems by soil microbial uptake than taken up directly by the leaves, a possible assimilation pathway for NO, NO_2, and HNO_3 (Monson and Holland 2001).

Leaf uptake of NO, NO_2 and HNO_3, PAN, isoprene nitrates, and gaseous amines is a neglected but important pathway for N uptake to ecosystems. It provides plants with a direct pathway of N uptake, which short-circuits plant microbial competition for N uptake and increases the likelihood of N incorporation into plant biomass (Wildt et al. 1997; Amman et al. 1999; Lerdau, Munger, and Jacob 2000; Monson and Holland 2001; Siegwolf et al. 2001; Sparks et al. 2001). Direct leaf uptake of reactive N is only occasionally represented in physiological or dry deposition models. More process studies of the underlying biochemical, physiological, and ecological controls are needed before realistic models can be developed.

Non-linearities in biotic regulation of the terrestrial nitrogen cycle result from nitric and sulfuric acid deposition, which has substantial deleterious impacts including soil acidification, O_3 uptake, base cation depletion, NO_2, and aluminum toxicity (Aber et al. 1989; Schulze et al. 1989; Aber et al. 1998; Schulze 2000). These processes collectively contribute to N saturation, and they may slow or stop the growth of forests altogether. A number of studies have examined the effect of N deposition on terrestrial ecosystem C uptake (Peterson and Melillo 1985; Schindler and Bayley 1993; Townsend et al. 1996; Wedin and Tilman 1996; Holland et al. 1997; Nadelhoffer et al. 1999a,b; Nadelhoffer, Downs, and Fry 1999). At the community scale plant species shifts may be induced by N additions, leading to species distribution changes that often favor introduced species, including weeds over native plants, which can reduce species diversity and later ecosystem function (Aerts, Berendse, and Klerk 1989; Aerts, Berendse, and de Caluwe 1990; Berendse, Elberse, and Geertz 1992; Bobbink, Heil, and Raeseen 1992; Berendse, Aerts, and Bobbink 1993; Sutton, Pitcairn, and Fowler 1993; Chapin et al. 1997; Hooper and Vitousek 1997; Tilman et al. 1997). There is good evidence that shifts in below:aboveground allocation in grasslands contribute to non-linearities in N mineralization and loss (Holland et al. 1992), but it remains unclear how changes in plant community composition may contribute to the non-linear response of the biotic system.

Different vegetation types retain differing proportions of the deposited nitrogen, independent of the amount of nitrogen deposited (Holland et al. 2000). Across the spectrum of response, Century model simulations show that those biomes with the largest woody components also retain the greatest proportion of nitrogen (Table 15.3; Parton et al. 1987; VEMAP 1995). In a comparison of grassland and forested biomes, a substantially greater proportion of N added through N deposition is retained in forest biomes (90 versus 70 percent N retention) because of the wide C:N ratio of wood and its long turnover time (Table 15.3; Holland, Braswell, Bossdorf, pers. comm.). Nitrogen stored in compartments with shorter turnover times, such as some soil organic matter compartments, is more likely to be lost via volatilization and trace gas emissions or via nitrate leaching. Models of nitrogen retention are remarkably consistent across vegetation

Table 15.3. Average loss rate of N for representative vegetation types in the continental United States for current-day N and a tenfold increase in N deposition as a sensitivity analysis

Vegetation type/indicator	*N loss*
Coniferous forests	
Current N deposition	4.71
Tenfold increase	29.78
Deciduous forests	
Current N deposition	5.21
Tenfold increase	32.14
Mixed forests	
Current N deposition	4.80
Tenfold increase	30.82
Shrublands	
Current N deposition	1.50
Tenfold increase	10.75
Savannas	
Current N deposition	8.78
Tenfold increase	55.94
Grasslands	
Current N deposition	5.77
Tenfold increase	39.14

Note: All simulations were done using the Century simulation model using VEMAP climate and land use over the past 100 years (Braswell, Bossdorf, pers. comm.). Maps of N deposition inputs are described in Holland et al., submitted). All units are in kg N ha^{-1} accumulated over the ninety-nine-year simulation. Output is the sum of all simulated N losses = gaseous losses (NH_3 volatilization [plant, animal waste, soil]) + soil N gas losses (NO, N_2O, N_2) + leaching (NO_3^-, dissolved organic N).

types, however, with an approximate fivefold increase in N inputs for a tenfold increase in deposition (Table 15.3; range is a 5.68- to 6.93-fold increase across the vegetation types). These models' results conflict with observations that suggest that losses become disproportionately larger as inputs grow, showing that the model fails to capture measured indices of N saturation (Aber et al. 1989; Schulze et al. 1989; Aber et al. 1998).

The fate of N deposited onto terrestrial ecosystems is key to understanding its impact. A series of ^{15}N studies summarized by Nadelhoffer et al. 1999 (a,b) shows that nitrogen applied to the soil is immobilized in soil organic matter rather than taken up directly by plants. The method used for N application limits the relevance of these ^{15}N studies to total reactive N exchange. This is because it bypasses canopy processing of the deposited N. Canopy interception can account for more than half of the N deposited onto terrestrial ecosystems (Lovett and Lindbergh 1993; Lovett 1994; Lovett, Weathers, and Sobczak 2000; Sievering et al. 2000; Ganzeveld et al. 2002; Holland et al. submitted). Because N availability limits plant growth, the balance between microbial and plant uptake of the added nitrogen is a key determinant of its ultimate fate. When plant and microbial uptake are tightly coupled as occurs under natural conditions, the opportunity for nitrogen loss is significantly reduced. Under conditions of high nitrogen inputs through nitrogen deposition or fertilization, however, the plant and microbial N uptake can become imbalanced. Nitrogen losses grow, with significant losses to the atmosphere via NH_3, NO, and N_2O emissions and hydrologic systems (Aber et al. 1989; Durka et al. 1994; Howarth et al. 1996; Aber et al. 1998; Townsend 1999). Our current models of biogeochemistry fail to capture these non-linear responses to increasing N inputs and their effect on C storage (Table 15.3). It has been hypothesized that important chemical feedbacks, such as O_3 formation, provide a key mechanism to limit C uptake, thus decoupling the carbon and nitrogen cycles, allowing the measured non-linear increase in N outputs (Harrison et al. 2000).

In addition to regulation of ecosystem losses of nitrogen, plant versus microbial competition determines the residence time and the ability of the nitrogen deposition to stimulate carbon uptake and storage. The deposited nitrogen, which is immobilized by microbes to form soil organic matter, is stored at a C:N ratio close to that of microbial biomass, between 4 and 15. If the nitrogen is taken up by plants, the nitrogen is stored at a much wider C:N ratio of between 30 and 90 for leaves and a C:N of 150–300 for wood (Townsend et al. 1996). Microbial N has a residence time of less than a year compared with the nitrogen in wood, which can have a residence time of decades to centuries.

Deposition and gaseous uptake through stomata provide a pathway for direct plant uptake of nitrogen. Dry deposition of nitrogen containing gases accounts for an estimated 40–60 percent of total N deposition (Whelpdale and Kaiser 1997; Holland et al. 2000; Holland et al. submitted). The dry deposited nitrogen can be washed off of the leaf and deposited to the soil surface, or a substantial portion of the nitrogen can be taken up directly through the stomata and incorporated directly into plant biomass. Each pathway for uptake results in different carbon uptake rates in response to the added N and results in different N turnover times. Much of the experimental work has focused on the fate of nitrogen added to the soil rather than the plant canopy because the experiments are more tractable (Sievering et al. 2000). So far, the mechanism of direct leaf uptake of N is not included in current models of terrestrial nitrogen and carbon cycling.

The partitioning and speciation of deposited N can determine its fate, residence time, and impact, but the body of knowledge about the uptake and metabolism of specific gaseous N compounds by plant canopies is limited (Sparks et al. 2001; Colorplate 3). Chemical processing within the atmosphere determines the partitioning of N between its various forms: soluble and insoluble, inorganic and organic, oxidized and reduced, and gas-phase and particulate. This partitioning, together with surface processes and precipitation patterns within the atmospheric model, determine the speciation and type (wet or dry) of deposition of reactive nitrogen. Inputs of N through wet deposition are more likely to be delivered to microbes in the soil. N deposited as a gas will pass through the stomata or be deposited onto the leaf surface. The chemical species exchanged through the stomata, NO(g), NH_3(g), HNO_3(g), PAN(g), and NO_2(g), will be taken up directly by the plants. Furthermore, measurements of HONO(g) (nitrous acid) provide evidence that deposited nitrogen may undergo photolysis on canopy surfaces. In dew, the result is re-emission of gaseous reactive oxidized nitrogen species to the atmosphere (Munger et al. 1998).

The partitioning among the species is less certain than is suggested by the pie charts shown in Colorplate 3 and is largely driven by uncertainties in estimating dry deposition. For example, the range of percentages of particulate NH_4^+ is between 17 and 22 in the United States. In Europe the range is greater for particulate N fluxes, varying between 6.5 and 20.3 for dry deposition of NO_y and between 4 and 12 for particulate NH_4^+.

Nitrogen-Carbon Interactions

Nitrogen limitation of terrestrial productivity is widespread (Vitousek and Howarth 1991). Nitrogen deposition is an important external source of nitrogen to ecosystems and represents a significant contribution of plant-available nitrogen, particularly over areas of western Europe where nitrogen deposition inputs are similar to agricultural fertilizer application rates (Holland et al. 1999). The effect of nitrogen deposition on terrestrial ecosystems covers the spectrum of "beneficial" with the stimulation of carbon uptake by plants to "deleterious" with the onset of nitrogen saturation. The effects operate along a spectrum of time scales from the "instantaneous" influence of nitrogen availability on stomatal conductance of leaves to the influence of nitrogen deposition on carbon sequestration in wood and soil organic matter, which operates on annual to centennial time scales. As well, biotic regulation of the nitrogen cycle and the influence of nitrogen deposition on the carbon cycle include potential non-linear feedbacks. For example, NO_x and N_2O emissions by soils can increase disproportionately to nitrogen inputs (Hall and Matson 1999). The interaction of O_3 and N content of leaves may produce a collapse in the regu-

Box 15.1. Missing elements and interactions in current models of C and N cycles

1. Canopy interactions.
2. Soil chemistry-pH effects.
3. Direct toxic and nutrient impacts of reactive oxidized N and reduced N, O_3, HO_2, per-oxy radical damage, NO_2 and PAN phyto-toxicity.
4. Atmospheric chemistry feedbacks on N chemical speciation.
5. Biotic impacts of N deposition on plant and microbial species interactions and trophic relationships.

lation of stomatal conductance. Furthermore, at rates of N deposition above a critical threshold, carbon uptake may stop or slow and forest dieback ensue, and export of nitrogen to the hydrologic system can increase disproportionately relative to inputs. Current terrestrial biogeochemistry models fail to capture these important non-linearities in N and C cycling (Table 15.3; Box 15.1). Models that include land use change, CO_2, and O_3 effects demonstrate key interactions among these environmental changes (Figures 15.3 and 15.4; Ollinger and Aber 2002).

Atmospheric Chemistry and N Speciation

The bioatmospheric N cycle is intricately linked to atmospheric chemistry through tropospheric ozone formation, nitrogen deposition, aerosol formation, stratospheric ozone depletion, and the chemical speciation of N-determined hydrocarbon, NO_x, and O_3 interactions. Nitrogen oxides are of central importance in tropospheric photochemistry. They catalyze the production of ozone during the oxidation of carbon monoxide, methane, and nonmethane hydrocarbons (NMHCs). In most regions of the troposphere, ozone production is limited primarily by the abundance of NO_x (Chameides et al. 1992). The dependence of ozone production on NO_x, however, is highly non-linear (Liu et al. 1987; Ehhalt, Rohrer, and Wahner 1992). Nitrogen oxides indirectly control the oxidizing capacity of the atmosphere by regulating the abundance of ozone, which in turn regulates the abundance of the hydroxyl (OH) radical, the primary atmospheric oxidant. Sources of NO_x to the troposphere have increased considerably over the past century as a result of increases in fossil fuel combustion, biomass burning, and fertilizer use (Galloway et al. 1995). The short lifetime (typically less than a day) of NO_x and the heterogeneous distribution of its emission sources pose a challenge for simulating the global distribution of NO_x in the troposphere. A three-dimen-

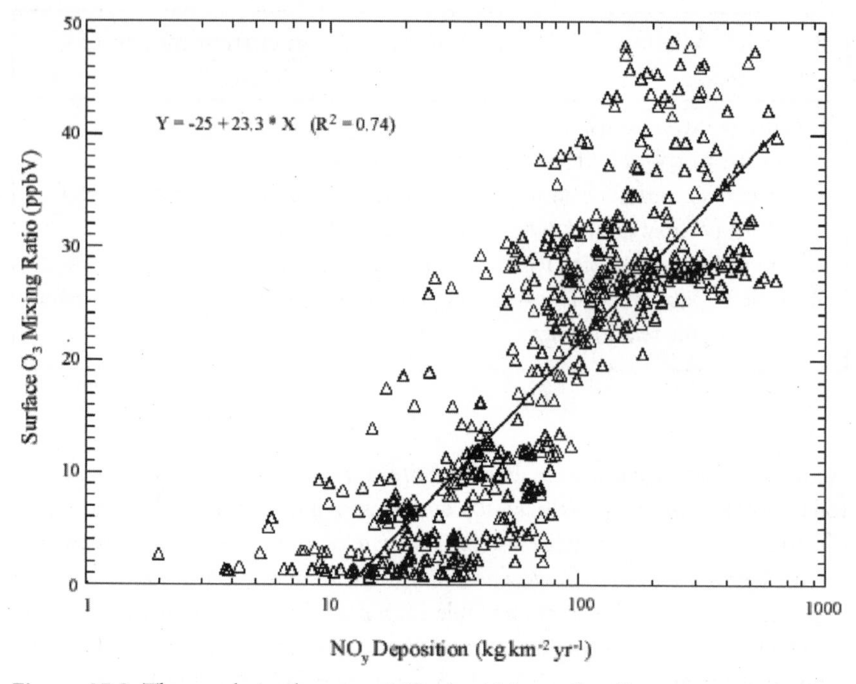

Figure 15.3. The correlation beetween NO_y deposition and surface ozone concentrations predicted by IMAGES, a 3-D chemical transport model. The correlation occurs because both depend on the same sets of chemical reactions, precursors (from Holland et al. 1997).

sional model, with accurate representations of the sources and sinks of NO_x, is necessary, as is a detailed description of the chemical interactions of NO_x with ozone and NMHCs (Horowitz et al. 1998).

Once emitted to the atmosphere, NO_x interacts with many other chemical species and undergoes rapid photochemical conversion to other forms of reactive nitrogen. It can be converted to more stable "reservoir species" such as organic nitrates (including peroxyacetyl nitrate [PAN]), which can be transported over large distances. The primary mechanism for NO_x removal from the atmosphere is by oxidation to nitric acid (HNO_3), either through gas-phase reaction with OH or through heterogeneous reaction with ozone. Nitric acid is removed by wet and dry deposition or can be taken up by aqueous aerosols (as nitrate ion). This uptake by aerosols depends strongly on temperature and the relative abundance of ammonium and sulfate ions in the aerosol particles, since un-neutralized nitrate aerosols are thermodynamically unstable at the warm temperatures found in the lower troposphere (Adams, Seinfeld, and Koch 1999).

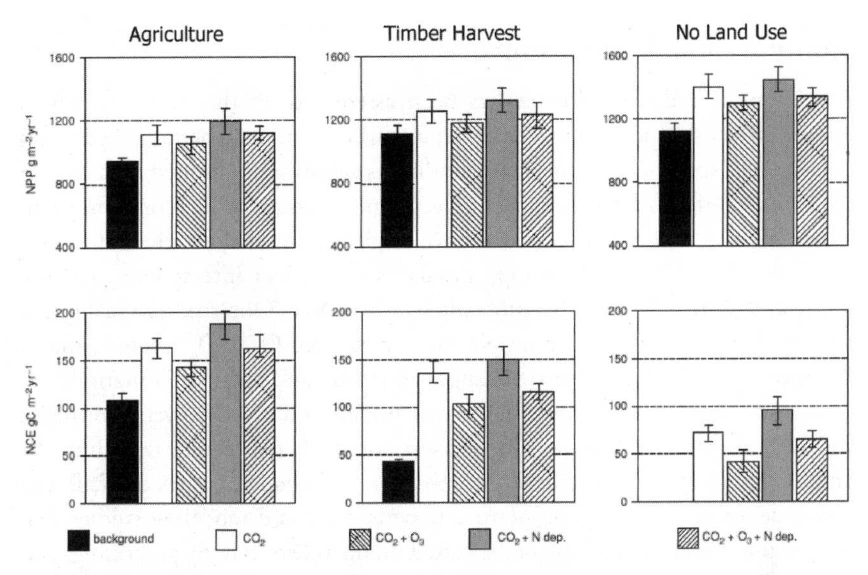

Figure 15.4. Carbon uptake and fluxes under different land use scenarios with different combinations of CO_2, O_3, and N deposition (from Ollinger and Aber, 2002).

Reactive nitrogen, NO_y and NH_x, are removed from the atmosphere primarily by wet and dry deposition. Species such as HNO_3, NH_3, some organic nitrates, and nitrate and ammonium aerosols are highly soluble and are readily removed by precipitation. This wet deposition typically limits the lifetimes of these species in the troposphere to about a week. Oxidants and soluble species are also removed from the atmosphere by dry deposition onto the surfaces of vegetation, soil, ocean, or other surfaces. The rate at which a given species deposits to the surface depends on chemical properties of the species, atmospheric conditions, and the surface type. For many species (including NO and NO_2 , as for O_3) the deposition velocity depends strongly on vegetation type, with uptake occurring through the stomata or the surface of leaves (Wesely 1989). For other species, such as HNO_3, deposition is limited primarily by the rate of turbulent mixing from the atmospheric boundary layer down to the surface. In order to realistically represent dry deposition of trace gases in a chemical transport model, it is necessary to include couplings to boundary layer and land surface models. Previous studies (e.g., Ganzeveld and Lelieveld 1995; Ganzeveld, Lelieveld, and Roelofs 1998; Ganzeveld et al. 2002) indicate that including a more realistic representation of dry deposition can considerably improve the quality of a simulation in a chemical transport model, and such a representation is critical to modeling intersystem nitrogen cycling and addressing questions regarding the impact of deposited N on carbon cycling and sequestration.

Summary and Conclusions

A substantial body of evidence has been assembled on the extent to which humans affect the global nitrogen cycle, in particular the impact of added N on ecosystems. Nonetheless, our understanding is significantly limited by a lack of results of the effects of deposited N and its chemical speciation on ecosystem properties. This is in part because we have neglected to include the important processes, consideration of canopy uptake, soil nutrient interactions, and the direct and indirect effects of atmospheric chemistry. Our studies are limited because they fail to consider multiple stresses: elevated O_3, CO_2 aerosols, and N deposition together with climate changes and water and nutrient limitations.

A more comprehensive and quantitative understanding of ecosystem nitrogen and carbon cycling is required to develop our predictive ability regarding the impacts of changing atmospheric composition and changing rates of N, P, and cation deposition on these properties. Experimental and modeling studies that address the speciation and fate of deposited N, the role of canopy processing, and subsequent change with changing atmospheric composition and atmosphere-biosphere feedbacks are needed. This data plus an ability to assess the effects of multiple stresses must be high priorities for the atmosphere and biospheric science communities.

Literature Cited

Aber, J. D., J. K. Nadelhoffer, P. A. Steudler, and J. M. Melillo. 1989. Nitrogen saturation in northern forest ecosystems: Hypotheses and implications. *BioScience* 39:378–386.

Aber, J., W. McDowell, J. K. Nadelhoffer, G. Bernston, M. Kamakea, S. McNulty, W. Currie, L. Rustad, and I. Fernandez. 1998. Nitrogen saturation in temperate forest ecosystems. *BioScience* 11:921–934.

Adams, P. J., J. H. Seinfeld, and D. M. Koch. 1999. Global concentrations of tropospheric sulfate, nitrate, and ammonium aerosol simulated in a general circulation model. *Journal of Geophysical Research* 104:13,791–13,823.

Aerts, R., F. Berendse, and N. M. Klerk. 1989. Root production and root turnover in two dominant species of wet heathlands. *Oecologia* 81:374.

Aerts, R., F. Berendse, and H. de Caluwe. 1990. Competition in heathland along an experimental gradient of nutrient availability. *Oikos* 57:310.

AGU (American Geophysical Union). 2000. Nitrogen loading in coastal water bodies: An atmospheric perspective, edited by R. Valigura, R. Alexander, M. S. Castro, T. Meyers, H. Paerl, P. E. Stacey, and R. E. Runder. Coastal and Estuarine Studies 57. Washington, D.C.

Albritton, D. L., M. R. Allen, A. P. M. Baede, J. A. Church, U. Cubasch, D. Xiaosu, D. Yihui, D. H. Ehhalt, C. K. Folland, F. Giorgi, J. M. Gregory, D. J. Griggs, J. M. Haywood, B. Hewitson, J. T. Houghton, J. I. House, M. Hulme, I. Isaksen, V. J.

Jaramillo, A. Jayaraman, C. A. Johnson, F. Joos, S. Joussaume, T. Karl, D. J. Karoly, H. S. Kheshgi, C. Le Quéré, K. Maskell, L. J. Mata, B. J. McAvaney, M. McFarland, L. O. Mearns, G. A. Meehl, L. G. Meira-Filho, V. P. Meleshko, J. F. B. Mitchell, B. Moore, R. K. Mugara, M. Noguer, B. S. Nyenzi, M. Oppenheimer, J. E. Penner, S. Pollonais, M. Prather, I. C. Prentice, V. Ramaswamy, A. Ramirez-Rojas, S. C. B. Raper, M. J. Salinger, R. J. Scholes, S. Solomon, T. F. Stocker, J. M. R. Stone, R. J. Stouffer, K. E. Trenberth, M. X. Wang, R. T. Watson, K. S. Yap, and J. Zillman. 2001. *Summary for policy makers.* A report of Working Group 1 of the Intergovernmental Panel on Climate Change. London: Cambridge University Press.

Amman, M., R. Siegwolf, F. Pichlmayer, M. Suter, M. Sauerer, and C. Brunold. 1999. Estimating the uptake of traffic-derived NO_2 from ^{15}N abundance in Norway spruce needles. *Oecologia* 118:124–131.

Berendse, F., W. T. Elberse, and R. H. M. E. Geertz. 1992. Competition and nitrogen loss from plants in grassland ecosystems. *Ecology* 73:46.

Berendse, F., R. Aerts, and R. Bobbink. 1993. Atmospheric N deposition and its impact on terrestrial ecosystems. In *Landscape ecology of a stressed environment,* edited by C. C. Vos and P. Opdam. London: Chapman and Hall.

Bobbink, R., G. W. Heil, and M. B. A. G. Raeseen. 1992. Atmospheric deposition and canopy exchange processes in heathland ecosystems. In *Critical loads for nitrogen.* Report Nord. Copenhagen: Nordic Council of Ministers.

Bolin, B., and R. B. Cook, eds. 1983. *The major biogeochemical cycles and their interactions.* SCOPE 21. New York: John Wiley.

Boyer, E. W., and R. W. Howarth, eds. 2002. *The nitrogen cycle at regional to global scales.* Dordrecht, Netherlands: Kluwer.

Broecker, W. S., and G. M. Henderson. 1998. The sequence of events surrounding Termination II and their implications for the cause of glacial-interglacial CO_2 changes. *Paleooceanography* 13:352–364.

Chameides, W. L., F. Fehsenfeld, O. Rodgers, C.. Cardelino, J. Martinez, D. Parrish, W. Lonneman, D. R. Lawson, R. A. Rasmussen, P. Zimmerman, J. Greenberg, P. Middleton, and T. Wang. 1992. Ozone precursor relationships in the ambient atmosphere. *Journal of Geophysical Research* 97:6037–6055.

Chameides, W. L., P. S. Kasibhatla, J. Yienger, and H. Levy II. 1994. Growth of continental-scale metro-agro-plexes, regional ozone pollution, and world food production. *Science* 264:74–77.

Chapin III, F. S., B. H. Walker, R. J. Hobbs, D. H. Hooper, J. H. Lawton, O. E. Sala, and D. Tilman. 1997. Biotic control over the functioning of ecosystems. *Science* 277:500–503.

Dickinson, R. E., J. A. Berry, G. B. Bonon, G. J. Collatz, C. B. Field, I. Y. Fung, M. Goulden, W. A. Hoffman, R. B. Jackson, R. Myneni, P. J. Sellers, and M. Shaikh. 2002. Nitrogen controls on climate model evapotranspiration. *Journal of Climate* 15:278–295.

Durka, W. E., D. Schulze, G. Gebauer, and S. Voerkelius. 1994. Effects of forest decline and leaching of deposited nitrate determined from ^{15}N and ^{18}O measurements. *Nature* 372:765–767.

Edgerton, E. S., T. F. Lavery, and R. P. Boksleitner. 1992. Preliminary data from the USEPA dry deposition network: 1989. *Environmental Pollution* 75: 145–155.

EMEP (Co-operative programme for monitoring and evaluation of the long-range transmissions of air pollutants in Europe). 1995. *Data report 1995.* http://www.emep.int/index.html. Oslo.

Ehhalt, D. H., F. Rohrer, and A. Wahner. 1992. Sources and distribution of NO_x in the upper troposphere at northern and mid-latitudes. *Journal of Geophysical Research* 97:3725–3738.

Galloway, J. N., L. I. Hiram, and P. S. Kasibhatla. 1994. Year 2020: Consequences of population growth and development on deposition of oxidized nitrogen. *Ambio* 23:120–123.

Galloway, J. N., W. H. Schlesinger, H. Levy II, A. Michaels, and J. L. Schnoor. 1995. Nitrogen fixation: Anthropogenic enhancement-environmental response. *Global Biogeochemical Cycles* 9:235–252.

Ganzeveld, L., and J. Lelieveld. 1995. Dry deposition parameterization in a chemistry general circulation model and its influence on the distribution of reactive trace gases. *Journal of Geophysical Research* 100D:20,999–21,012.

Ganzeveld, L., H. Lelieveld, and G.-J. Roelofs. 1998. A dry deposition parameterization for sulfur oxides in a chemistry and general circulation model. *Journal of Geophysical Research* 103:5679–5694.

Ganzeveld, L., J. Lelieveld, F. Dentener, M. Krol, L. Bouwman, and G.-J. Roelofs. 2002. Global soil-biogenic NO_x emissions and the role of canopy processes. *Journal of Geophysical Research* 107:4298.

Hall, S. J., and P. A. Matson. 1999. Nitrogen oxide emissions after N additions in tropical forests. *Nature* 400:152–155.

Hansen, J. E., M. Sato, A. Lacis, R. Ruedy, I. Tegen, and E. Matthews. 1998. Climate forcings in the industrial era. *Proceedings of the National Academy of Sciences USA* 95:12,753–12,758.

Hansen, J. E., ed. 2002. *Air pollution as a climate forcing: A workshop, April 29–May 3, 2002.* New York: Goddard Institute for Space Studies Library.

Harrison, A. F., E. D. Schulze, G. Gebauer, and G. Bruckner. 2000. Canopy uptake and utilization of atmospheric pollutant nitrogen. In *Carbon and nitrogen cycling in European forest ecosystems,* edited by E. D. Schulze. Ecological Studies 142. New York: Springer. CD-ROM.

Holland, E. A., W. J. Parton, J. K. Detling, and D. L. Coppock. 1992. Physiological responses of plant populations to herbivory and their consequences for ecosystem nutrient flow. *American Naturalist* 140:685–706.

Holland, E. A., B. H. Braswell, J.-F. Lamarque, A. Townsend, J. M. Sulzman, J.-F. Müller, F. Dentener, G. Brasseur, H. Levy II, J. E. Penner, and G. Roelofs. 1997. Variations in the predicted spatial distribution of atmospheric nitrogen deposition and their impact on carbon uptake by terrestrial ecosystems. *Journal of Geophysical Research* 102:15,849.

Holland, E. A., F. J. Dentener, B. H. Braswell, and J. M. Sulzman. 1999. Contemporary and pre-industrial global reactive nitrogen budgets *Biogeochemistry* 46:7–43.

Holland, E. A., B. H. Braswell, O. Bossdorf, and J. C. Neff. 2000. Changes in the nitrogen and carbon budgets of Eastern U.S. forest ecosystems. Abstract of a paper presented at the annual meeting of the Ecological Society of America, August 6–10, Snowbird, Utah.

Holland, E. A., B. H. Braswell, J. Sulzman, and J.-F. Lamarque. 2003. Nitrogen deposition onto the United States and Western Europe: A synthesis of observations and models. *Global Biogeochemical Cycles,* submitted.

Hooper, D. U., and P. M. Vitousek. 1997. The effects of plant composition and diversity on ecosystem processes. *Science* 277:1302–1305.

Horowitz, L. W., J. Liang, G. M. Gardner, and D. J. Jacob. 1998. Export of reactive nitrogen from North America during summertime: Sensitivity to hydrocarbon chemistry. *Journal of Geophysical Research* 103:13,451–13,476.

Howarth, R. W., G. Billen, D. Swaney, A. Townsend, N. Jaeorski, K. Lajtha, J. A. Downing, R. Elmgren, N. Caraco, T. Jordan, F. Berendse, J. Freney, V. Kudeyarov, P. Murdoch, and Z. Zhao-liang. 1996. Regional nitrogen budgets and riverine N & P fluxes for the drainages to the North Atlantic Ocean: Natural and human influences. *Biogeochemistry* 35:75–139.

Howarth, R. W., ed. 1996. Nitrogen cycling in the North Atlantic Ocean and its watersheds. Dordrecht, Netherlands: Kluwer.

Husar, R. B. 1986. Emissions of sulfur dioxide and nitrogen oxides and trends for Eastern North America. In *Acid deposition: Long-term trends.* Washington, D.C.: National Academy Press.

International Global Atmospheric Chemistry Project. 2003. *Atmospheric chemistry in a changing world: An integration and synthesis of a decade of tropospheric chemistry research.* New York: Springer Verlag.

IPCC (Intergovermental Panel on Climate Change). 2001. *Climate change 2001: The scientific basis.* The Working Group I contribution to the IPCC. Cambridge: Cambridge University Press.

Jacob, D. J., J. A. Logan, and P. P. Murti. 1999. Effect of rising Asian emissions on surface ozone in the United States. *Geophysical Research Letters* 26:2175–2178.

Jacob, D. J. 2000. Heterogeneous chemistry and tropospheric ozone. *Atmospheric Environment* 34:2131–2159.

Lamb, D., and V. C. Bowersox. 2000. National Atmospheric Deposition Program (NADP). *Atmospheric Environment* 34:1661–1738.

Lerdau, M.T., J. W. Munger, and D. Jacob. 2000. The NO_2 flux conundrum. *Science* 289:2291–2293.

Liu, S. C., M. Trainer, F. C. Fehsenfeld, D. D. Parrish, E. J. Williams, D. W. Fahey, G. Hübler, and P. C. Murphy. 1987. Ozone production in the rural troposphere and the implications for regional and global ozone distributions. *Journal of Geophysical Research* 92:4191–4207.

Lovett, G. M., and S. E. Lindbergh. 1993. Atmospheric deposition and canopy interactions of nitrogen in forests. *Canadian Journal of Forest Research* 23:1,603–1,616.

Lovett, G. M. 1994. Atmospheric deposition of nutrients and pollutants in North America: An ecological perspective. *Ecological Applications* 4:629–650.

Lovett, G. M., K. C. Weathers, and W. V. Sobczak. 2000. Nitrogen saturation and retention in forested watersheds of the Catskill Mountains, New York. *Ecological Applications* 9:73–84.

Mayewski, P. A., W. B. Lyons, and M. J. Spencer. 1990. An ice-core record of atmospheric response to anthropogenic sulfate and nitrate. *Nature* 346:554–556.

Monson, R. K., and E. A. Holland. 2001. Biospheric trace gas fluxes and their control over tropospheric chemistry. *Annual Reviews in Ecology and Systematics* 32:547–576.

Mosier, A. R., M. A. Bleken, P. Chaiwanakupt, E. C. Ellis, J. R. Freney, R. B. Howarth,

P. A. Matson, K. Minami, R. Naylor, K. N. Weeks, and Z. L. Zhu. 2002. Policy implications of human-accelerated nitrogen cycling. *Biogeochemistry* 57/58:477–516.
Munger, J. W., S.-M. Fan, P. S. Bakwin, M. L. Goulden, A. H. Goldstein, A. S. Colman, and S. C. Wofsy. 1998. Regional budgets for nitrogen oxides from continental sources: Variations of rates for oxidation and deposition with season and distance from source regions. *Journal of Geophysical Research* 103:8355–8368.
Nadelhoffer, K. J., B. A. Emmett, P. Gundersen, O. J. Kjonass, C. J. Koopmans, P. Schleppi, A. Tietema, and R. F. Wright. 1999a. Nitrogen deposition makes a minor contribution to carbon seqestration in temperate forests, reply. *Nature* 400:630.
———. 1999b. Scientific correspondence. *Nature* 400:630.
Nadelhoffer, K. J., M. R. Downs, and B. Fry. 1999c. Sinks for ^{15}N-enriched additions to an oak forest and a red pine plantation. *Ecological Applications* 9:72–86.
NADP/NTN (National Atmospheric Deposition Program (NRSP-3)/National Trends Network). 1995. http://nadp.sws.uiuc.edu/. NADP/NTN Coordination Office, Natural Resource Ecology Laboratory, Colorado State University, Fort Collins, Colorado.
NARSTO (North American Research Strategy on Tropospheric Ozone). 2000. *An assessment of tropospheric ozone pollution: A North American perspective.* (http://cdiad/esd.ornl.gov/programs/NARSTO). Washington, D.C.: Commission on Geosciences Environment and Resources, National Academy Press.
Neff, J. C., E. A. Holland, F. J. Dentener, W. H. McDowell, and K. M. Russel. 2002a. Atmospheric organic nitrogen: Implications for the global N cycle. *Biogeochemistry* 57/58:99–136.
Neff, J. C., A. R. Townsend, G. Gleixner, S. J. Lehman, J. Turnbull, and W. D. Bowman. 2002b. Variable effects of nitrogen additions on the stability and turnover of soil carbon. *Nature* 419:915–917.
Olivier, J. G. J., A. F. Bouwman, K. W. van der Hoek, and J .J. M. Berdowski. 1999. Global air emission inventories for anthropogenic sources of NO_X, NH_3 and N_2O in 1990. *Environmental Pollution* 102:135–148.
Ollinger, S. V., J. D. Aber, and P. B. Reich. 1997. Simulating ozone effects on forest productivity: Interactions among leaf-, canopy-, and stand-level processes. *Ecological Applications* 7:1237–1251.
Ollinger, S. V., and J. D. Aber. 2002. The interactive effects of land use, carbon dioxide, ozone, and N deposition. *Global Change Biology* 8:545–562.
Parton, W. J., D. S. Schimel, C. V. Cole, and D. S. Ojima. 1987. Analysis of factors controlling soil organic matter levels in Great Plains grasslands. *Soil Science Society of America Journal* 51:1173–1179.
Paerl, J. W. 1997. Coastal eutrophication and harmful algal blooms: Importance of atmospheric deposition and groundwater as "new" nitrogen and other nutrient sources. *Limnology and Oceanography* 42:1154–1165.
Peterson, B. J., and J. M. Melillo. 1985. The potential storage of carbon caused by eutrophication of the biosphere. *Tellus* Series B 37:117–127.
Prather, M., D. Ehhalt, F. J. Dentener, R. Derwent, E. Dlugokencky, E. Holland, I. Isaksen, J. Katima, V. Kirchhoff, P. Matson, P. Midgley, T. Wang, T. Berntsen, I. Bey, G. Brasseur, L. Buja, W. Collins, J. J. Daniel, D. W. B. N. Derek, R. Dickerson, D. Etheridge, J. Feichter, P. Fraser, R. Friedl, J. Fuglestvedt, M. N. Gauss, L. U. Grenfell, A. A. Grübler, N. U. Harris, D. F. Hauglustaine, L. U. Horowitz, C. U. Jackman, D. U. Jacob, L. U. Jaeglé, A. U. Jain, M. G. Kanakidou, S. N. Karlsdottir, M. U. Ko, R.

G. von Kuhlmann, M. U. Kurylo, M. U. Lawrence, J. A. U. Logan, M. N. Z. Manning, D. U. Mauzerall, J. C. McConnell, L. U. Mickley, S. U. Montzka, J. F. B. Müller, J. N. Oliver, K. U. Pickering, G. I. Pitari, G. J. N. Foelofs, H. U. Rogers, B. N. Rognerud, S. U. Solomon, J. S. Staehelin, P. A. Steele, D. U. Stevenson, J. N. Sundet, S. U. Smith, A. U. Thompson, M. N. van Weele, Y. U. Wang, D. U. Weisenstein, T. U. Wigley, O. U. Wild, D. U. Wuebbles, and R. U. Yantosca. 2001. Atmospheric chemistry and greenhouse gases. Contribution of Working Group 1. *Third Annual Assessment Report of the Intergovernmental Panel on Climate Change.* Cambridge: Cambridge University Press.

Reich, P. B. 1987. Quantifying plant response to ozone: A unifying theory. *Tree Physiology* 3:63–91.

Reich, P. B., D. S. Ellsworth, B. D. Koeppel, J. H. Fownes, and S. T. Gower. 1990. Vertical variation in canopy structure. *Tree Physiology* 7:349–345.

Schindler, D. W., and S. E. Bayley. 1993. The biosphere as an increasing sink for atmospheric carbon: Estimates from increased nitrogen deposition. *Global Biogeochemical Cycles* 7:717–734.

Schulze, E. D., W. De Vries, M. Hauhs, K. Rosén, L. Rasmussen, O.-C. Tann, and J. Nilsson. 1989. Critical loads for nitrogen deposition in forest ecosystems. *Water, Air, and Soil Pollution* 48:451–456.

Schulze E. D. 2000. Carbon and nitrogen cycling in European forest ecosystems. *Ecological Studies* 142. New York: Springer.

Siegwolf, R. T., R. Mayssek, M. Saurer, S. Maurer, M. S. Guenthardt-Goerg, P. Schmutz, and J. B. Bucher. 2001. Stable isotope analysis reveals differential effects of soil nitrogen and nitrogen dioxide on the water use efficiency of hybrid poplar leaves. *New Phytologist* 149:233–246.

Sievering, H. I., J. Fernandes, J. Lee, J. Hom, and L. Rustad. 2000. Forest canopy uptake of atmospheric N deposition at eastern U.S. conifer sites: Carbon storage implications? *Global Biogeochemical Cycles* 14:1153–1160.

Sparks, J. P., R. K. Monson, K. L. Sparks, M. Lerdau. 2001. Leaf uptake of nitrogen dioxide (NO_2) in a tropical wet forest: Implications for tropospheric chemistry. *Oecologia* 127:214–221.

Sutton, M. A., C. E. R. Pitcairn, and D. Fowler. 1993. The exchange of ammonia between the atmosphere and plant communities. *Advances in Ecological Research* 24:301.

Tilman, D., J. Knops, D. Wedin, P. Reich, M. Ritchie, and E. Siemann. 1997. The influence of functional diversity and composition on ecosystem processes. *Science* 277:1300–1302.

Townsend, A. R., B. H. Braswell, E. A. Holland, and J. E. Penner. 1996. Spatial and temporal patterns in potential terrestrial carbon storage resulting from deposition of fossil fuel derived nitrogen. *Ecological Applications* 6:806–814.

Townsend, A. R., ed. 1999. New perspectives on nitrogen cycling in the temperate and tropical Americas. Report of the International SCOPE Nitrogen Project. *Biogeochemistry* 46:1–293.

U.S. Environmental Protection Agency. 2003. Clean Air Status and Trends Network. http://www.epa.gov/castnet/. Washington, D.C.

VEMAP (Vegetation/Ecosystem Modeling and Analysis Project). 1995. Vegetation/ Ecosystem Modeling and Analysis Project: Comparing biogeography and biogeochemistry models in a continental-scale study of terrestrial ecosystem responses to climate change and CO_2 doubling. *Global Biogeochemical Cycles* 9:407–437.

Vitousek, P. M., and R. W. Howarth. 1991. Nitrogen limitation on land and in the sea: How can it occur? *Biogeochemistry* 13:87–115.

Vitousek, P. M. 1994. Beyond global warming: Ecology and global change. *Ecology* 75:1861–1876.

Vitousek, P. M., J. D. Aber, R. W. Howarth, G. E. Likens, P. A. Matson, D. W. Schindler, W. H. Schlesinger, and D. G. Tilman. 1997. Human alteration of the global nitrogen cycle: Sources and consequences. *Ecological Applications* 7:737–750.

Wedin, D. A., and D. Tilman. 1996. Influence of nitrogen loading and species composition on the carbon balance of grasslands. *Science* 274:1720–1723.

Wesely, M. L. 1989. Parameterization of surface resistances to gaseous dry deposition in regional-scale numerical models. *Atmospheric Environment* 23:1293–1304.

Wesely, M. L., and B. B. Hicks. 1999. A review of the current status of knowledge on dry deposition. *Atmospheric Environment* 34:2261–2282.

Whelpdale, D. M., and M. S. Kaiser. 1997. *Global acid deposition assessment.* Geneva: World Meterological Organization.

Wildt, J., D. Kley, A. Roeckel, P. Roeckel, and J. J. Segschneider. 1997. Emission of NO from several higher plant species. *Journal of Geophysical Research* 102:5919–5927.

PART V
Hydrosphere

16

The Role of Iron as a Limiting Nutrient for Marine Plankton Processes

Hugh W. Ducklow, Jacques L. Oliver, and Walker O. Smith Jr.

Our understanding of elemental cycling in the ocean, its mechanisms and implications for ecology and climate, has undergone revolutionary change since the previous SCOPE volume was published in 1983 (Bolin and Cook 1983). The most notable of the advances in ocean biogeochemistry concern the role of the oceans in the global carbon cycle (Hanson, Ducklow, and Field 2000) and recognition of iron as a key nutrient limiting ocean photosynthesis, carbon storage, and glacial-interglacial variations in atmospheric carbon dioxide concentrations. De Baar (1994) elegantly and comprehensively reviewed iron limitation up to the advent of the *in situ* iron fertilization studies of the past decade (see below). In this chapter we outline biological control of the ocean carbon cycle, focusing on the role of iron and its interactions with carbon, nitrogen, and phosphorus.

The Ocean Carbon Cycle, the Biological Pump, and Iron Limitation

Atmospheric CO_2 exchanges with the surface ocean by thermodynamic processes, as modulated by ocean circulation and biological cycling (Wallace 2001; Takahashi et al. 2002). Variations in temperature, wind, regional to basin-scale circulation, and ocean ecology result in a characteristic geographical distribution of the partial pressure of CO_2 (pCO_2) in the ocean surface layer, with strong atmospheric sources in the equatorial zone and strong sinks (areas of net ocean uptake) at higher latitudes, particularly in the North Atlantic and Southern Oceans (Takahashi et al. 2002). Primary productivity in the sea is dominated by unicellular phytoplankton including species less than one micron in diameter (picoplankton; Chisholm et al. 1988; Karl 1999). These autotrophs convert CO_2 into reduced organic matter, contributing about half the total global photosynthesis of 105 petagrams C (10^{15} gC; PgC) per year (Field et al. 1998) and drawing down pCO_2 below the atmospheric equilibrium. Carbon fixed in organic tissues is either respired in the surface layer or else transported to depth. Only about 0.1 percent of the organic carbon exported from the surface layer is buried in deep ocean sediments; most is oxidized back to CO_2 by bacteria in the water column. The net transport of organic carbon into the deep sea and its remineralization there are the result of complex interactions between trophodynamic processes, including ingestion by zooplankton, excretion of dissolved organic compounds and oxidation by bacteria, particle aggregation, gravitational settling, and advection. These processes combine to pump atmospheric CO_2 into the deep ocean (Volk and Hoffert 1985; Longhurst and Harrison 1989). This biological pump is responsible for maintaining about 80–90 percent of the vertical gradient in dissolved inorganic carbon (DIC) between the ocean surface and deep water, with the rest being due to physical solubility and circulation (Gruber and Sarmiento 2002).

The ecological process most responsible for regulating the biological pump is new production, or that fraction of the marine primary production supported by inputs of "new" or externally supplied limiting nutrients. Over sufficiently large time and space scales, the new production balances losses, or export of organic matter from the euphotic zone into the deep sea (Eppley and Peterson 1979; Ducklow 1995), and is about 10 percent of the total oceanic production. Traditionally, the most common limiting nutrient for marine productivity was assumed to be nitrogen, supplied as nitrate from deep water (Dugdale and Goering 1967). This traditional view is accurate for the oligotrophic gyres and the highly seasonal North Atlantic, where N and P are depleted in the surface layer by phytoplankton utilization, either seasonally or persistently. But in vast expanses of the Southern Ocean and the equatorial and subarctic North Pacific, these macronutrients remain at high concentrations and are never fully utilized by plant growth (Ruud 1930; McAllister, Parsons, and Strickland 1960). Moreover, and paradoxically, chlorophyll *a* remains persistently low in these regions of nutrient plenitude (Miller et al. 1991). These puzzling regimes are termed "high-nutrient–low-chlorophyll" (HNLC) systems (Colorplate 4). In these regions some other factor besides N or P must be limiting phytoplankton growth. Gran (1931) first hypothesized iron limitation in place of N or P. John Martin recast the modern debate over the maintenance of the HNLC condition by hypothesizing that iron limits primary production in the HNLC regimes (Martin and Fitzwater 1988).

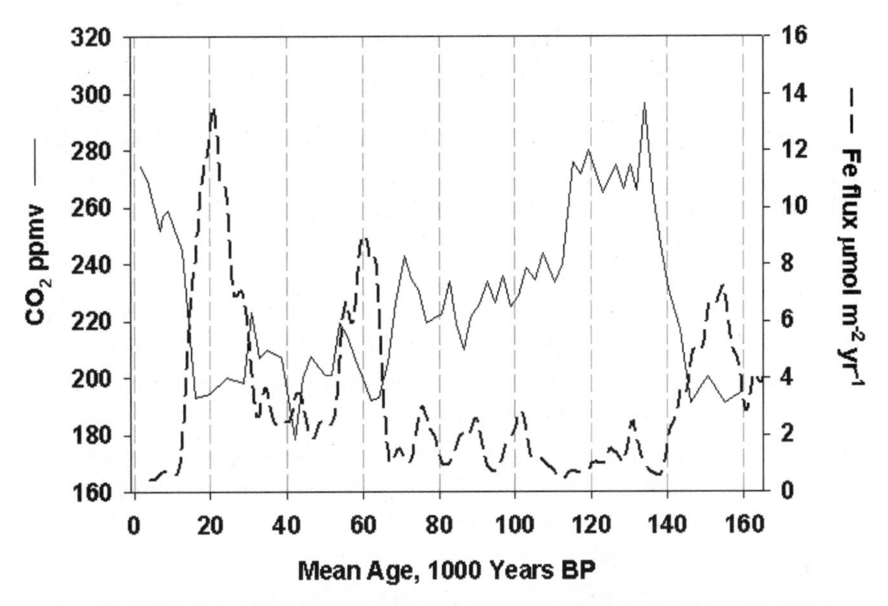

Figure 16.1. The Iron Hypothesis. Mean CO_2 concentrations from air entrapped in the Vostok ice core (Barnola et al. 1987) and mean iron flux in atmospheric dust (after Petit et. al. 1990). The data are from http://www.ngdc.noaa.gov/paleo/icecore/antarctica/vostok/. Iron flux was calculated from Petit's dust flux using a crustal iron abundance of 5.63×10^4 mg kg^{-1}. Graph revised from Martin (1990).

Solving the puzzle of HNLC systems is a key to a deeper understanding of climate variation. Three seminal papers (Knox and McElroy 1984; Sarmiento and Toggweiler 1984; Siegenthaler and Wenk 1984) linked the ocean carbon cycle to glacial-interglacial variations in atmospheric CO_2 and focused new attention on ocean biogeochemistry as a key driver of global climate. In the persistently nutrient-rich oceans DIC is not drawn down to the levels it would be if the deepwater macronutrients were fully utilized. In other words, the biological pump is not working at full efficiency in HNLC regions. If it were, atmospheric CO_2 would be 165 parts per million (ppm) in the absence of anthropogenic forcing, compared with the preindustrial value of 280 or the current, anthropogenically forced level of 370 ppm. Martin showed a striking correlation between glacial-interglacial iron supply rates to the ocean and atmospheric CO_2 (Figure 16.1) and boldly claimed that variations in iron regulated global ocean productivity and the efficiency of the biological pump, thus controlling the atmospheric CO_2 (Martin 1990). This is the Iron Hypothesis.

Iron is the fourth most abundant element in the Earth's crust and thus plentiful in windborne dust blown over the oceans (Colorplate 5), but it exists in dissolved (biologically available) form in only vanishingly low concentrations in the ocean. The mean concentration in the upper 200 meters is just 0.07 nanomoles per kilogram of seawater (nmol kg^{-1}) (Johnson, Gordon, and Cole 1997), a level comparable to the half-saturation coefficient for iron uptake by phytoplankton (Coale et al. 1996a). Iron concentration is kept low in surface waters by biological uptake and the ultralow solubility of Fe(III), the principal form in oxygenated solution. The rest stays in the particulate phase and is rapidly removed by sedimentation. Atmospheric deposition is the principal source of support for iron-based new production (Duce and Tindale 1991), supplying 96×10^9 mol per year to the open sea (Fung et al. 2000) away from continental margins, where river outflow, sediments, and upwelling are additional sources. This flux is well in excess of the model-estimated phytoplankton assimilation rate (12×10^9 mol yr^{-1} [Fung et al. 2000]), but only 1–10 percent of the iron is soluble. Thus the transport of dust in the atmosphere is a vector ultimately linking together ocean productivity and climate with the terrestrial biosphere.

Confirming Iron Limitation of Phytoplankton Growth

Using trace-metal clean sampling techniques, Martin and Fitzwater (1988) added trace amounts of iron (<2.5 nanomolar) to bottles containing natural phytoplankton assemblages and found marked increases in biomass through time. Because bottles do not contain copepod grazers and because microzooplankton often suffer declines under containment, the applicability of these results to *in situ* processes was questioned (Banse 1994). At the same time it became possible to add an inert tracer to the ocean in extremely low quantities (sulfur hexaflouride; SF_6) to follow a patch of the ocean through time. By adding iron and SF_6 simultaneously, it was possible to investigate the effects of trace metal additions to natural plankton communities without enclosures. Thus, oceanographers were able to manipulate the ocean on a scale never before possible and to follow the effects of the perturbation unambiguously.

The first iron addition experiment (IronEx-I) added iron and SF_6 once at the start of the experiment off the Galapagos. Kolber et al. (1994) convincingly showed a dramatic increase in photochemical efficiency upon iron addition for all components and sizes of phytoplankton, suggesting that photosynthesis was indeed limited by iron concentrations. Lindley and Barber (1998) also found that the quantum yield (a measure of photochemical efficiency) increased within twenty-four hours of iron addition and remained elevated for at least seven days. Although photochemical and physiological enhancements of phytoplankton photosynthesis were noted, no nutrient or CO_2 reduction was observed (Watson et al. 1994). Banse (1994) argued that this was because grazing and removal processes

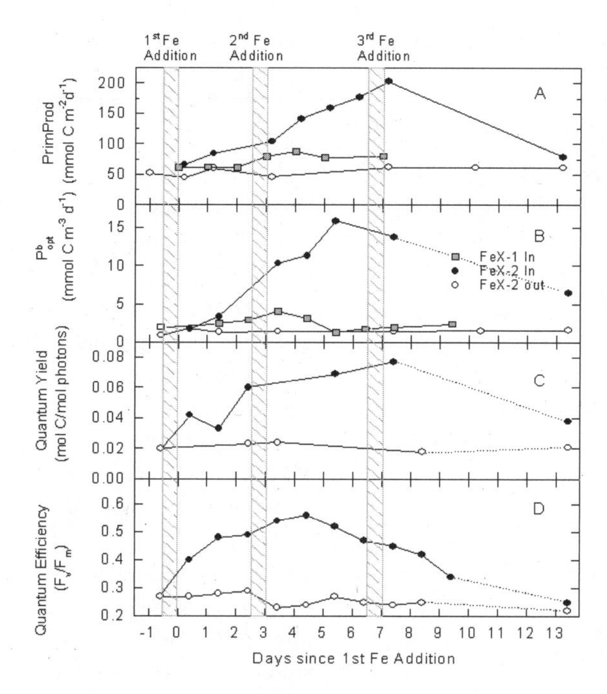

Figure 16.2. Response of phytoplankton to *in situ* iron enrichment in the IronEx-II iron fertilization experiment in the eastern Pacific (see Table 16.2 for details). The responses include increases in primary productivity (A), the maximum rate of photosynthesis within the water column (P^b_{opt}) (B), quantum yield (C), and photosynthetic efficiency (D).

balanced CO_2 uptake, but it also was suggested that the water mass was subducted to depth and that the iron was rapidly lost from the surface layer (Coale et al. 1996b).

A second experiment (IronEx-II) was conducted in 1995, and the revised experimental strategy included refertilizing the same patch with iron approximately every three days. In this experiment a clear and unambiguous response to iron enrichment was noted (Coale et al. 1996b): nutrients and CO_2 were removed, phytoplankton biomass increased, and primary productivity was stimulated (Figure 16.2 and Colorplate 6). All components of the phytoplankton responded, but diatom biomass responded most markedly (an eighty-five-fold increase over non-enriched waters). Photochemical efficiency again increased (Behrenfeld et al. 1996). Picoplankton doubled, but their increase was largely balanced by increased losses due to microzooplankton grazing. Mesozooplankton ingestion of larger phytoplankton was insignificant, suggesting a minor role for larger zooplankton on the time scales of the induced bloom. Hence, the iron hypothesis was proven, albeit slightly modified. The iron paradigm thus became that iron limits the growth of diatoms, whereas smaller forms, while iron-stressed, were limited by their removal by microzooplankton (Landry et al. 1997).

Since the IronEx-II experiment additional trace-metal fertilizations have been

Table 16.1. Mesoscale iron enrichment experiments conducted in the open sea

Study	*Dates*	*Location*	*Result*
Iron-Ex I[a]	October 1993	Equatorial Pacific Galapagos Islands, 5°S; 93°W	Chl 3× increase, photochemical response, small pCO_2 decrease, 80% increase in DMSP
Iron-Ex II[b]	May–June 1995	Equatorial Pacific Galapagos Islands, 6°S; 108°W	Chl 20×; diatoms 80×, NO_3 and DIC utilization, and pCO_2 drawdown
SOIREE[c]	February 1999	Southern Ocean, 61°S; 140°E	Chl 5×, NO_3, $Si(OH)_4$ decrease, pCO_2 drawdown, DMS increase, no increase in particle export, long-lived bloom
EISENEX[d]	November 2000	Southern Ocean, 48°S; 21°E	Chl ≥4× increase, NO_3 down 10%, $Si(OH)_4$ down 30%, pCO_2 drawdown, storms and mixing obscured results
SOFEX	February 2002	Southern Ocean, 56°S; 172°W and 66°S; 171°W	2 experiments in areas with differing NO_3:Si ratios, N and S of Polar Front; Chl increases and diatom growth in both areas

[a]Martin et al. (1994). [b]Coale et al. (1996b). [c]Boyd et al. (2000). [d]Smetacek (2001).

conducted in the Southern Ocean (Table 16.1), since that region is the largest reservoir of unutilized nutrients in the ocean, and the region which potentially could have the largest impact on the marine carbon cycle (Sarmiento et al. 1998). The locations of the cruises were quite different, however. SOIREE (Boyd et al. 2000) was conducted in February at 62°S in a region with about 20 micromolar μM nitrate and 5 μM silicate, whereas EISENEX was conducted in November at about 55°S (initial nitrate ≈ 30 μM, initial silicate ≈ 60 μM). SOFEX, conducted in January-February, fertilized two regions, one at about 65°S (initial nitrate ≈ 28 μM, initial silicate ≈ 60 μM) and one at approximately 57°S, where nitrate and silicic acid were about 5 and 1 μM at the experiment's start, respectively. All of these experiments showed similar responses (large phytoplankton are selectively stimulated, with a concomitant increase in photosynthetic efficiency, biomass, and

Table 16.2. Inorganic nutrient limitation of phytoplankton functional groups by various elements

Group	*Limiting elements*
Diatoms	Si, Fe, N
Diazotrophs (N fixers)	Fe, P
Other groups (Coccolithophorids, dinoflagellates, cyanobacteria, etc.)	Fe, N, P

Note: The most commonly limiting element is listed first. Other likely limiting nutrients listed subsequently. Importance varies regionally and seasonally.

productivity, paralleled by a reduction in CO_2 and macronutrients), although variations in some details have been noted (e.g., the surprising increase in Fe^{+2} in SOIREE; Boyd et al. 2000). None of the experiments conclusively demonstrated that iron enrichment stimulated organic matter export below 300 meters, but the duration of the experiments was limited and the fate of the increased organic matter load was difficult to quantify. On the other hand, there is no example anywhere in the ocean where diatoms have been stimulated to bloom (by any nutrient) and biogenic matter has *not* been exported to depth. Regardless of the relationship of flux and iron-stimulated growth, the paramount role of iron to the ocean's plankton processes and in maintaining HNLC regions is now firmly established. Research has turned to defining the inputs of iron, clarifying the cycling of trace metals in surface waters, and understanding the controls of iron on plankton community composition.

Iron, Elemental, and Organic Matter Interactions

Phytoplankton may be limited by one or more inorganic nutrients (Table 16.2). Iron not only influences the different phytoplankton taxa differentially, but it also modifies the biogeochemistry of other elements during phytoplankton growth. Silicic acid (a component of diatom cell walls) uptake is greatly enhanced relative to carbon or nitrogen when iron is limiting (Hutchins and Bruland 1998; Takeda 1998), so that the Si:N ratio increased from about 1 (iron replete growth) to ratios of 5 and more under severe iron limitation. Carbon:nitrogen ratios appear to be unaffected by iron concentrations. Hence, diatoms that grow under iron limitation are highly enriched with silica, and the relative fluxes of carbon, nitrogen and silica to depth are also altered. Such an alteration might explain the highly siliceous (and relatively carbon depauperate) deposits that are found in the Polar Front region in the Antarctic.

Nitrogen-fixing cyanobacteria apparently have elevated iron demands that are approximately tenfold greater than those of non-N_2-fixing phytoplankton (iron replete cultures have Fe:C molar ratios of about 50×10^{-6}; Berman-Frank et al. 2001). Indeed, the Fe:C ratios of pelagic plankton are highly variable and can range over an order of magnitude (from 5×10^{-6} to 0.2×10^{-6}); furthermore, they appear to respond dramatically to reduced iron concentrations and allow phytoplankton to maintain cellular processes at an optimal rate.

Models are now being run that include significantly more detail in the biological processes than were previously available. For example, Moore et al. (2001) simulated the growth of three functional components of phytoplankton (diatoms, small phytoplankton, and diazotrophs) in a global circulation model and predicted the areas where iron limitation would occur. In general, mid-ocean gyres are nitrogen-limited for diatoms and small phytoplankton, whereas the subarctic and equatorial Pacific and the Southern Ocean are iron-limited. Other regions (equatorial waters of the Indian and Atlantic Oceans) and the North Atlantic also become iron-limited during summer months, largely because of reduced atmospheric inputs during this time.

Within the past decade oceanographers have recognized the significance of microbial transformations of organic matter as a dominant process in ocean biogeochemistry (Colorplate 7; Azam 1998). This dominance extends to controlling the concentrations and availability of iron in low-iron waters. Apart from marine viruses (Fuhrman 1999), heterotrophic bacteria (Whitman, Coleman, and Wiebe 1998), autotrophic bacteria (Chisholm et al. 1988; Beja et al. 2002), and *Archaea* (Karner, DeLong, and Karl 2001) are the numerically dominant organisms in the ocean. Their biomass can equal or exceed that of phytoplankton (Ducklow 1999), and no other member of the plankton community exerts as much influence on the fate of elements in the ocean (Williams 2000; Carlson 2002). The principal ecological role of heterotrophic bacteria in the plankton foodweb is remineralizing nitrogen-, phosphorus-, and iron-containing organic matter (Pomeroy 1974; Cotner and Biddanda 2002).

Dissolved organic matter (DOM) is the largest pool of organic carbon in the ocean (685 Pg C) and its export is a component of the biological pump (Hansell and Carlson 1998a,b; Williams 2000; Carlson 2002). The majority of the removal of DOM is through consumption of carbon by heterotrophic bacteria. Consumption is a variable process as DOM has a wide range of composition and lability, which is determined by the production process and the biogeochemical transformations occurring within the ocean foodweb. Much of the DOM is refractory owing to extensive microbial transformation (Carlson 2002), allowing it to serve as a carbon reservoir on the time scale of the ocean circulation (Hansell and Carlson 1998b). Heterotrophic bacteria are major remineralizers not only of the carbon in DOM, but also of its phosphorus (Clark, Ingall, and Benner 1998), nitrogen (Zehr and Ward 2002), and biogenic silica (Bidle and Azam 1999).

Concurrent with heterotrophic growth and remineralization of DOM are the uptake and recycling of inorganic nutrients (N and P) and trace elements such as iron. Bacteria possess the ability to compete effectively with phytoplankton for these nutrients, particularly in the regions of the global ocean that are most influential in the carbon cycle (HNLC regions) where iron is the probable limiting nutrient. Low atmospheric flux of iron to HNLC waters and its low solubility in seawater are initial constraints on phytoplankton and possibly heterotrophic bacterial growth. The numerous small-scale and mesoscale iron enrichment experiments have confirmed this for phytoplankton but not for heterotrophic bacteria. This is surprising given the fact that heterotrophic bacteria have a higher per cell iron demand than phytoplankton (Tortell, Maldonado, and Price 1996). The bacterial response within the fertilized patches during IronEx II (Cochlan 2001) and SOIREE (Hall and Safi 2001) was modest with little to no accumulation of biomass and threefold increases in bacterial production in both experiments. Direct stimulation of bacterial growth by iron could not be discounted entirely in either experiment, but multiple lines of evidence point toward growth enhancement by phytoplankton-derived DOM (Church, Hutchins, and Ducklow 2000; Kirchman et al. 2000). Scenarios may exist where iron can directly stimulate growth (Pakulski et al. 1996); however, the sum of the evidence suggests that iron in HNLC waters is rarely in low enough concentration to limit heterotrophic bacterial growth.

Siderophores: A Competitive Edge

Low atmospheric flux of iron coupled with low solubility in seawater explains the paradox of the HNLC condition. The findings that DOM primarily limits heterotrophic bacterial growth rather than iron suggests an additional mechanism that might allow the HNLC condition to persist. Such a mechanism might be mediated through the extracellular release of iron-binding ligands called siderophores. Siderophores are low molecular weight organic molecules (500–1,000 Daltons) secreted by heterotrophic and autotrophic bacteria generally under conditions of low dissolved iron concentrations (Reid et al. 1993; Wilhelm and Trick 1994; Wilhelm, Maxwell, and Trick 1996). They have an extremely high affinity for ferric (Fe^{3+}) iron (stability constant, $K = 10^{49\text{-}53}$; Reid et al. 1993), as well as other trace elements (Butler 1998). Once the siderophore binds iron, the siderophore-iron complex is acquired by the cell through cell-surface receptors and internalized where the iron is reduced and catabolized (Raymond, Muller, and Matzanke

1984). The strategy of siderophore production has enabled marine bacteria to be highly competitive and influential in the biogeochemical cycling of iron (Tortell et al. 1999). This biochemical strategy, however, has been usurped by some siderophore-producing and non-siderophore-producing organisms (cyanobacteria and phytoplankton), which have employed their own strategies to utilize siderophore-iron complexes as a source of iron (Trick 1989; Granger and Price 1999; Hutchins et al. 1999; Maldonado and Price 1999). Utilization of siderophores as a carbon source has been demonstrated for terrestrial bacteria (Warren and Neilands 1964; DeAngelis, Forsyth, and Castignetti 1993) but not for any marine organism. The interactions between iron, siderophores, and the plankton community are fundamentally important because the majority of the dissolved inorganic iron in the oceans (more than 99 percent) is complexed to low-molecular-weight organic ligands. These ligands appear to be ubiquitous; however, they vary slightly in their concentrations and their affinity for iron. In the North Pacific these ligands fall into two classes (L1 and L2) based on their conditional stability constants for inorganic ferric iron (Rue and Bruland 1995). Organic ligands have also been found in the northwest Atlantic Ocean (Wu and Luther 1995), South and equatorial Atlantic Ocean (Powell and Donat 2001), and the Southern Ocean (Nolting et al. 1998). Rue and Bruland (1997) observed the production of these iron-binding ligands during IronEx II but stopped short of calling them siderophores. Presently, the chemical composition of these organic ligands is unknown. Identifying the origin of siderophores and other organic ligands within the milieu of DOM, characterizing their composition, and understanding their role in iron-plankton dynamics represents a formidable challenge to biological and chemical oceanographers (Colorplate 7).

Geoengineering with Iron

The postulated link between iron limitation, increased iron supply during glacial periods, and atmospheric CO_2 raised the possibility of slowing the increase in anthropogenic CO_2 accumulation in the atmosphere and ameliorating global warming by stimulating carbon export to depth through large-scale iron fertilization (Chisholm and Morel 1991; Chisholm 1995). Modeling studies suggest the practical effect on atmospheric CO_2 would be small even with prolonged, intensive fertilization (Sarmiento and Orr 1991), as well as practically impossible to document for carbon accounting purposes. The same studies also suggest other effects on the ocean might be profound and far reaching. For example, increased organic matter production will stimulate respiration, leading to hypoxia in some areas (Sarmiento and Orr 1991) and increased nitrous oxide production (Fuhrman and Capone 1991). Unintended consequences like release of N_2O and methane, powerful greenhouse gases, suggest the complexity of such interventions in global biogeochemical cycles. Although the issue remains controversial a decade after it was first raised (Chisholm, Falkowski, and Cullen 2001; Johnson and Karl 2002), several patents have been applied for, and various commercial efforts aimed at global trading of carbon credits gained through iron fertilization seem likely (see, e.g., http://www.planktos.com/eco-solutions.htm).

The manifold and pervasive roles of iron in ocean ecology and biogeochemistry have only begun to be understood. Differential responses to iron additions by diatoms modulate element ratios in the nearshore and deep ocean. Bacterial production of and interactions with organic ligands regulate iron availability for primary producers. Terrestrial soil fungi as well as ultraviolet radiation may also mobilize iron in airborne dust (Saydam and Senyuva 2002). Although our focus on iron has been its role as a limiting nutrient in the sea, a wider focus on its role as a critical linking mechanism in the Earth system (Falkowski 1997) seems justified.

Acknowledgments

Preparation of this chapter was supported by National Science Foundation Grants OPP-0000329 and 0217282 and Department of Energy Grant ER63082. We thank Keith Moore (National Center for Atmospheric Research, Boulder CO) and Kenneth Coale (Moss Landing) for providing original figures for this chapter.

Literature Cited

Azam, F. 1998. Microbial control of oceanic carbon flux: The plot thickens. *Science* 280:694–696.

Banse, K. 1994. Grazing and zooplankton production as key controls of phytoplankton production in the open ocean. *Oceanography* 7:13–20.

Barnola, J. M., D. Raynaud, Y. S. Korotkevich, and C. Lorius. 1987. Vostok ice core provides 160,000 year record of atmospheric CO_2. *Nature* 329:408–414.

Behrenfeld, M. J., A. J. Bale, Z. S. Kolber, J. Aiken, and P. G. Falkowski. 1996. Confirmation of iron limitation of phytoplankton photosynthesis in the equatorial Pacific Ocean. *Nature* 383:508–511.

Beja, O., M. T. Suzuki, J. F. Heidelberg, W. C. Nelson, C. M. Preston, T. Hamada, J. A. Eisen, C. M. Fraser, and E. F. DeLong. 2002. Unsuspected diversity among marine aerobic anoxygenic phototrophs. *Nature* 415:630–633.

Berman-Frank, I., P. Lundgren, Y.-B. Chen, H. Kupper, Z. Kolber, B. Bergman, and P. Falkowski. 2001. Segregation of nitrogen fixation and oxygenic photosynthesis in the marine Cyanobacterium Trichodesmium. *Science* 294:1534–1537.

Bidle, K. D., and F. Azam. 1999. Accelerated dissolution of diatom silica by marine bacterial assemblages. *Nature* 397:508–512.

Bolin, B., and R. B. Cook, eds. 1983. *The major biogeochemical cycles and their interactions.* SCOPE 21. Chichester, U.K.: John Wiley and Sons.

Boyd, P. W., A. J. Watson, C. S. Law, E. R. Abraham, T. Trull, R. Murdoch, D. C. E. Bakker, A. R. Bowie, K. O. Buesseler, H. Chang, M. Charette, P. Croot, K. Downing, R. Frew, M. Gall, M. Hadfield, J. Hall, M. Harvey, G. Jameson, J. Laroche, M. Liddicoat, R. Ling, M. T. Maldonado, R. M. McKay, S. Nodder, S. Pickmere, R. Pridmore, S. Rintoul, K. Safi, P. Sutton, R. Strzepek, K. Tanneberger, S. Turner, A. Waite, and J. Zeldis. 2000. A mesoscale phytoplankton bloom in the polar Southern Ocean stimulated by iron fertilization. *Nature* 407:695–702.

Butler, A. 1998. Acquisition and utilization of transition metal ions by marine organisms. *Science* 281:207–210.

Carlson, C. A. 2002. Production and removal processes. Pp. 91–151 in *Biogeochemistry of marine dissolved organic matter*, edited by D. A. Hansell and C. A. Carlson. New York: Elsevier Science.

Chisholm, S. W., R. J. Olson, E. R. Zettler, R. Goericke, J. B. Waterbury, and N. A. Welschmeyer. 1988. A novel free-living prochlorophyte abundant in the oceanic euphotic zone. *Nature* 334:340–43.

Chisholm, S. W. 1995. The iron hypothesis: Basic research meets environmental policy. *Reviews of Geophysics*, Supplement: 1277–1286.

Chisholm, S. W., and F. M. M. Morel. 1991. Preface. *Limnology and Oceanography* 36:ii–vi.

Chisholm, S. W., P. G. Falkowski, and J. J. Cullen. 2001. Oceans: Discrediting ocean fertilization. *Science* 294:309–310.

Church, M., D. A. Hutchins, and H. W. Ducklow. 2000. Limitation of bacterial growth by dissolved organic matter and iron in the Southern Ocean. *Applied and Environmental Microbiology* 66:455–466.

Clark, L. L., E. D. Ingall, and R. Benner. 1998. Marine phosphorus is selectively remineralized. *Nature* 393:426.

Coale, K. H., S. E. Fitzwater, R. M. Gordon, K. S. Johnson, and R. T. Barber. 1996a. Control of community growth and export production by upwelled iron in the Equatorial Pacific Ocean. *Nature* 379:621–624.

Coale, K. H., K. S. Johnson, S. E. Fitzwater, R. M. Gordon, S. Tanner, F. P. Chavez, L. Ferioli, C. Sakamoto, P. Rogers, F. Millero, P. Steinberg, P. Nightingale, D. Cooper, W. P. Cochlan, and R. Kudela. 1996b. A massive phytoplankton bloom induced by an ecosystem-scale iron fertilization experiment in the Equatorial Pacific Ocean. *Nature* 383:495–501.

Cochlan, W. P. 2001. The heterotrophic bacterial response during a mesoscale iron enrichment experiment (IronEx II) in the eastern Equatorial Pacific Ocean. *Limnology and Oceanography* 46:428–435.

Cotner, J. B., and B. A. Biddanda. 2002. Small players, large role: Microbial influence on biogeochemical processes in pelagic aquatic ecosystems. *Ecosystems* 5:105–121.

De Baar, H. J. W. 1994. Von Liebig's Law of the Minimum and plankton ecology (1899–1991). *Progress in Oceanography* 33:347–386.

DeAngelis, R., M. Forsyth, and D. Castignetti. 1993. The nutritional selectivity of a siderophore-catabolizing bacterium. *Biometals* 6:234–238.

Duce, R. A., and N. W. Tindale. 1991. Atmospheric transport of iron and its deposition in the ocean. *Limnology and Oceanography* 36:1715–1726.

Ducklow, H. W. 1995. Ocean biogeochemical fluxes: New production and export of organic matter from the upper ocean. *Reviews of Geophysics.* Supplement (U.S. National Report to IUGG, Contributions in Ocean Sciences):1271–1276.

Ducklow, H. W. 1999. The bacterial content of the oceanic euphotic zone. *FEMS Microbiology-Ecology* 30:1–10.

Dugdale, R. C., and J. J. Goering. 1967. Uptake of new and regenerated forms of nitrogen in primary production. *Limnology and Oceanography* 12:196–206.

Eppley, R. W., and B. J. Peterson. 1979. Particulate organic matter flux and planktonic new production in the deep ocean. *Nature* 282:677–680.

Falkowski, P. G. 1997. Evolution of the nitrogen cycle and its influence on the biological sequestration of CO_2 in the ocean. *Nature* 387:272–275.

Field, C. B., M. J. Behrenfeld, J. T. Randerson, and P. Falkowski. 1998. Primary production of the biosphere: Integrating terrestrial and oceanic components. *Science* 281:237–240.

Fuhrman, J. A. 1999. Marine viruses and their biogeochemical and ecological effects. *Nature* 399:541–548.

Fuhrman, J. A., and D. G. Capone. 1991. Possible biogeochemical consequences of ocean fertilization. *Limnology and Oceanography* 36:1951–1959.

Fung, I., S. Meyn, I. Tegen, S. C. Doney, J. John, and J. K. B. Bishop. 2000. Iron supply and demand in the upper ocean. *Global Biogeochemical Cycles* 14:281–296.

Gran, H. H. 1931. On the conditions for the production of phytoplankton in the sea. *Rapports et Procès Verbaux des Réunions, Conseil International pour l'Exploration de la Mer* 75:37–46.

Granger, J., and N. M. Price. 1999. The importance of siderophores in iron nutrition of heterotrophic marine bacteria. *Limnology and Oceanography* 44:541–555.

Gruber, N., and J. L. Sarmiento. 2002. Large-scale biogeochemical /physical interactions in elemental cycles. In *The Sea: Biological-physical interactions in the sea,* edited by A. R. Robinson, J. J. McCarthy, and B. J. Rothschild. New York: John Wiley and Sons.

Hall, J. A., and K. Safi. 2001. The impact of *in situ* Fe fertilisation on the microbial food web in the Southern Ocean. *Deep Sea Research* 48:11–12.

Hansell, D. A., and C. A. Carlson. 1998a. Net community production of dissolved organic carbon. *Global Biogeochemical Cycles* 12:443–453.

Hansell, D. A., and C. A. Carlson. 1998b. Deep-ocean gradients in the concentration of dissolved organic carbon. *Nature* 395:263–266.

Hanson, R., H. W. Ducklow, and J. G. Field. 2000. *The changing carbon cycle in the oceans.* Cambridge: Cambridge University Press.

Hutchins, D. A., and K. W. Bruland. 1998. Iron-limited diatom growth and Si:N uptake ratios in a coastal upwelling regime. *Nature* 393:561–564.

Hutchins, D. A., V. M. Franck, M. A. Brzezinski, and K. W. Bruland. 1999. Inducing phytoplankton iron limitation in iron-replete coastal waters with a strong chelating ligand. *Limnology and Oceanography* 44:1009–1018.

Johnson, K. S., and D. M. Karl. 2002. Is ocean fertilization credible and creditable? *Science* 296:467–468.

Johnson, K. S., R. M. Gordon, and K. H. Coale. 1997. What controls dissolved iron concentrations in the world ocean? *Marine Chemistry* 57:137–161.

Karl, D. M. 1999. A sea of change: Biogeochemical variability in the North Pacific subtropical gyre. *Ecosystems* 2:181–214.

Karner, M. B., E. F. DeLong, and D. M. Karl. 2001. Archaeal dominance in the mesopelagic zone of the Pacific Ocean. *Nature* 409:507–510.
Kirchman, D. L., B. Meon, M. T. Cottrell, D. A. Hutchins, D. Weeks, and K. W. Bruland. 2000. Carbon versus iron limitation of bacterial growth in the California upwelling regime. *Limnology and Oceanography* 45:1681–1688.
Knox, F., and M. McElroy. 1984. Changes in atmospheric CO_2: Influence of the marine biota at high latitudes. *Journal Geophysical Research* 89:4629–4637.
Kolber, Z. S., R. T. Barber, K. H. Coale, S. E. Fitzwater, R. M. Greene, K. S. Johnson, S. Lindley, and P. G. Falkowski. 1994. Iron limitation of phytoplankton photosynthesis in the Equatorial Pacific Ocean. *Nature* 371:145–149.
Landry, M. R., R. T. Barber, R. R. Bidigare, F. Chai, K. H. Coale, H. G. Dam, M. R. Lewis, S. T. Lindley, J. J. McCarthy, M. R. Roman, D. K. Stoecker, P. G. Verity, and J. R. White. 1997. Iron and grazing constraints on primary production in the central equatorial Pacific: An EqPac synthesis. *Limnology and Oceanography* 42:405–418.
Lindley, S. T., and R. T. Barber. 1998. Phytoplankton response to natural and experimental iron addition. *Deep Sea Research* (Part II: Topical Studies in Oceanography) 45:1135–1150.
Longhurst, A. R., and W. G. Harrison. 1989. The biological pump: Profiles of plankton production and consumption in the upper ocean. *Progress in Oceanography* 22:47–123.
Maldonado, M. T., and N. M. Price. 1999. Utilization of iron bound to strong organic ligands by plankton communities in the subarctic Pacific Ocean. *Deep-Sea Research* (Part 2, Topical Studies in Oceanography) 46:2447–2473.
Martin, J. H. 1990. Glacial-interglacial CO_2 change: the iron hypothesis. *Paleoceanography* 5:1–13.
Martin, J. H., and S. Fitzwater. 1988. Iron deficiency limits phytoplankton growth in the northeast Pacific subarctic. *Nature* 331:341–343.
Martin, J. H., K. H. Coale, K. S. Johnson, S. E. Fitzwater, R. M. Gordon, S. J. Tanner, C. N. Hunter, V. A. Elrod, J. L. Nowicki, T. L. Coley, R. T. Barber, S. Lindley, A. J. Watson, K. Van Scoy, and C. S. Law. 1994. Testing the iron hypothesis in ecosystems of the Equatorial Pacific Ocean. *Nature* 371:123–129.
McAllister, C. D., T. R. Parsons, and J. D. H. Strickland. 1960. Primary productivity and fertility at Station "P" in the north-east Pacific Ocean. *Journal de Conseil International pour l'Exploration de la Mer* 25:240–259.
Miller, C. B., B. W. Frost, P. A. Wheeler, M. R. Landry, N. Welschmeyer, and T. M. Powell. 1991. Ecological dynamics in the subarctic Pacific, a possibly iron-limited ecosystem. *Limnology and Oceanography* 36:1600–1615.
Moore, J. K., S. C. Doney, D. M. Glover, and I. Y. Fung. 2001. Iron cycling and nutrient-limitation patterns in surface waters of the World Ocean. *Deep Sea Research* (Part II: Topical Studies in Oceanography) 49:463–507.
Nolting, R. F., L. J. A. Gerringa, M. J. W. Swagerman, K. R. Timmermans, and H. J. W. de Baar. 1998. Fe(III) speciation in the high current, low chlorophyll Pacific region of the Southern Ocean. *Marine Chemistry* 62:335–352.
Pakulski, J. D., R. B. Coffin, C. A. Kelley, S. L. Holder, R. Downer, P. Aas, M. M. Lyons, and W. H. Jeffrey. 1996. Iron stimulation of Antarctic bacteria. *Nature* 383:133–134.
Petit, J. R., L. Mounier, J. Jouzel, Y. Korotkevitch, V. Kotlyakov, and C. Lorius. 1990. Paleoclimatological implications of the Vostok core dust record. *Nature* 343:56–58.

Pomeroy, L. R. 1974. The oceans food web, a changing paradigm. *BioScience* 24:499–504.

Powell, R. T., and J. R. Donat. 2001. Organic complexation and speciation of iron in the South and Equatorial Atlantic. *Deep-Sea Research* (Part 2, Topical Studies in Oceanography) 48:2877–93.

Raymond, K. N., G. Muller, and B. F. Matzanke. 1984. Complexation of iron by siderophores: A review of their solution and structural chemistry and biological function. Pp. 49–102 in *Topics in current chemistry*, edited by F. L. Boscheke. New York: Springer.

Reid, R. T., D. H. Live, D. J. Faulkner, and A. Butler. 1993. A siderophore from a marine bacterium with an exceptional ferric ion affinity constant. *Nature* 366:455–458.

Rue, E. L., and K. W. Bruland. 1995. Complexation of iron(III) by natural organic ligands in the Central North Pacific as determined by a new competitive ligand equilibration/adsorptive cathodic stripping voltammetric method. *Marine Chemistry* 50:117–138.

Rue, E. L., and K. W. Bruland. 1997. The role of organic complexation on ambient iron chemistry in the equatorial Pacific Ocean and the response of a mesoscale iron addition experiment. *Limnology and Oceanography* 42:901–910.

Ruud, J. T. 1930. Nitrates and phosphates in the Southern seas. *Rapports et Procès Verbaux des Réunions, Conseil International pour l'Exploration de la Mer* 5:347–360.

Sarmiento, J. L., and J. R. Toggweiler. 1984. A new model for the role of the oceans in determining atmospheric pCO_2. *Nature* 308:621–624.

Sarmiento, J. L., and J. C. Orr. 1991. Three-dimensional simulations of the impact of Southern Ocean nutrient depletion on atmospheric CO sub(2) and ocean chemistry. *Limnology and Oceanography* 36:1928–1950.

Sarmiento, J. L., T. M. C. Hughes, R. J. Stouffer, and S. Manabe. 1998. Simulated response of the ocean carbon cycle to anthropogenic climate warming. *Nature* 393:245–249.

Saydam, A. C., and H. Z. Senyuva. 2002. Deserts: Can they be the potential suppliers of bioavailable iron? *Geophysical Research Letters* 29:1–3.

Siegenthaler, U., and T. Wenk. 1984. Rapid atmospheric CO_2 variations and oceanic circulation. *Nature* 308:624–625.

Smetacek, V. 2001. EisenEx: International team conducts iron experiment in the Southern Ocean. *US JGOFS News* 11:11–12,14.

Takahashi, T., S. C. Sutherland, C. Sweeney, A. Poisson, N. Metzl, B. Tilbrook, N. Bates, R. Wanninkhof, R. A. Feely, C. Sabine, J. Olafsson, and Y. Nojiri. 2002. Global sea-air CO_2 flux based on climatological surface ocean pCO_2, and seasonal biological and temperature effects. *Deep-Sea Research* (Part II: Topical Studies in Oceanography) 49:1601–1622.

Takeda, S. 1998. Influence of iron availability on nutrient consumption ratio of diatoms in oceanic waters. *Nature* 393:774–777.

Tortell, P. D., M. T. Maldonado, and N. M. Price. 1996. The role of heterotrophic bacteria in iron-limited ocean ecosystems. *Nature* 383:330–332.

Tortell, P. D., M. T. Maldonado, J. Granger, and N. M. Price. 1999. Marine bacteria and biogeochemical cycling of iron in the oceans. *FEMS Microbiology Ecology* 29:1–11.

Trick, C. G. 1989. Hydroxymate-siderophore production and utilization by marine eubacteria. *Current Microbiology* 18:375–378.

Turner, D. R., and K. A. Hunter, eds. 2001. *The biogeochemistry of iron in seawater.* New York: John Wiley and Sons.

Volk, T., and M. I. Hoffert. 1985. Ocean carbon pumps: Analysis of relative strengths and efficiencies in ocean-driven atmospheric CO_2 changes. Pp. 99–110 in *The carbon cycle and atmospheric* CO_2*: Natural variations Archean to present,* edited by E. T. Sundquist and W. S. Broecker. Washington, D.C.: American Geophysical Union.

Wallace, D. W. R. 2001. Storage and transport of excess CO_2 in the oceans: The JGOFS/WOCE global CO_2 survey. Pp. 489–524 in *Ocean circulation and climate: Observing and modeling the global ocean,* edited by G. Siedler, J. Church, and J. Gould. San Diego: Academic Press.

Warren, R. A. J., and J. B. Neilands. 1964. Microbial degradation of the ferrichrome compounds. *Journal of General Microbiology* 64:459–470.

Watson, A. J., C. S. Law, K. A. Van Scoy, F. J. Millero, W. Yao, G. E. Friedderich, M. I. Liddicoat, R. H. Wanninkhof, R. T. Barber, and K. H. Coale. 1994. Minimal effect of iron fertilization on sea-surface carbon dioxide concentrations. *Nature* 371:143–145.

Whitman, W. B., D. C. Coleman, and W. J. Wiebe. 1998. Prokaryotes: The unseen majority. *Proceedings National Academy of Sciences* 95:6578–6583.

Wilhelm, S. W., and C. G. Trick. 1994. Iron-limited growth of cyanobacteria: Multiple siderophore production is a common response. *Limnology and Oceanography* 39:1979–1984.

Wilhelm, S. W., D. P. Maxwell, and C. G. Trick. 1996. Growth, iron requirements, and siderophore production in iron-limited Synechococcus PCC 7002. *Limnology and Oceanography* 41:89–97.

Williams, P. J. L. 2000. Heterotrophic bacteria and the dynamics of dissolved organic material. Pp. 153– 200 in *Microbial ecology of the oceans,* edited by D. L. Kirchman. New York: Wiley-Liss.

Wu, J., and G. W. I. Luther. 1995. Evidence for the existence of Fe(III) organic complexation in the surface water of the Northwest Atlantic Ocean by a competitive ligand equilibrium method and a kinetic approach. *Marine Chemistry* 50:159–177.

Zehr, J. P., and B. B. Ward. 2002. Nitrogen cycling in the ocean: New perspectives on processes and paradigms. *Applied and Environmental Microbiology* 68:1015–1024.

17

Carbon-Silicon Interactions

Venugopalan Ittekkot, Christoph Humborg, Lars Rahm, and Nguyen Tac An

Silicon dioxide is the most abundant component of the Earth's crust. It occurs as silicate minerals in association with igneous, metamorphic, and sedimentary rocks, which are continuously subjected to physical and chemical weathering. The processes associated with weathering and the products released form the basis of silicon biogeochemistry and its interactions with other elements. This chapter looks at the interactions between silicon and carbon, the cycles of which have been intimately linked through geological time. Silicate minerals (clay minerals) are implicated in the concentration of simple organic molecules to form complex ones early in the development of life on Earth because of their surface charge and repeating crystal structure (Cairns-Smith 1985). Organic carbon compounds, such as proteins and carbohydrates, synthesized by organisms mediated the formation of Si-containing minerals (biogenic silica-opal) in frustules of diatoms and radiolarians through the process commonly known as biomineralization (Hecky et al. 1973; Degens 1976). Further, the tight absorption of organic molecules to silicate minerals in clays is thought to be a major mechanism by which these organic molecules escape microbial attack and are ultimately buried in marine sediments (Mayer 1999). This burial of organic C is the key mechanism that maintains O_2 in the atmosphere (Garrels and MacKenzie 1971; Holland 1978). Of special interest in the context of carbon-silicon interactions is the modulation of atmospheric CO_2 contents through geological time through the chemical weathering of silicates and the sequestration of atmospheric CO_2 through the biological carbon pump in the sea. These and their perturbations by human activities are discussed in this chapter.

Silicate Weathering and Atmospheric CO_2

Weathering of silicate rocks on land is a sink for atmospheric CO_2, albeit on long time scales (Berner, Lasaga, and Garrels 1983; Wollast and Mackenzie 1986; Brady and Carrol 1994; White and Brantley 1995). Although the rate of CO_2 consumption by weathering reactions is much slower than its rate of cycling through organisms, weathering processes appear to have been important in controlling the long-term concentrations of CO_2 and O_2 in the atmosphere and the ocean.

In general, chemical weathering of carbonate and silicate rocks can be represented as follows:

for carbonates

$$CO_2 + H_2O + CaCO_3 \rightarrow Ca^{++} + 2HCO_3^-$$
$$2CO_2 + 2\,H_2O + CaMg\,(CO_3)_2 \rightarrow Ca^{++} + Mg^{++} + 4HCO_3^-$$

and for silicates

$$2\,CO_2 + 3\,H_2O + CaSiO_3 \rightarrow Ca^{++} + 2\,HCO_3^- + H_4SiO_4$$
$$2\,CO_2 + 3\,H_2O + MgSiO_3 \rightarrow Mg^{++} + 2\,HCO_3^- + H_4SiO_4$$

CO_2 is supplied either via the high CO_2 partial pressure in the atmosphere as has occurred during the Phanerozoic (Berner 1998) or via degradation products of plant biomass (e.g., Humborg et al. 2002). The link between vegetation and increased dissolved silicate loads of rivers is indicated by the positive correlation between silicate and total organic carbon (TOC) concentrations; the latter can be seen as a proxy for vegetation coverage, found in pristine northern Swedish rivers (Figure 17.1).

The weathering products are transferred via the rivers to the oceans, where they accumulate. Oceans' capacity to hold the weathering products is, however, limited,

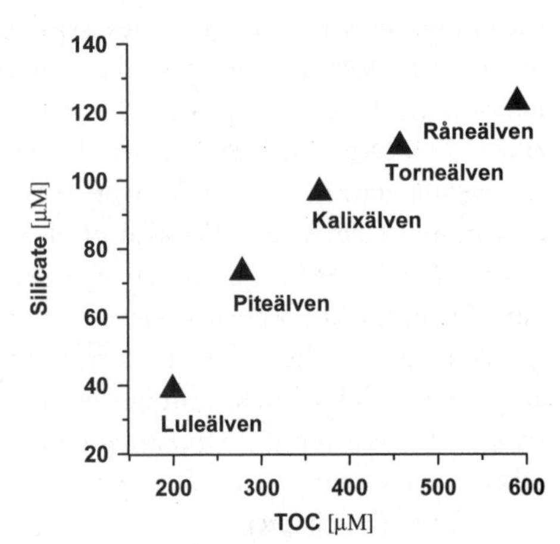

Figure 17.1. Silicate concentrations versus total organic carbon (TOC) concentration in the northernmost Swedish rivers (long-term mean values).

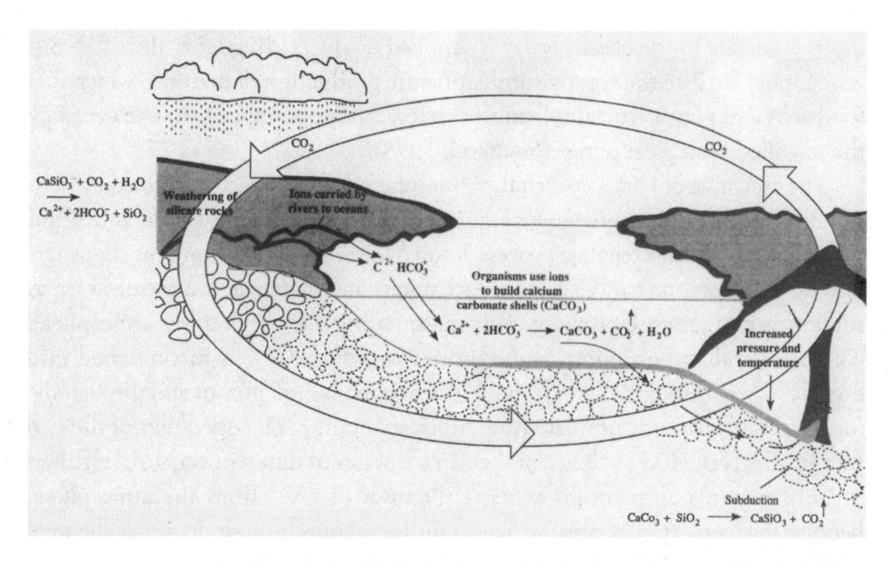

Figure 17.2. The interaction between the carbonate and silicate cycles in the surface of the Earth (modified from Schlesinger 1997). Long-term control of atmospheric CO_2 is achieved by dissolution in surface waters and its participation in the weathering of rocks. Eventually carbon is buried as part of carbonate rocks in the oceanic crust and is released to the atmosphere when these rocks undergo metamorphism at high temperature and pressure in the Earth's crust.

and processes such as biological extraction of carbonate and silicate in the tissues and skeleton of marine plankton and their subsequent sedimentation lead to their permanent removal into sediments. These reactions involving carbonate and silicates are of importance in the removal of atmospheric CO_2 (Figure 17.2). In the case of carbonate weathering, half the HCO_3 transported is of atmospheric origin; in the case of silicate weathering, all of it is. When silicate minerals are formed inorganically, the reaction is basically the reverse of weathering on land. Thus for each mole of Ca/Mg SiO_3 formed, a mole of CO_2 is generated. Accumulation of silicates in sediments at longer time scales leads to an increase of the CO_2 levels in the atmosphere; a decrease occurs when there is preferential accumulation of carbonates in sediments. Modeling results have shown that the changes in the intensity of weathering could have caused the changes in the atmospheric CO_2 contents recorded in the Phanerozoic record (Berner 1998).

Silicate-Driven Biological Carbon Pump in the Ocean

Chemical weathering on land is the process that supplies dissolved and particulate silicate to rivers and ultimately to oceans. More than 80 percent of the total input of silicate to the oceans today is supplied by rivers (Treguer et al. 1995). Silicate inputs fertilize the seas by stimulating the production of diatoms, which fuel foodwebs and play a crucial role in the biological uptake of CO_2 by the ocean by the so-called biological pump (Smetacek 1998).

The exchange of CO_2 between the atmosphere and the ocean is affected by both physical and biological processes (Figure 17.3; Heinze, Maier-Reimer, and Winn 1995). The physical exchange process, known as the solubility pump, is characterized by the dissolution of CO_2 in surface waters and its transfer to the deep sea in sinking water masses. Upwelling water masses bring CO_2 back to the atmosphere. The biological carbon pump is the process by which CO_2 is incorporated into organic matter through photosynthesis (the organic carbon pump) and through the formation of calcium carbonate (the carbonate pump). The operation of the two pumps affects the CO_2 balance of the surface ocean in different ways. An efficient organic carbon pump means a net withdrawal of CO_2 from the atmosphere, because the formation of organic matter during photosynthesis decreases the total carbon content and partial pressure of CO_2 in the surface layers:

$$CO_2 + H_2O \rightarrow CH_2O + O_2$$

Although carbonate production reduces the total dissolved inorganic carbon in the surface layers, the reaction increases the partial pressure of CO_2 at the surface ocean, thus driving CO_2 from the ocean to the atmosphere. For each mole of carbonate formed, a mole of CO_2 is released:

$$Ca^{2+} + 2HCO_3 \rightarrow CaCO_3 + CO_2 + H_2O$$

Diatoms (silica-secreting organisms) and coccolithophorids (carbonate-secreting organisms) are among the major players in the working of the biological carbon pump (Figure 17.3). Because of the difference in the ways the two reactions affect the CO_2 system in the surface ocean, the efficiency of the biological pump in the short term is determined by the relative abundance of the two species: diatoms are more efficient at carbon sequestration than coccolithophorids. This may seem counterintuitive at first, but note that the formation of $CaCO_3$ (in coccolithophorids) releases CO_2 to the surface water; the formation of a diatoms silica shell does not. So for an equal sedimentation of organic C, the net effect of coccolithophorids is to sequester less C into the deep oceans than diatoms.

The efficiency of the biological pump is also reflected in the nature of material settling out of the surface layers of the oceans. In areas where the biological CO_2 pump works efficiently to remove CO_2, the material settling to the ocean's interior exhibits higher ratios of biogenic silica to carbonate and organic carbon to carbonate carbon (C_{org}:C_{carb}; also called rain ratios), which are indicative of the efficiency of CO_2 sequestration at the sea surface (Berger and Keir 1984).

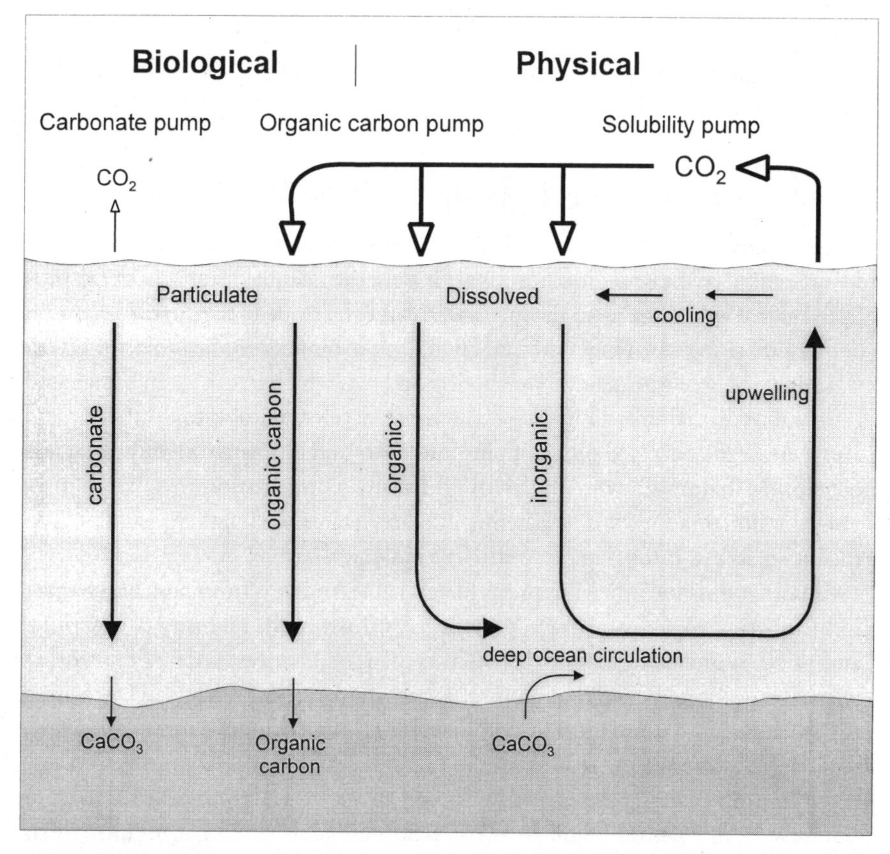

Figure 17.3. Carbon pumps in the ocean. Both physical and biological processes affect the exchange of CO_2 between the atmosphere and the ocean. The biological processes that are highlighted on the left side of the diagram are the formation of carbonate (the carbonate pump) and the formation of particulate matter during photosynthesis (the organic carbon pump). The right side of the diagram highlights the physical exchange processes, namely the dissolution of CO_2 in surface waters and its transfer to the deep sea in sinking water masses (known as the solubility pump); CO_2 is brought back to the atmosphere via upwelling of water masses. See text for explanation. Figure modified from SCOR (1990).

Short-term fluctuations in the atmospheric CO_2 contents (e.g., glacial-interglacial) have recently been suggested to result from changes in the relative abundance of diatoms and coccolithophorids in the surface ocean biological community structure, which in turn is brought about by the variations in the silicate delivery to the sea surface, for example, by windborne dust deposition (Treguer and Pondavan 2000) or in the possible changes in the vegetation coverage on land (Humborg et al. 2002). Transfer of the newly fixed CO_2 at sea surface to the deep ocean occurs in high-density aggregates formed by the interaction of organic matter and freshly formed organic matter (Haake and Ittekkot 1990; Ittekkot 1993).

Human Perturbations in Carbon-Silicon Interactions

Human activities in the past decades have altered the nature and magnitude of C-Si interactions with wider implications. First are the possible feedbacks of the projected global warming due to the increase in carbon dioxide in the atmosphere from deforestation and fossil fuel burning. High CO_2 contents and the projected global warming will accelerate silicate weathering on land, which in turn is an accelerated transient terrestrial sink for CO_2. Furthermore, accelerated silicate weathering can also increase the dissolved silicate inputs into rivers and ultimately to the coastal seas promoting diatom growth. By favoring diatom sedimentation over coccolithophorid sedimentation, this weathering also accelerates the long term sink of C in the ocean as well through a more efficient biological carbon pump.

Another possible impact of projected global warming is from possible changes in the physical regime of the oceans, especially those in the intensity of upwelling and shifts in upwelling centers (Ittekkot et al. 1996). In vast areas of the world's oceans, the silicate demand of diatoms is met by the silicate-rich waters reaching the sea surface by upwelling. In such areas (e.g., the equatorial upwelling zone), dissolved silicate concentrations set the upper limit on the total possible biological utilization of inorganic carbon. Diatoms appear to be responsible for most of the new primary production in these areas (Dugdale, Wilkerson, and Minas 1995; Dugdale and Wilkerson 1998).

Diatom populations in coastal waters are also sustained by silicate inputs from rivers. Although coastal seas represent ≤10 percent of the oceanic area and <0.5 percent of oceanic volume, they make a disproportionately high contribution to global marine primary production (Mantoura, Martin, and Wollast 1991). Recent estimates show that the coastal seas contribute 30–50 percent to the new primary production of global oceans (Pearl 1995). Moreover, deltaic and shelf sediments incorporate about 80 percent of the organic carbon being sequestered in modern marine sediments (Berner 1982). Studies over the past few years have indicated a decrease in dissolved silicate in many coastal marine regions of the world (Conley, Schelske, and Stoermer 1993). This decrease has been explained by increased growth of silicate-utilizing diatoms as a result of nitrate- and phosphate-induced eutrophicaton, and subsequent removal of fixed biogenic silica via sedimentation out of the water column (Billen, Lancelot, and Meybeck 1991; Rahm et al. 1996). The resulting changes in the ratios of nutrient elements (e.g., Si:N:P ratios) have caused shifts in phytoplankton populations in water bodies (Admiral et al. 1990; Turner and Rabalais 1994). Shifts from diatoms to non-siliceous phytoplankton have been observed much earlier in the season in several estuarine and coastal regions, such as in the receiving marine waters of the Rhine River.

It is now known that large-scale hydrological alterations on land such as river damming and river diversion could directly cause reductions of silicate inputs to the sea as have been observed for the Black Sea and the Baltic Sea (Humborg et al. 1997, 2000). One explanation is that hydrological alterations decrease the rate of weathering (Humborg et al. 2002), and the second is that diatom blooms in the reservoirs sequester silicate. By contrast, although all nutrients (nitrogen, phosphorus, and silicon) get trapped in reservoirs behind dams, nitrate and phosphate discharged from human activities downstream of the dams more than make up for what is trapped in reservoirs; for silicate there is no such compensation (Ittekkot, Humborg, and Schäfer 2000). It is estimated that 160 teragrams (Tg) of carbon accumulate in reservoirs compared with about 60–130 Tg in the open ocean or 60 Tg in natural lakes (Dean and Graham 1998). It is of vital interest to investigate how much of this carbon is from autochthonous production—that is, diatom blooms and thus an additional global carbon sink—or from allochthonous sources. The resulting alteration in the nutrient mix reaching the sea, however, could also exacerbate the effect of eutrophication. That is, silicate limitation in perturbed water bodies can set in much more rapidly than under pristine conditions, leading to changes in the composition of phytoplankton in coastal waters.

Recent studies on material fluxes into the ocean's interior in the Bay of Bengal, an area that receives sediment and freshwater inputs from some of the largest rivers of the world such as the Ganges, Brahmaputra, and the peninsular Indian rivers, suggest that the response of the biological carbon pump in the ocean to changing river inputs can be rapid. The rivers in question discharge large amounts of dissolved silicate accounting for about 5 percent of global river inputs. The study involved the collection of settling particles using sediment traps (devices that allow continuous collection from months to years and that can be moored at various depths in the ocean) at two locations off the mouths of the Ganges and Brahmaputra separated by about 200 kilometers in a north-south direction (locations North and South in Figure 17.4). Results from location North showed that fluxes to deep sea varied with inputs from the Ganges-Brahmaputra River system (Figure 17.4a; Ittekkot et al. 1985, 1991). Most of the material reaches the ocean's interior within the three months of the southwest monsoon. More important, the prevailing biological community structure showed an increase in the silica-secreting diatoms relative to the carbonate-secreting coccolithophorids. The signals of these changes are recorded in the material collected in sediment traps, especially in their high biogenic silica:biogenic carbonate and the C_{org}:C_{carb} ratios. The location South, on the other hand, was characterized by low fluxes and an absence of pronounced seasonal signals (Figure 17.4b; Schäfer et al. 1996). Fluxes of bio-

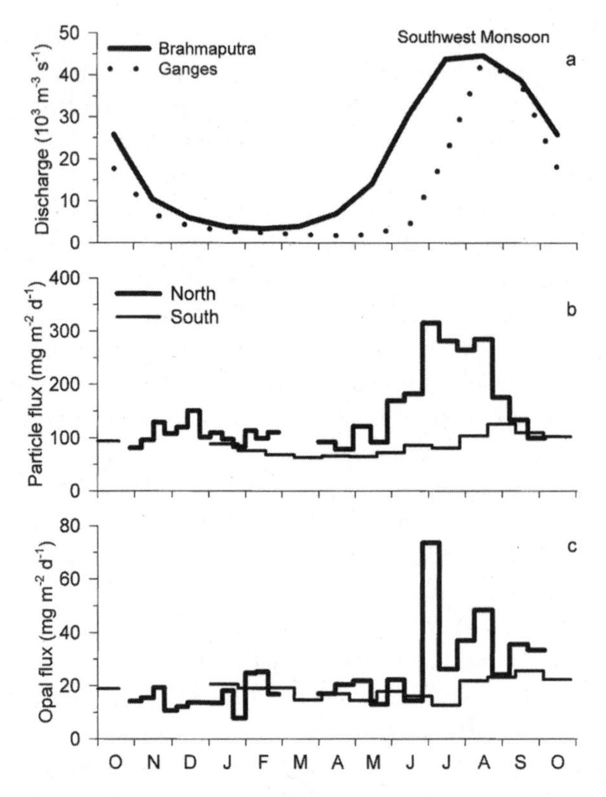

Figure 17.4. River inputs and marine fluxes. Seasonal fluctuations in (a) inputs from the Ganges and Brahmaputra Rivers, (b) total fluxes of particulate material, and (c) biogenic opal-silica fluxes from stations North and South.

genic silica are also low and uniform (Figure 17.4c), indicating lower contribution from diatoms. The difference in fluxes and their quality at the two sediment trap locations basically reflects the lateral extent of the river plumes (with location South being away from the plume), with diatom blooms at plume fronts acting as barriers both for dissolved nutrients and for river-derived lithogenic (mineral) material. This material acts as ballast in the formation of rapidly sinking aggregates, which remove the freshly formed biogenic silica from diatom blooms and organic matter to the sea bottom (Figure 17.5; Ittekkot et al. 1992). An increase in the efficiency of the biological carbon pump is the result.

Thus, alterations in the supply of silicate inputs to the productive surface layers of the oceans from either changing river inputs or upwelling intensities, and the accompanying potential shifts in species composition, could have a significant impact on carbon cycling in the ocean. Many important regulatory and socio-

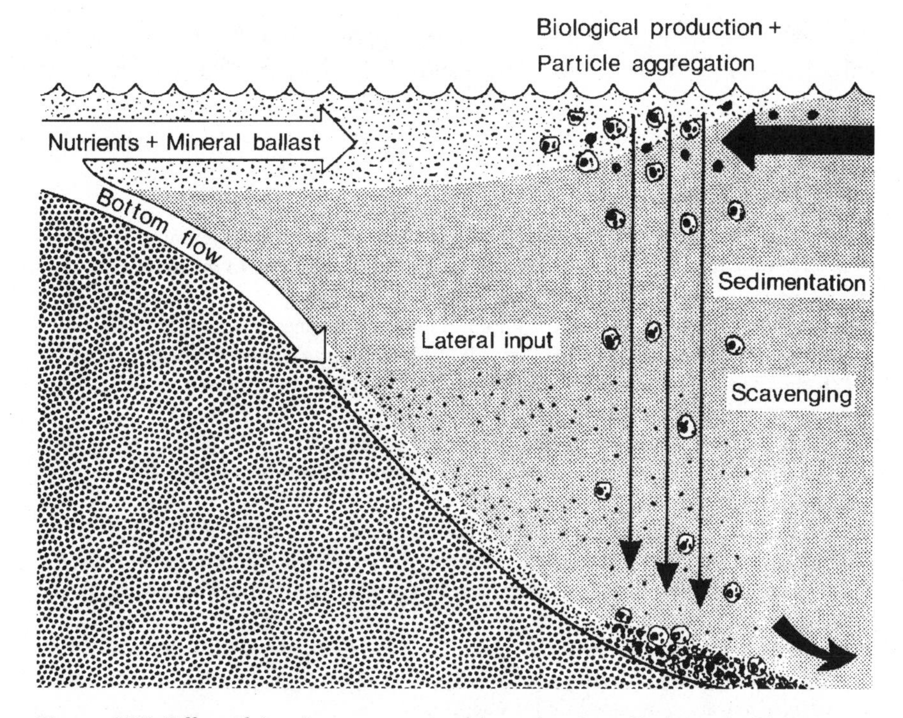

Figure 17.5. Effect of river input on marine biogeochemistry. Both nutrient inputs (in this case high silica inputs) and mineral matter input from rivers have an effect on the production and removal of biogenic silica and organic matter. The biological material (biogenic silica and organic matter) interacts with mineral matter, leading to the formation of high-density aggregates that settle rapidly to the sea bottom (from Ittekkot et al. 1992).

economic functions of water bodies will be affected. The ability of these water bodies to sustain economically important fisheries will be reduced; severe perturbations can be expected in the biogeochemical cycling of elements, with consequences for the role of coastal seas to act as sinks for anthropogenic gases such as CO_2. Since silica is also implicated in the long-term sequestration of organic C into marine sediments, the largest long-term sink on earth, human alterations of silica delivery to the oceans could have profound effects on carbon burial and ultimately atmospheric oxygen.

Acknowledgments

We are grateful to Dr. Jonathan Cole for his review comments. We also thank the Deutsche Forschungsgemeinschaft (DFG, Bonn) and the Bundesministerium für

Bildung und Forschung (BMBF, Bonn) for supporting our work. This is a contribution to the SCOPE Project on Land-Ocean Nutrient Fluxes: The Silica Cycle.

Literature Cited

Admiral, W., P. Breugem, D. M. Jacobs, and E. D. de Ruyter van Steveninck. 1990. Fixation of dissolved silicate and sedimentation of biogenic silicate in the lower river Rhine during diatom blooms. *Biogeochemistry* 9:175–185.

Berger, W. H., and R. S. Keir. 1984. Glacial-Holocene changes in atmospheric CO_2 and the deep-sea record. Pp. 337–351 in *Climate processes and climate sensitivity,* edited by J. K. Hansen and T. Takahashi. Geophysical Monograph 29. Washington, D.C.: American Geophysical Union.

Berner, R. A. 1982. Burial of organic carbon and pyrite sulfur in the modern ocean: Its geochemical and environmental significance. *American Journal of Science* 282:451–473.

Berner, R. A. 1998. Chemical weathering and its effect on atmospheric CO_2 and climate. Pp. 565–583 in *Chemical weathering rates of silicate minerals,* edited by A. F. White and S. L. Brantley. Reviews in Mineralogy 31. Washington, D.C.: Mineralogical Society of America.

Berner, R. A., A. C. Lasaga, and R. M. Garrels. 1983. The carbonate-silicate geochemical cycle and its effect on atmospheric carbon dioxide over the past 100 million years. *American Journal of Science* 283:641–683.

Billen, G., C. Lancelot, and M. Meybeck. 1991. N, P, Si retention along aquatic continuum from land to ocean. Pp. 19–44 in *Ocean margin processes in global change,* edited by R. F. C. Mantoura, J. M. Martin, and R. Wollast. Chichester, U.K.: John Wiley and Sons.

Brady, P. V., and S. A. Q. Carrol. 1994. Direct effects of CO_2 and temperature on silicate weathering: Possible implications for climate control. *Geochimica Cosmochimica Acta* 58:1853–1856.

Cairns-Smith, A. G. 1985. The first organisms. *Scientific American* 252:90–100.

Conley, D. J., C. L. Schelske, and E. F. Stoermer. 1993. Modification of the biogeochemical cycle of silica with eutrophication. *Marine Ecology Progress Series* 101:179–192.

Dean, W. E, and E. Graham. 1998. Magnitude and significance of carbon burial in lakes, reservoirs, and peatlands. *Geology* 26:535–538.

Degens, E. T. 1976. Molecular mechanisms on carbonate, phosphate, and silica deposition in the living cell. *Topics in Current Chemistry* 64:1–112.

Dugdale, R. C., and F. P. Wilkerson. 1998. Silicate regulation of new production in the equatorial Pacific upwelling. *Nature* 391:270–273.

Dugdale, R. C., F. P. Wilkerson, and H. J. Minas. 1995. The role of a silicate pump in driving new production. *Deep-Sea Research* 42:697–719.

Garrels, R. M., and F. T. Mackenzie 1971. *Evolution of sedimentary rocks.* New York: Norton.

Haake, B., and V. Ittekkot. 1990. Die wind-getriebene "biologische Pumpe" und der Kohlenstoffentzug im Ozean. *Naturwissenschaften* 77:75–79.

Hecky, R. E., K. Mopper, P. Kilham, and E. T. Degens. 1973. The amino acids and sugar composition of diatom cell walls. *Marine Biology* 19:323–331.

Heinze, C., E. Maier-Reimer, and K. Winn. 1995. Glacial pCO_2 reduction by the world ocean: Experiments with the Hamburg Carbon Cycle model. *Paleoceanography* 6:395–430.

Holland, H. D. 1978. *Chemistry of the atmosphere and the oceans.* New York: Wiley.

Humborg, C., V. Ittekkot, A. Cociasu, and B. von Bodungen. 1997. Effect of Danube river dam on Black Sea biogeochemistry and ecosystem structure. *Nature* 386:385–388.

Humborg, C., D. L. Conley, L. Rahm, F. Wulff, A. Cociasu, and V. Ittekkot. 2000. Silicon retention in river basins: Far reaching effects on biogeochemistry and aquatic food webs in coastal marine environments. *Ambio* 29:45–50.

Humborg, C., S. Blomqvist, E. Avsan, Y. Bergensund, and E. Smedberg. 2002. Hydrological alterations with river damming in northern Sweden: Implications for weathering and river biogeochemistry. *Global Biogeochemical Cycles* 16:10.1029/2000GB001369.

Ittekkot, V. 1993. An abiotically driven biological pump in the ocean and short term fluctuations in atmospheric CO_2 contents. *Global and Planetary Change* 8:17–25.

Ittekkot, V., B. Haake, M. Bartsch, R. R. Nair, and V. Ramaswamy. 1992. Organic carbon removal in the sea: The continental connection. Pp. 167–176 in *Upwelling systems: Evolution since the early Miocene,* edited by C. P. Summerhayes, W. L. Prell, and K. C. Emeis. Publication No. 64. London: Geological Society.

Ittekkot, V., C. Humborg, and P. Schäfer. 2000. Large-scale hydrological alterations and marine biogeochemistry: A silicate issue. *BioScience* 50:776–782.

Ittekkot, V., R. R. Nair, S. Honjo, V. Ramaswamy, M. Bartsch, S. Manganini, and B. N. Desai. 1991. Enhanced particle fluxes in Bay of Bengal induced by injection of freshwater. *Nature* 351:385–387.

Ittekkot, V., S. Safiullah, B. Mycke, and R. Seifert. 1985. Organic matter in the river Ganges, Bangladesh: Seasonal variability and geochemical significance. *Nature* 317:800–803.

Ittekkot, V., J. Su, E. Miles, E. Desa, B. N. Desai, J. T. Everett, J. J. Magnuson, A. Tsyban, and S. Zuta. 1996. Oceans. Pp. 267–288 in *Climate change 1995: Impacts, adaptations, and mitigation of climate change: Scientific-technical analyses,* edited by R. T. Watson, M. C. Zinyowera, and R. H. Moss. Contribution of Working Group II to the Second Assessment Report of the Intergovernmental Panel on Climate Change. Cambridge: Cambridge University Press.

Mantoura, R. F. C., J. M. Martin, and R. Wollast. 1991. *Ocean margin processes in global change.* Chichester, U.K.: John Wiley and Sons.

Mayer, L. 1999. Surface area control of organic carbon accumulation in continental shelf sediments. *Geochimica Cosmochimica Acta* 58:1271–1284.

Pearl, H. 1995. Coastal eutrophication in relation to atmospheric nitrogen deposition: current perspectives. *Ophelia* 41:237–259.

Rahm, L., D. Conley, P. Sanden, F. Wulff, and P. Stalnacke. 1996. Time series analysis of nutrient inputs to the Baltic Sea and changing DSi:DIN ratios. *Marine Ecology Progress Series* 130:221–228.

Schäfer, P., V. Ittekkot, M. Bartsch, R. R. Nair, and J. Tiemann. 1996. Freshwater influx and particle flux variability in the Bay of Bengal. Pp. 271–292 in *Particle flux in the ocean,* edited by V. Ittekkot, P. Schäfer, S. Honjo, and P.-J. Depetris. SCOPE 57. Chichester, U.K.: John Wiley and Sons.

Schlesinger, W. H. 1997. *Biogeochemistry: An analysis of global change.* 2nd ed. San Diego: Academic Press.

SCOR (Scientific Committee on Ocean Research). 1990. *Oceans, carbon, and climate change.* An introduction to the Joint Global Ocean Flux Study. Halifax: SCOR ICSU (International Council of Scientific Unions) Publication.

Smetacek, V. 1998. Diatoms and the silicate factor. *Nature* 391:224–225.

Treguer, P., D. M. Nelson, A. J. van Bennekom, D. J. DeMaster, A. Leynart, and B. Queguiner. 1995. The silica balance in the world ocean: A reestimate. *Science* 268:375–379.

Treguer, P., and P. Pondavan. 2000. Silica control of carbon dioxide. *Nature* 406:358–359.

Turner, R. E., and N. N. Rabalais. 1994. Coastal eutrophication near the Mississippi River delta. *Nature* 368:619–621.

White, A. F., and S. L. Brantley. 1995. Chemical weathering rates of silicate minerals. P. 583 in *Chemical weathering rates of silicate minerals,* edited by A. F. White and S. L. Brantley. Reviews in Mineralogy 31. Washington, D.C.: Mineralogical Society of America.

Wollast, R., and F. T. Mackenzie. 1986. The global cycle of silica. Pp. 39–76 in *Silicon geochemistry and biogeochemistry,* edited by R. Astonm. London: Academic Press.

18

Interactions among Carbon, Sulfur, and Nitrogen Cycles in Anoxic and Extreme Marine Environments

Mikhail V. Ivanov and Alla Y. Lein

More than two decades have passed since the publication of the series of SCOPE monographs devoted to biogeochemical cycles of elements and to their interaction (Bolin et al. 1979b; Bolin and Cook 1983; Ivanov and Freney 1983). New data have accumulated in the world's scientific literature that necessitate the revision of some former estimates of fluxes and reservoirs of biogenic elements. In addition, detailed investigations have been carried out on the previously poorly studied cycles of carbon, sulfur, and nitrogen in anaerobic and extreme ecosystems of the oceans of our planet, in particular, investigations of the sulfur and carbon cycles in the rift valleys of the ocean.

The main processes of the biogeochemical cycles of carbon, sulfur, and nitrogen and their interactions are shown in Figure 18.1. The figure shows that all anaerobic processes (sulfate reduction, methanogenesis, and denitrification) are coupled to consumption of organic matter. Therefore, quantitative evaluation of all anaerobic processes occurring in bottom sediments of the shelf, continental slope, and deep-sea troughs, as well as in water columns of fjords and some seas (the Black Sea, the Banda Sea, the Cariaco trench in the Caribbean Sea, and deep-sea basins in the eastern part of the Mediterranean Sea) requires refined estimates of the primary production in the ocean and of the fluxes of organic carbon to the oceanic bottom sediments.

Modern estimates of the primary production of phytoplankton in the ocean vary from 48,500 teragrams (Tg) per year (Field et al. 1998) to 60,000 Tg per year (Romankevich and Vetrov 1997) and do not differ considerably from the estimates made in the 1970s (Bolin et al. 1979a). The situation is quite differ-

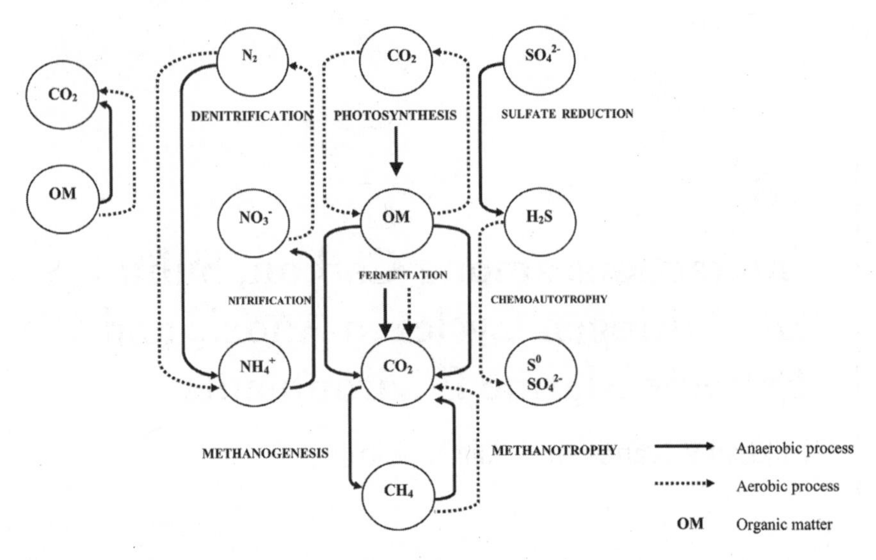

Figure 18.1. Linkage of carbon, sulfur, and nitrogen cycles

ent with the estimates of the flux of organic carbon from the zone of photosynthesis to bottom sediments. The model of the global biogeochemical cycle published in 1979 estimated this flux to be less than 100 Tg per year (Bolin et al. 1979a). As soon as the early 1980s it became clear that this value was underestimated, since sulfate reduction alone in bottom sediments of the continental margin results in the consumption of at least 369 Tg of carbon per year (Table 18.1). A still greater amount of organic carbon (more than 1,900 Tg per year; Table 18.1) is consumed in the course of aerobic processes in the upper horizons of sediments (Jorgensen 1983). Both of the values given are based on experimental data on the rates of sulfate reduction and oxygen consumption under *in situ* conditions.

In the early 1980s sediment traps came into use for assessing the flux of organic matter from the photic zone. Generalization of the data obtained by this method shows that 5 percent of the organic carbon fixed by phytoplankton reach the ocean bottom; this amounts to 1,000–3,000 Tg of carbon per year (Romankevich and Vetrov 1997).

Table 18.1 shows the values of aerobic and anaerobic consumption of organic carbon in bottom sediments of various zones of the ocean. The aerobic consumption is calculated from the data of Jorgensen (1983), who summarized data on oxygen consumption by bottom sediments; these data were recalculated to carbon according to the formula:

Table 18.1. Consumption of organic carbon as a result of aerobic oxidation and anaerobic sulfate reduction in sediments of various zones of the world ocean

			Consumption of C_{org}			
			Aerobic		*Anaerobic*	
Zone	*Depth (meters)*	*Area (10^{12} m^2)*	*mg m^{-2} day^{-1}*	*Tg $year^{-1}$ $area^{-1}$*	*mg m^{-2} day^{-1}*	*Tg $year^{-1}$ $area^{-1}$*
Inner shelf	0–50	13 (2.12)[a]	240	940	28.5	59.5
Outer shelf	50–200	18 (2.85)[a]	120	750	14.8	40.0
Upper slope	200–1,000	15	36	164	10.0	151.0
Lower slope	1,000–4,000	106	3.6	117	1.1	119.2
Deep sea	>4,000	208	0.6	36	—[b]	—
TOTAL		360		1,907		369.7

[a]Figures in parentheses are areas of reduced sediments where sulfate reduction is observed.
[b]Reduced sediments are observed only in deepwater trenches.

$$C_{org} + O_2 = CO_2.$$

The consumption of C_{org} in the course of sulfate reduction was calculated by Lein (1983) according to the formula:

$$SO_4^{2-} + 2\,C_{org} = S^{2-} + 2CO_2.$$

According to the data given in Table 18.1, the maximum rates of aerobic and anaerobic consumption of organic carbon at one square meter per day have been seen in shelf sediments. This was precisely the ocean zone where the maximum values of the primary production of organic carbon from the water column to bottom sediments have been recorded (Jorgensen 1983). As primary production moves away from the continents it tends to decrease, and as the depths increase most of the organic matter is mineralized under the aerobic conditions of the water column. Both of these circumstances lead to a decrease in the fluxes of organic matter to bottom sediments and the reduction of activity of aerobic as well as anaerobic processes of mineralization in the continental slope sediments and the deepwater ocean zone.

The rates of sulfate reduction under one square meter were determined experimentally by using radiolabeled sulfate (Ivanov et al. 1976) and used to calculate the values for various zones of the ocean (Lein 1983; Ivanov et al. 1989). The total con-

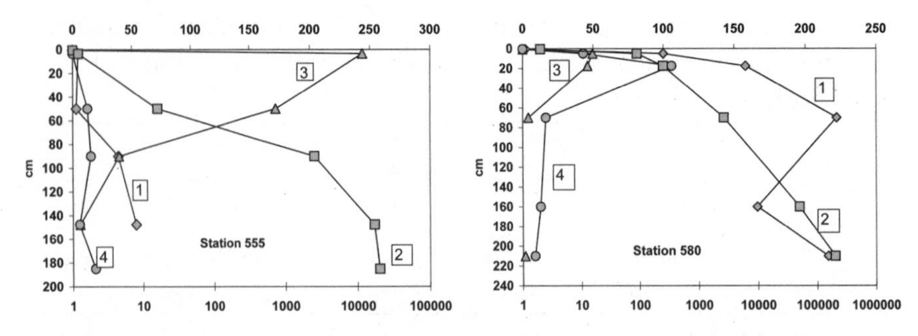

Figure 18.2. Distribution of reduced sulfur compounds, methane, rate of sulfate reduction (SR), and rate of methane production (MG) in bottom sediments of western part of the Black Sea (Ivanov at al. 1983). 1, ΣS_{red} (10^{-2} %); 2, $lgCH_4$ (10^{-5} cm^3 kg^{-1}); 3, SR (μg kg^{-1} day^{-1}); 4, MG (ng kg^{-1} day^{-1})

sumption of organic carbon in the processes of sulfate reduction in ocean sediments amounts to 369.7 Tg per year, or 12 percent of organic carbon, whose flux to the bottom sediments is about 3,000 Tg per year (Romankevich and Vetrov 1997).

Between the two other anaerobic processes of organic matter consumption, namely methanogenesis and denitrification, a considerable consumption of organic carbon may occur only via methanogenesis, since anaerobic sediments lack nitrates and nitrites.

It is possible to derive an approximate estimate of the consumption of organic carbon as a result of methanogenesis in reduced oceanic sediments from our experimental data obtained under *in situ* conditions using radiolabeled substrates of methanogenesis, $H^{14}CO_3$, and $^{14}CH_3COONa$. In upper horizons of sediments, the amount of C_{org} consumed by methanogenesis is usually no more than 5 percent of the amount of C_{org} consumed by sulfate reduction (Belyaev, Lein, and Ivanov 1980). Taking into account that methanogenesis actively proceeds in the lower horizons of sediments, however, where sulfate reduction does not occur because of the absence of sulfate (Figure 18.2), we may assume that the amount of C_{org} consumed during methanogenesis is equivalent to 10 percent of the amount of C_{org} consumed during sulfate reduction.

The value thus calculated (about 37 Tg per year) is indirectly supported by data of Bange et al. (1994), who summarized their own data and data from the literature and estimated that methane flux from shallow sediments to the atmosphere makes up about 11–18 Tg C per year. Since a notable part of methane formed in bottom sediments is oxidized by methanotrophs during its migration through the water column, it is evident that the methane production is higher than the emission to the atmosphere.

Thus, the minimum value of anaerobic decomposition of organic matter during two basic anaerobic microbial processes, sulfate reduction and methanogenesis, in bottom sediments of the ocean is 400–406 Tg of carbon per year. Proceeding from this value and the Redfield equation, we can calculate the amount of ammonium nitrogen released in the course of anaerobic decomposition of organic matter. The calculated flux of ammonium nitrogen from bottom sediments to the water column is ≈7.0 Tg per year. The distribution pattern of NH_4^+ in bottom sediments shows that the process of nitrogen release occurs throughout the entire thickness of the anaerobic sediments investigated.

Another interesting example of interaction of nitrogen, carbon, and sulfur cycles is observed in coastal zones of inland and marginal seas. In the 1970s and 1980s the northwestern Black Sea shelf became one of the most contaminated areas of the world ocean. An increased flux of nitrogen and phosphorus compounds from the agricultural regions of Bulgaria, Romania, and Ukraine resulted in a considerable increase of primary production, which may have reached 1,000–1,500 milligrams (mg) C per square meter per day in shallow waters (Vedernikov 1991).

The increase of the organic matter flux from the photosynthesis zone has led to hypoxia in prebottom waters (Zaitsev 1991), activation of sulfate reduction process in bottom sediments (Ivanov and Lein 1992), and mass mortality of benthic organisms. Over several years oxygen-deficient areas have covered up to 20,000 square kilometers (km^2) of the northwestern Black Sea shelf (Zaitsev 1991).

Following the economic collapse in socialist countries, the use of fertilizers in agriculture in these countries has drastically decreased, with a consequent remarkable reduction of nitrogen flowing to the Black Sea (Figure 18.3). This reduction, in turn, has led to the decrease of the primary production of organic matter and the reduction of sulfate reduction rates in bottom sediments of the northwestern shelf (Lein et al. 2002).

The use of manned submersibles and highly sensitive locators made it possible to reveal an earlier unknown large-scale flux of carbon from the bottom sediments to the water column and further to the atmosphere, namely cold methane and oil seeps. Localization of large fields of cold seeps is shown in Figure 18.4. Some of the seeps, for example, in the Gulf of Mexico, occur in oil-bearing provinces; other seeps are located in subduction zones along continental coasts. The third group of methane seeps is associated with submarine mud volcanoes.

As can be seen from data presented in Table 18.2 (page 330), the carbon isotopic composition of methane from different seeps varies widely from typical biogenic values (Black Sea seeps) to values characteristic of the methane of high-temperature genesis (seeps of the Gulf of Cadiz).

No quantitative estimation of the global flux of methane from cold seeps has

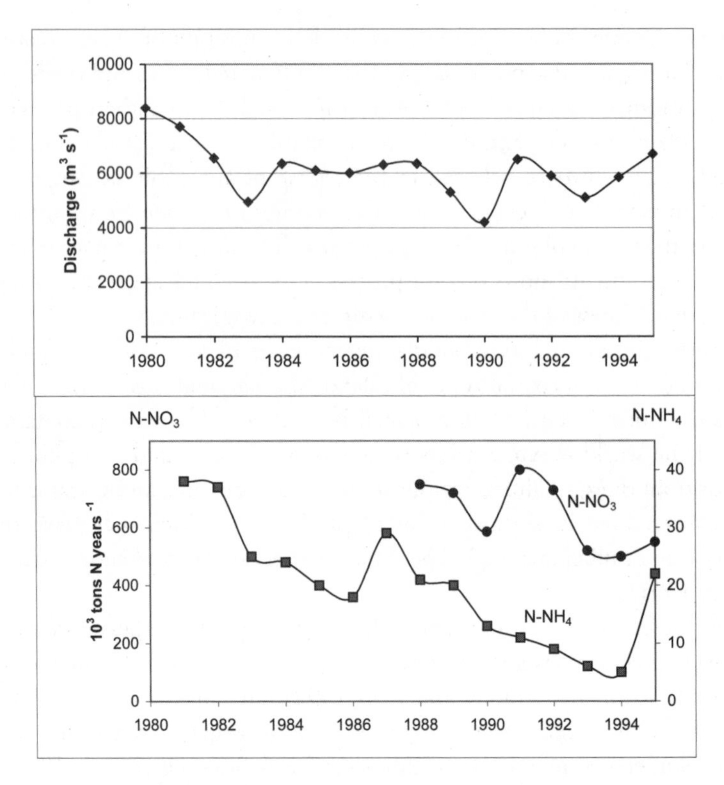

Figure 18.3. Interannual variations in water discharge and nitrogen ($N\text{-}NO_3$ and $N\text{-}NH_4$) fluxes at the outlet of the Danube Basin (Cociasu et al. 1997)

yet been made owing to technical difficulties; however, the values characterizing methane fluxes from certain areas are rather impressive. Thus, the minimum emission of deep methane in the caldera of the Haakon Mosby mud volcano (Norwegian Sea) is 56,000 kilograms (kg) C as methane per year (Lein et al. 2000a); the numerous cold seeps of the southeastern part of the Black Sea release more than 3,414,000 kg C per year.

The idea that methane seeps release a large amount of methane is supported by the occurrence of numerous carbonate constructions formed as a result of aerobic and anaerobic oxidation of methane. It is important to emphasize that, despite the presence of the oxidative microbial filter, a significant portion of methane slips through the water column in the form of bubbles and enters the atmosphere, contributing to the pool of atmospheric greenhouse gases. It is a pity that this large-scale flux of carbon from the lithosphere to the atmosphere has not been taken into account in any of the models of the global carbon cycle.

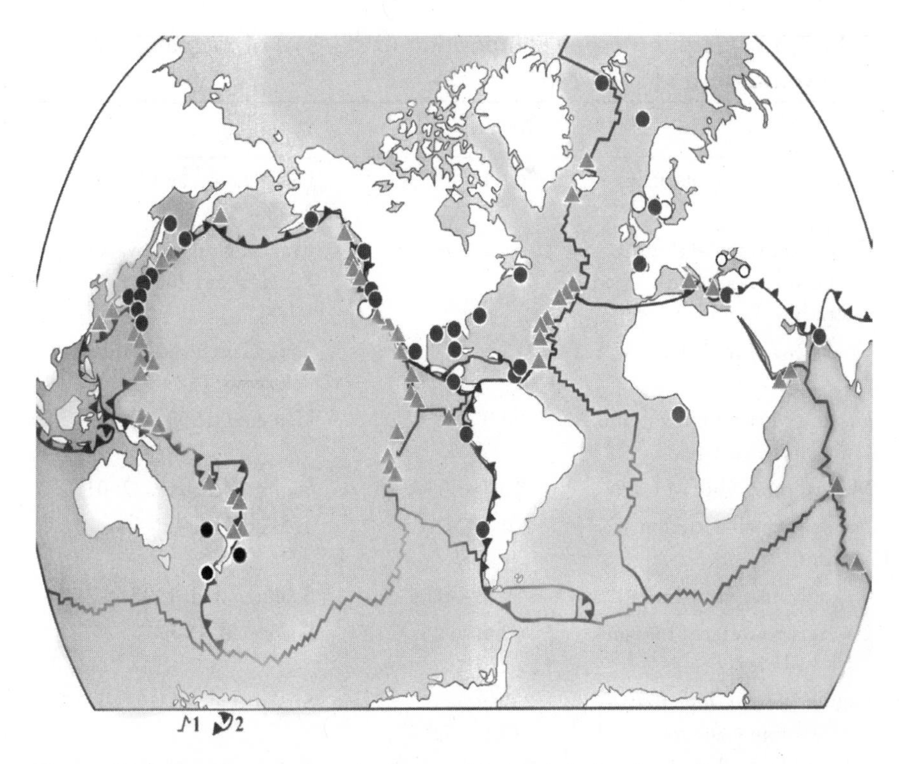

Figure 18.4. Distribution of mud volcanoes (closed circles), cold methane seeps (open circles), and submarine hydrothermal fields (triangles) in the world ocean. 1—spreading zones; 2—subduction zones (Gebruk 2002)

An enormous amount of data on the geochemistry of biogenic elements and metals has accumulated over the past twenty years for such extreme ecosystems as the submarine hydrothermal fields of rift valleys of mid-oceanic ridges. In the late 1970s the study of such ecosystems had just started, and therefore they went virtually unmentioned in SCOPE reports on biogeochemical cycles, with the exception of one report edited by the SCOPE-UNEP (United Nations Environment Programme) Sulphur Unit (Brimblecombe and Lein 1989).

The main element of these extreme ecosystems is ultra-acidic hydrothermal fluids overheated to 360–390 degrees Celsius; these fluids are formed during the interaction of oceanic water with hot subsurface rocks of oceanic crust. The key diagram of the formation of such solutions is shown in Figure 18.5, and the chemical composition of water released from black and white smokers into seawater is given in Table 18.3. On the surface of gigantic sulfide constructions of the smokers and in the hot fluid, which is saturated with gases and metal ions, occur

Table 18.2. Carbon isotope composition ($\partial^{13}C$, ‰) of methane from mud volcanoes and cold seeps

Site	*$\partial^{13}C$, ‰*	*Reference*
Cold seeps of the Black Sea	-62.4 ÷ -70.5	Lein, Ivanov, and Pimenov 2002
Crimea continental slope, mud volcanoes MGU, acad. Strachov, Bezimjany	-61.8 ÷ -63.55	Byakov, Kruglyakova, and Kruglyakova 2002
Cold seeps of the Okchotsk Sea	-56.4	Lein, Galtchenko, and Pokrovsky 1989
Haakon Mosby mud volcano, the Norwegian Sea	-60.0	Lein et al. 1999
Mud volcano, Gulf of Cadiz	-32.0 ÷ -63.0	Stadnitaskaia et al. 2001
Naukai Trough accretionary prism	-70	Toki et al. 2001
Cascadia margin hydrates	-59.5 ÷ -62.4	Schlüter et al. 1998
Cascadia convergent margin gas hydrates	-62.4; -68.5; -71.5	Suess et al. 1999
Gulf of Mexico slope from Mississippi Canyon: Bush Hill, Jolliet	Vent gas: -44.1 ÷ -49.0 Gas hydrates: -42.5 ÷ -48.3	Sassen et al. 2001

intense chemical and biological processes associated with the oxidation of reduced compounds from depth with the oxygen dissolved in seawater.

In this chapter we are primarily interested in the biogeochemical processes of the cycles of sulfur, carbon, and nitrogen. The distribution of the population density of key groups of microorganisms involved in the oxidation of reduced compounds of carbon (methane), sulfur (hydrogen sulfide, suspended metal sulfides, elemental sulfur), and nitrogen (ammonium) and the rates of the processes of chemosynthesis and methane oxidation in the plumes of the black smokers of the Guaymas trench are shown in Figure 18.6.

The maximum concentrations of suspended and dissolved reduced compounds are determined from the value of maximum turbidity and the content of dissolved manganese and methane. These horizons contain the maximum numbers of bacterial cells recorded by direct microscopic count and maximum numbers of chemoautotrophic thionic bacteria, nitrifiers, and methanotrophs. In the

Table 18.3. Chemical composition of hydrothermal fluid compared with seawater

Element	*Unit*	*Hydrothermal fluid*	*Seawater*	*Enrichment factor (minimum)*
H_2S	mM kg^{-1}	3–12	0	∞
H_2*	mM kg^{-1}	0.02–13	0	∞
CH_4*	mM kg^{-1}	0.04–2.3	0	∞
NH_4*	mM kg^{-1}	0.01–3.65	<0.01	∞
CO_2*	mM kg^{-1}	3–8	2	2.5
Mn	μM kg^{-1}	360–1,140	0	∞
Fe	μM kg^{-1}	750–6,500	0	∞
Be	nM kg^{-1}	10–40	0	∞
Zn	μM kg^{-1}	40–00	0.01	4,000
Cu	μM kg^{-1}	10–40	0.007	1,500
Ag	nM kg^{-1}	25–0	0.02	1,250
Pb	nM kg^{-1}	10–60	0.01	1,000
Co	nM kg^{-1}	20–220	0.03	650
Si	mM kg^{-1}	15–20	0.05	300
Al	μM kg^{-1}	5–20	0.02	250
Ba	μM kg^{-1}	10–40	0.15	66
Ca	mM kg^{-1}	10–5	10	1
Sr	μM kg^{-1}	90	85	1
P	μM kg^{-1}	0.5	2	0.25
Mg	mM kg^{-1}	0	50	0
SO_4	mM kg^{-1}	0–1	30	0
Alk	mM kg^{-1}	(-0.1)–(-1)	2	0

Sources: Data for elements marked with an asterisk (*) are from Lein et al. (2000b); other data are from van Dover (2000).

same horizons, the maximum rates of organic matter chemosynthesis and methane oxidation are observed.

The estimates of the fluxes of hydrogen sulfide and methane with the high-temperature hydrothermal solutions in the oceanic rift zones increase virtually every year, because virtually every year new sites of deep-sea hydrothermal fields are discovered. According to some data (McColom 2000), the total flux of hydrothermal solutions is as large as about 10^{11} liters per year. Proceeding from this value and the data on the content of hydrogen sulfide and methane in hydrothermal solutions (Table 18.3), the flux of H_2S ranges from 1.02 to 4.02 Tg per year, and the flux of methane is 1.6 Tg per year.

These should be considered minimum values, because not all mid-oceanic

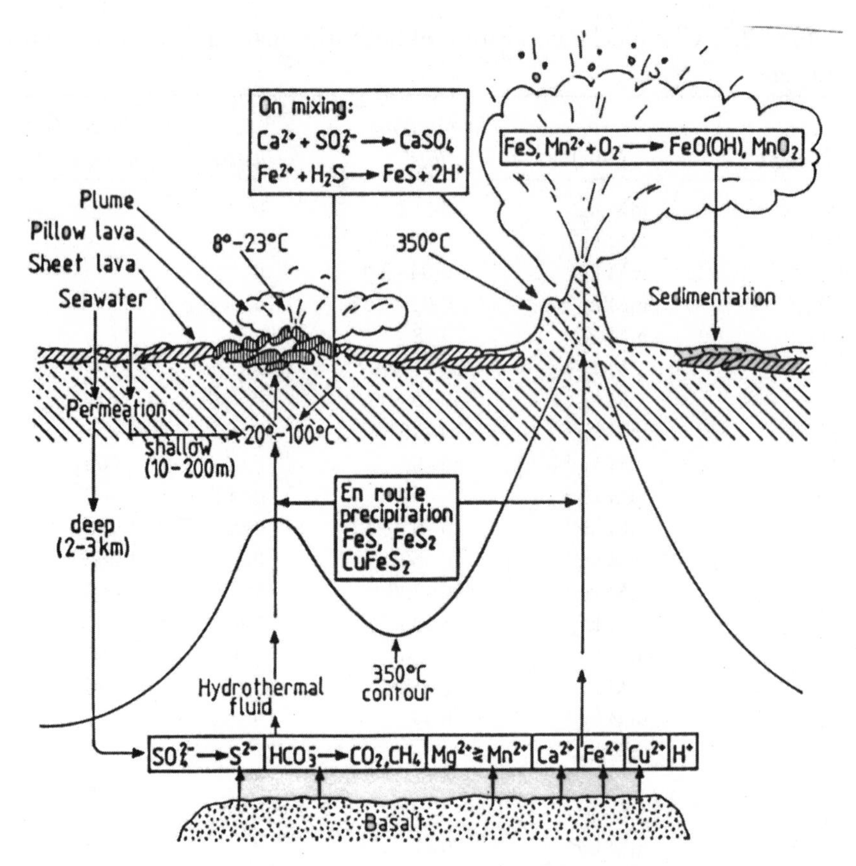

Figure 18.5. The major geochemical reactions occurring during the hydrothermal cycling of seawater through the Earth's crust at ocean floor spreading zones, indicating the two commonly observed types of vents (Jannash 1989)

ridges have been studied for the occurrence of hydrothermal fields. Even at the already known hydrothermal fields, only part of the hydrothermal solution is released through the craters of the smokers, where the flux can be quantitatively assessed. Large amounts of the solutions enter the oceanic water through small fractures of the seafloor or soak through the sediments surrounding the hydrothermal constructions. The reduced compounds contained in the solutions released in these ways still escape quantitative assessment, although just these dispersed sources, and not the overheated hydrothermal jets, provide for the existence of chemoautotrophic microflora inhabiting specialized tissues (trophosomes) of abundant symbiotic animals thriving at the hydrothermal fields (Van Dover 2000).

The most important result of the geochemical activity of methanotrophic and

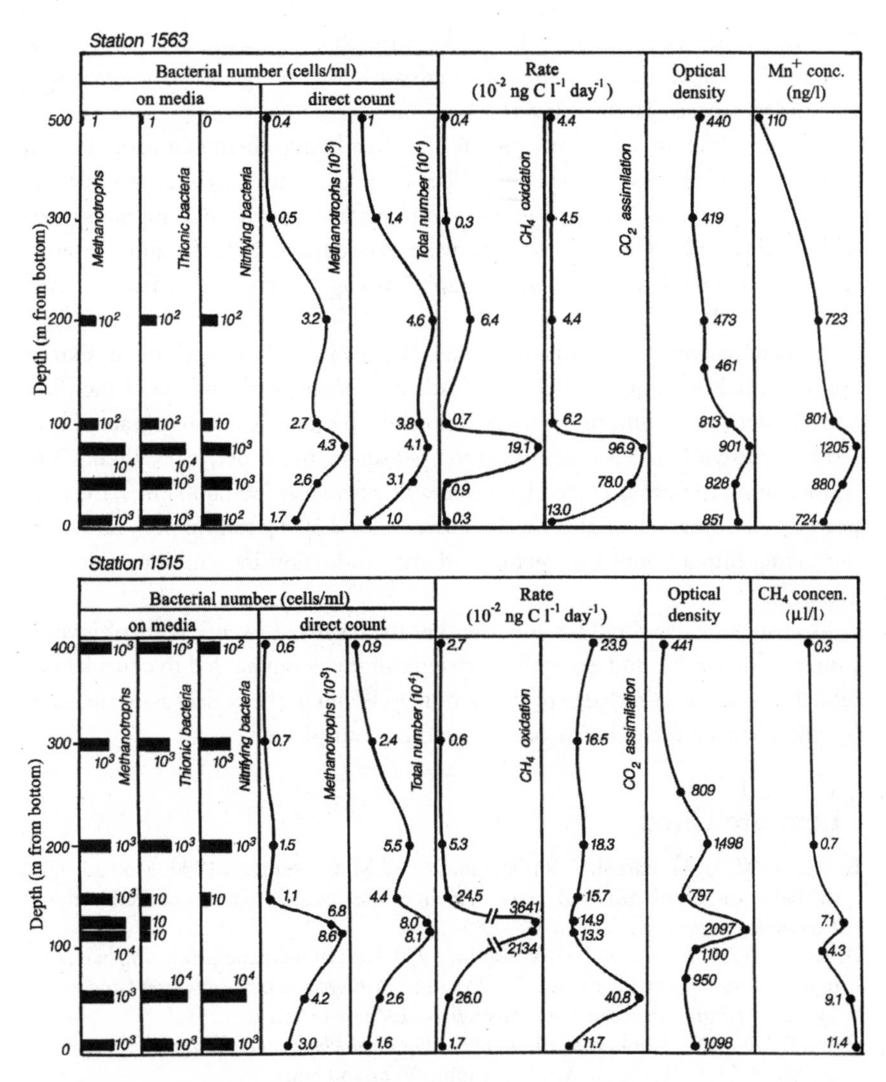

Figure 18.6. Microbial and biogeochemical characteristics of the water column at the Guaymas Basin hydrothermal site (Galtchenko, Ivanov, and Lein 1989)

chemoautotrophic bacteria developing at hydrothermal fields is the resynthesis of the organic matter of microbial biomass, which is the main (and in the deepest regions the single) source of organic matter for the entire community of heterotrophic microorganisms inhabiting these deep-sea oases of life. The scale of the activity of symbiotrophic microorganisms inhabiting the trophosomes of hydrothermal animals may be estimated from the biomass of these organisms. On some hydrothermal fields it reaches 10 kg per square meter, which is comparable with the ecosystems whose trophic pyramids are based on organic matter produced via photosynthesis (Jannash 1984).

The production of organic matter by free-living chemoautotrophic and methanotrophic bacteria grazed on by filter-feeding animals can be experimentally determined in short-term experiments with isolated water samples supplemented with $^{14}CH_4$ and $^{14}CO_2$. These data were summarized by Lein and Pimenov (2002) and published in the monograph "Biology of Hydrothermal Systems" (Gebruk 2002).

Generalization of experimental data obtained by these authors at sixteen hydrothermal fields of the Pacific and Atlantic Oceans with the use of the Pices and Mir deep-sea submersibles made it possible to calculate the mean rate of bacterial chemosynthesis (100 grams carbon per square meter per year). Taking into consideration the area occupied by the oceanic rift valleys (600,000 km^2), the total production of organic matter via chemosynthesis is 60 Tg carbon per year (without taking into account the chemosynthetic production by symbiotic microorganisms).

To conclude, we should emphasize that the results of twenty years of investigations of anaerobic and extreme ecosystems of the ocean has led to considerable refinement of our knowledge of the carbon cycle and of the closely associated sulfur and nitrogen cycles in the oceanic sector of our planet.

Literature Cited

Bange, H. W., U. H. Bartell, S. Rapsomanicis, and M. O. Andreae. 1994. Methane in the Baltic and North Seas and a reassessment of the marine emissions of methane. *Global Biogeochemical Cycles* 8:465–480.

Belyaev, S. S., A. Y. Lein, and M. V. Ivanov. 1980. Role of methane-producing bacteria in destruction of organic matter. Pp. 235–242 in *Biogeochemistry of ancient and modern environments.* Canberrra: Australian Academy of Sciences.

Bolin, B., and R. B. Cook, eds. 1983. *The major biogeochemical cycles and their interactions*, SCOPE 21. Chichester, U.K.: John Wiley and Sons.

Bolin, B., E. T. Degens, P. Duvigneaud, and S. Kempe. 1979a. The global biogeochemical carbon cycle. In *The global carbon cycle,* edited by B. Bolin, E. T. Degens, S. Kempe, and P. Ketner. SCOPE 13. Chichester, U.K.: John Wiley and Sons.

Bolin, B., E. T. Degens, S. Kempe, and P. Ketner, eds. 1979b. *The global carbon cycle.* SCOPE 13. Chichester, U.K.: John Wiley and Sons.

Brimblecombe, P., and A. Y. Lein, eds. 1989. *Evolution of the global biogeochemical sulphur cycle.* SCOPE 39. Chichester, U.K.: John Wiley and Sons.

Byakov, Y. A., R. P. Kruglyakova, and M. V. Kruglyakova. 2002. Gas hydrates of the Black Sea sediment section: Genesis, geophysical methods for their discovery and mapping. Minerals of the Ocean, edited by L.S. Granberg. Intern. Conference. St. Petersburg: VNII Okeangeologia. Pp. 159–160.

Cociasu, A., V. Diaconu, L. Popa, I. Nae, I. Derogem, and V. Maclin. 1997. The nutrient stock of the Romanian shelf of the Black Sea during the last three decades. Pp. 49–65 in *Sensitivity of change: Black Sea, Baltic Sea, and North Sea,* edited by E. Ozsoy and A. Mikaelyan. NATO ASI Series 2, Environment 27. Dordrecht, the Netherlands: Kluwer.

Field, C. B., M. J. Behrenfeld, J. T. Randerson, and P. Falkowski. 1998. Primary production in the biosphere: Integrating terrestrial and oceanic components. *Science* 281:237–240.

Galtchenko, V. F., M. V. Ivanov, and A. Y. Lein. 1989. Microbial and geochemical processes in water columns of the ocean as tracers of activity of submarine hydrothermal vents. *Geochemistry International* 8:1075–1088.

Gebruk, A. V. 2002. *Biology of hydrothermal systems* (in Russian). Moscow: KMK Press.

Ivanov, M. V., and J. R. Freney, eds. 1983. *The global biogeochemical sulphur cycle.* SCOPE 19. Chichester, U.K.: John Wiley and Sons.

Ivanov, M. V., A. Y. Lein, and E. V. Kashparova. 1976. Intensity of formation and diagenetic transformation of reduced sulphur compounds in sediments of the Pacific Ocean. Pp. 171–181 in *The biogeochemistry of diagenesis of ocean sediments* (in Russian). Moscow: Nauka.

Ivanov, M. V., A. Y. Lein, W. S. Reeburg, and G. W. Skyring. 1989. Interaction of sulphur and carbon cycles in marine sediments. In *Evolution of the global biogeochemical sulphur cycle,* edited by P. Brimblecombe and A. Y. Lein. SCOPE 39. Chichester, U.K.: John Wiley and Sons.

Ivanov, M. V., M. B. Vainstein, V. F. Galtchenko, S. N. Gorlatov, and A. Y. Lein. 1983. Distribution and geochemical activity of bacteria in sediments of western part of the Black Sea. Pp. 150–180 in *Oil- and gas-genetic studies of the Bulgarian sector of the Black Sea* (in Russian). Sofia: Bulgarian Academy of Sciences.

Ivanov, M. V., and A. Y. Lein. 1992. Causes of appearance of anaerobic conditions in prebottom waters of the NW Shelf of Black Sea. *UNESCO Technical Papers in Marine Science* 64:232–245.

Jannash, H. W. 1984. Microbial processes at deep-sea hydrothermal vents. Pp. 677–709 in *Hydrothermal processes at seafloor spreading centers,* edited by D. A. Rona, K. Bostom, L. Lanbier, and K. Smith. NATO Conference Series IV, Marine Sciences Vol. 12. New York: Plenum.

Jannash, H. W. 1989. Sulphur emission and transformations at deep-sea hydrothermal vents. In *Evolution of the global biogeochemical sulphur cycle,* edited by P. Brimblecombe and A. Y. Lein. SCOPE 39. Chichester, U.K.: John Wiley and Sons.

Jorgensen, B. B. 1983. Processes at the sediment-water interface. In *The major biogeochemical cycles and their interactions,* edited by B. Bolin and K. B. Cook. SCOPE 21. Chichester, U.K.: John Wiley and Sons.

Lein, A. Y. 1983. Corg consumption by anaerobic processes of organic matter mineralization in modern oceanic sediments. *Geochemistry International* 11:1634–1639.

Lein, A. Y., and N. V. Pimenov. 2002. Role of bacterial production at active hydrothermal fields in the total balance of organic carbon in the ocean. Pp. 320–328 in *Biology of hydrothermal systems* (in Russian), edited by A. V. Gebruk. Moscow: KMK Press.

Lein, A. Y., V. F. Galtchenko, and B. G. Pokrovsky. 1989. Marine carbonate nodules as a result of microbial oxidation of gashydrate methane in the Okchotsk Sea. *Geochemistry International* 10:1396–1406.

Lein, A. Y., P. Vogt, K. Crane, A. Egorov, and M. V. Ivanov. 1999. Chemical and isotopeic evidences for the nature of fluid in CH^4-containing sediments of the Haakon Mosbi mud volcano. *Geo-Marine Letters* 19:76–83.

Lein, A. Y., N. V. Pimenov, A. S. Savvichev, G. A. Pavlova, P. Vogt, Y. A. Bogdanov, A. M. Sagalevich, and M. V. Ivanov. 2000a. Methane as a source of organic matter and carbon of carbonates at cold seep in the Norwegian Sea. *Geochemistry International* 4:268–281.

Lein, A. Y., D. V. Grichuk, E. G. Gurvich, and Y. A. Bogdanov. 2000b. New type of hydrothermal fluids enriched by hydrogen and methane in rift zone of Mid-Atlantic ridge (in Russian). *Doklady Academy of Sciences* 375:380–383.

Lein, A., N. Pimenov, C. Guillou, J.-M. Martin, C. Lancelot, I. Rusanov., S. Yusupov, Y. Miller, and M. V. Ivanov. 2002a. Seasonal dynamics of the sulfate reduction rate on the north-western Black Sea Shelf. *Estuarine, Coastal and Shelf Science* 54:385–401.

Lein, A. Y., M. V. Ivanov, and N. V. Pimenov. 2002b. Genesis of methane of cold seeps in Dniepr kanjone of the Black Sea (in Russian). *Doklady Academy of Sciences* 387:242–244.

McCollom, T. M. 2000. Geochemical constrains on primary vent plumes. *Deep-Sea Research* 47:85–101.

Romankevich, E. A., and A. A. Vetrov. 1997. Fluxes and masses of organic carbon in the ocean. *Geochemistry International* 9:829–836.

Sassen, R., S. T. Sweet, A. Milkov, D. A. D'Freitas, and M. C. Kennikutt. 2001. Thermogenic vent gas and gas hydrate in the Gulf of Mexico Slope: Is gas hydrate decomposition significant? *Geology* 29:107–110.

Schlüter, M., P. Linke, and E. Suess. 1998. Geochemistry of a sealed deep-sea borehole on the Cascadion Margin. *Marine Geology* 148:9–20.

Stadnitaskaia, A., M. K. Ivanov, T. C. E. van Weering, M. S. Sinnighe Damste, J. P. Werne, R. Krenlen, and V. Blinova. 2001. Molecular and isotopic characterization of hydrocarbon gas and organic matter from mud volcanoes of the Gulf of Cadiz, NE Atlantic. Pp. 90–91 in *Geological processes on deep-water European margins.* International conference and 10th anniversary training through research post-cruise meeting. Moscow: Moscow University .

Suess, E., M. E. Torres, G. Bohrman, R. W. Collier, J. Greinert, P. Linke, G. Rehder, A. Trehu, K. Wellmann, G. Winkler, and E. Zuleger. 1999. Gas hydrate destabilization: Enhanced dewatering, benthic material turnover, and large methane plumes at the Cascadia convergent margin. *Earth and Planetary Science Letters* 170:1–15.

Toki, T., T. Gamo, T. Yamanaka, J. Ishibashi, U. Tsunogai, and O. Matsubayashi. 2001. Methane migration from Naukai accretionary prism. *Bulletin of the Geological Survey of Japan* 52:1–8.

Van Dover, S. L. 2000. *The ecology of deep-sea hydrothermal vents.* Princeton: Princeton University Press.

Vedernikov, V. I. 1991. Distributions of primary production and chlorophyll in the Black Sea in spring and summer. Pp. 128–147 in *Variability of the Black Sea ecosystems* (in Russian). Moscow: Nauka.

Zaitsev, Y. P. 1991. Anthropogenic impact on the communities of biologically active zones of the Black Sea. Pp. 306–311 in *Variability of the Black Sea ecosystems* (in Russian), edited by M. E. Vinogradov. Moscow: Nauka.

List of Contributors

Göran Ågren
Department of Ecology and Environmental Research
Swedish University of Agricultural Sciences
Box 7072
SE 750 07 Uppsala, Sweden

Thomas R. Anderson
Southampton Oceanography Centre
Waterfront Campus, European Way
Southampton SO14 3ZH, UK

Mike Apps
Climate Change and Ecosystem Modelling
Canadian Forest Service
Pacific Forestry Centre
506 West Burnside Road
Victoria, BC, Canada V8Z 1M5

Amy T. Austin
Faculty of Agronomy
Department of Ecology
Universidad de Buenos Aires
Avenida San Martin 4453
1417 Buenos Aires, Argentina

Jill S. Baron
United States Geological Survey
Natural Resources Ecology Laboratory
NESB, B225, Colorado State University
Fort Collins, CO 80523-1499, USA

Bjorn Berg
Lehrstuhl für Bodenökologie
BITÖK, Postfach 101251
Uni Bayreuth
DE-95 448 Bayreuth, Germany

Mary Anne Carroll
Departments of Atmospheric, Oceanic, Space Sciences, and Chemistry
University of Michigan
Ann Arbor, MI 48109-1055, USA

F. Stuart Chapin III
Institute of Arctic Biology
University of Alaska-Fairbanks
Fairbanks, AK 99775-7480, USA

Torben R. Christensen
Department of Physical Geography and Ecosystem Analysis
GeoBiosphere Science Centre
Lund University, Box 118
SE-22100 Lund, Sweden

Jonathan J. Cole
Institute of Ecosystem Studies
P.O. Box AB
Millbrook, NY 12545-0129, USA

Eric Davidson
Woods Hole Research Center
P.O. Box AB
Woods Hole, MA 02543-0296, USA

Peter De Reuter
Department of Environmental Studies
University of Utrecht
P.O. Box 80115
Utrecht NL-3508 TC, Netherlands

Hugh W. Ducklow
The College of William and Mary
School of Marine Science
Route 1208, Greate Road, Box 1346
Gloucester Point, VA 23062, USA

James E. Elser
Department of Biology
Arizona State University
Tempe, AZ 85287, USA

Valerie Eviner
Institute of Ecosystem Studies
P.O. Box AB
Millbrook, NY 12545-0129, USA

Christopher B. Field
Department of Global Ecology
Carnegie Institution of Washington
260 Panama Street
Stanford, CA 94305, USA

James N. Galloway
Department of Environmental Sciences
University of Virginia, Clark Hall
291 McCormack Road
P.O. Box 400123
Charlottesville, VA 22904-4123, USA

Xingguo Han
Chinese Academy of Sciences
Institute of Botany
No. 20 Nan Xin Cun
Xiangshan, Haidian District
Beijing, 100093 P.R. China

Dag O. Hessen
University of Oslo
Department of Biology
P.O. Box 1027 Blindern
NO-0316 Oslo, Norway

Elisabeth A. Holland
National Center for Atmospheric Research
Boulder, CO 80305, USA

Robert Howarth
Department of Ecology and Evolutionary Biology
E311 Corson Hall
Cornell University
Ithaca, NY 14853, USA

Christoph Humborg
Department of Systems Ecology
Stockholm University
SE-10691 Stockholm, Sweden

Bruce A. Hungate
Northern Arizona University
Biological Sciences, Box 5640
Flagstaff, AZ 86011-5640, USA

Venugopalan Ittekkot
Zentrum für Marine Tropenökologie
Fahrenheitstrasse 6
28359 Bremen, Germany

Mikhail V. Ivanov
Institute of Microbiology
Russian Academy of Sciences
Prospekt 60-Letija Octyabrya 7
117811 Moscow B-312, Russia

Michael Keller
University of New Hampshire
Complex Systems Research Center
Durham, NH 03824, USA

Alla Y. Lein
Shirshov Institute of Oceanology
Russian Academy of Sciences
36, Nachimovsky
117851 Moscow, Russia

Manuel T. Lerdau
State University of New York (SUNY)
Ecology and Evolution Department
632 Life Science Building
Stony Brook, NY 11794-5245, USA

Luiz Antonio Martinelli
Centro de Energia na Agricultura (CENA)
Avenida Centenario 303
13416-000 Piracicaba
São Paulo, Brazil

Jerry Melillo
The Ecosystems Center
Marine Biological Laboratory
7 MBL Street
Woods Hole, MA 02543, USA

Bedrich Moldan
Environment Center
Charles University
U Krize 10
CZ-158 00 Praha 5, Czech Republic

Filip Moldan
IVL - POB 47086
Dagjämningsg. 1
Göteborg, Sweden

Robert J. Naiman
University of Washington
Fisheries Science Building
1122 NE Boat Street
Seattle, WA 98105, USA

Paolo Nannipieri
Dip. Scienza del Suolo e Nutrizione della Pianta
Università di Firenze
28 Piazzale delle Cascine
50144 Firenze, Italy

Jason Neff
Natural Resource Ecology Laboratory (NREL)
Natural and Environmental Science Building, B258
Colorado State University
Fort Collins, CO 80523-1499, USA

Jacques L. Oliver
The College of William and Mary
School of Marine Science
Route 1208, Greate Road, Box 1346
Gloucester Point, VA 23062, USA

Scott Ollinger
Complex Systems Research Center
University of New Hampshire
Morse Hall
Durham, NH 03824, USA

Lars Rahm
Department of Water and Environmental Studies
Linköping University
SE-58183 Linköping, Sweden

Osvaldo Sala
Faculty of Agronomy
Department of Ecology
Universidad de Buenos Aires
Avenida San Martin 4453
1417 Buenos Aires, Argentina

Mary Scholes
School of Animal, Plant and Environmental Sciences
University of the Witwatersrand
1 Jan Smuts Avenue
Private Bag 3
2050 Wits, Johannesburg 2000, South Africa

Sybil Seitzinger
Rutgers University
Institute of Marine and Coastal Sciences
Rutgers/NOAA CMER Program
71 Dudley Road
New Brunswick, NJ 08901, USA

Chao Shang
Department of Crop and Soil Management
Virginia Polytechnic Institute and State University
364 Smyth Hall
Blacksburg, VA 24061, USA

Walker O. Smith Jr.
The College of William and Mary
School of Marine Science
Route 1208, Greate Road, Box 1346
Gloucester Point, VA 23062, USA

Robert Sterner
Department of Ecology, Evolution, and Behavior
University of Minnesota
100 Ecology Building
1987 Upper Buford Circle
St. Paul, MN 55108, USA

John W. B. Stewart
University of Saskatchewan
118 Epron Road
Salt Spring Island, BC, V8K 1C7 Canada

Nguyen Tac An
Institute of Oceanography
Nha Trang, Vietnam

Holm Tiessen
Institute for Crop and Animal Production in the Tropics
Universität Göttingen
Grisebachstr. 6
37077 Göttingen, Germany

Reynaldo Victoria
Centro de Energia na Agricultura (CENA)
Avenida Centenario 303
13416-000 Piracicaba, São Paulo, Brazil

Peter M. Vitousek
Department of Biological Sciences
Stanford University
Stanford, CA 94305, USA

Andrew Woghiren
School of Animal, Plant, and Environmental Sciences
University of the Witwatersrand
1 Jan Smuts Avenue
Private Bag 3
2050 Wits, Johannesburg 2000, South Africa

SCOPE Series List

SCOPE 1–59 are now out of print. Selected titles from this series can be downloaded free of charge from the SCOPE Web site (http://www.icsu-scope.org).

SCOPE 1: *Global Environment Monitoring,* 1971, 68 pp
SCOPE 2: *Man-made Lakes as Modified Ecosystems,* 1972, 76 pp
SCOPE 3: *Global Environmental Monitoring Systems (GEMS): Action Plan for Phase I,* 1973, 132 pp
SCOPE 4: *Environmental Sciences in Developing Countries,* 1974, 72 pp
SCOPE 5: *Environmental Impact Assessment: Principles and Procedures,* Second Edition, 1979, 208 pp
SCOPE 6: *Environmental Pollutants: Selected Analytical Methods,* 1975, 277 pp
SCOPE 7: *Nitrogen, Phosphorus and Sulphur: Global Cycles,* 1975, 129 pp
SCOPE 8: *Risk Assessment of Environmental Hazard,* 1978, 132 pp
SCOPE 9: *Simulation Modelling of Environmental Problems,* 1978, 128 pp
SCOPE 10: *Environmental Issues,* 1977, 242 pp
SCOPE 11: *Shelter Provision in Developing Countries,* 1978, 112 pp
SCOPE 12: *Principles of Ecotoxicology,* 1978, 372 pp
SCOPE 13: *The Global Carbon Cycle,* 1979, 491 pp
SCOPE 14: *Saharan Dust: Mobilization, Transport, Deposition,* 1979, 320 pp
SCOPE 15: *Environmental Risk Assessment,* 1980, 176 pp
SCOPE 16: *Carbon Cycle Modelling,* 1981, 404 pp
SCOPE 17: *Some Perspectives of the Major Biogeochemical Cycles,* 1981, 175 pp
SCOPE 18: *The Role of Fire in Northern Circumpolar Ecosystems,* 1983, 344 pp
SCOPE 19: *The Global Biogeochemical Sulphur Cycle,* 1983, 495 pp
SCOPE 20: *Methods for Assessing the Effects of Chemicals on Reproductive Functions, SGOMSEC 1,* 1983, 568 pp
SCOPE 21: *The Major Biogeochemical Cycles and their Interactions,* 1983, 554 pp
SCOPE 22: *Effects of Pollutants at the Ecosystem Level,* 1984, 460 pp

SCOPE 46: *Methods for Assessing Exposure of Human and Non-human Biota SGOMSEC 5,* 1991, 448 pp
SCOPE 47: *Long-Term Ecological Research: An International Perspective,* 1991, 312 pp
SCOPE 48: *Sulphur Cycling on the Continents: Wetlands, Terrestrial Ecosystems and Associated Water Bodies,* 1992, 345 pp
SCOPE 49: *Methods to Assess Adverse Effects of Pesticides on Non-target Organisms, SGOMSEC 7,* 1992, 264 pp
SCOPE 50: *Radioecology after Chernobyl,* 1993, 367 pp
SCOPE 51: *Biogeochemistry of Small Catchments: A Tool for Environmental Research,* 1993, 432 pp
SCOPE 52: *Methods to Assess DNA Damage and Repair: Interspecies Comparisons, SGOMSEC 8,* 1994, 257 pp
SCOPE 53: *Methods to Assess the Effects of Chemicals on Ecosystems, SGOMSEC 10,* 1995, 440 pp
SCOPE 54: *Phosphorus in the Global Environment: Transfers, Cycles and Management,* 1995, 480 pp
SCOPE 55: *Functional Roles of Biodiversity: A Global Perspective,* 1996, 496 pp
SCOPE 56: *Global Change: Effects on Coniferous Forests and Grasslands,* 1996, 480 pp
SCOPE 57: *Particle Flux in the Ocean,* 1996, 396 pp
SCOPE 58: *Sustainability Indicators: A Report on the Project on Indicators of Sustainable Development,* 1997, 440 pp
SCOPE 59: *Nuclear Test Explosions: Environmental and Human Impacts,* 1999, 304 pp
SCOPE 60: *Resilience and the Behavior of Large-Scale Systems,* 2002, 287 pp

SCOPE Executive Committee 2001–2004

Index